Electrical Machine Drives

Electrical Machine Drives

Fundamental Basics and Practice

Claiton Moro Franchi

CRC Press is an imprint of the
Taylor & Francis Group, an **informa** business

CRC Press
Taylor & Francis Group
6000 Broken Sound Parkway NW, Suite 300
Boca Raton, FL 33487-2742

First issued in paperback 2022

ISBN-13: 978-1-138-09939-5 (hbk)
ISBN-13: 978-1-03-233862-0 (pbk)
DOI: 10.1201/b22314

Publisher's Note

The publisher has gone to great lengths to ensure the quality of this reprint but points out that some imperfections in the original copies may be apparent.

Library of Congress Cataloging-in-Publication Data

Names: Franchi, Claiton Moro, author.
Title: Electrical machine drives : fundamental basics and practice / Claiton Moro Franchi.
Description: Boca Raton : Taylor & Francis, a CRC title, part of the Taylor & Francis imprint, a member of the Taylor & Francis Group, the academic division of T&F Informa, plc, 2018. | Includes bibliographical references.
Identifiers: LCCN 2018042358 | ISBN 9781138099395 (hardback : alk. paper)
Subjects: LCSH: Electric driving. | Electric machines.
Classification: LCC TK4058 .F68 2018 | DDC 621.46--dc23
LC record available at https://lccn.loc.gov/2018042358

Visit the Taylor & Francis Web site at
http://www.taylorandfrancis.com

and the CRC Press Web site at
http://www.crcpress.com

I dedicate this book to my sons, Bruna and Renato, who are a source of inspiration and energy for the beginning of each day of my life and to my father, Calixtro, who cannot be present to see the finished book. Also, my wife, Eliane, for the companionship during these years and this work.

Contents

Preface

There are many good electrical machine books that present a rigorous and detailed analysis of its principle of operation meeting the needs of academia and research. However, there are many technologists, engineers, experimenters and others who are not interested in these aspects, but rather in practical aspects.

This book targets the large amount of people that know about electric concepts, but need practical knowledge related to electric inductions motors.

The motivation to elaborate the book is due to the difficulty of obtaining a didactic material that deals with electrical induction motor starters, with an objective and pleasant language and without losing the technical and formal essence.

This work content was developed based on the author experience of more than 10 years of working in research and industry in the areas of electrical drives and industrial automation, with the objective of treating the concepts in a practical way, seeking the connection between theory and its application.

It presents a detailed conceptual description with lots of figures and illustrative examples that harmonize the theoretical approach with the practice, which allows the full understanding of the content.

It is composed of 10 chapters and one appendix that describe in a dynamic and didactic way the fundamental concepts related to electric induction motor starters. At the end of each chapter is a set of exercises to ease the fixation of the presented content.

It also provides comprehensive coverage of electric motors and main relevant applications. Direct current (DC), synchronous, reluctance and permanent magnet motors are presented. The induction motor is treated in more depth as it is currently the most widely used motor in the industry.

It addresses the concepts related to single-phase electric motors from the description of their operating principle, electrical and constructive characteristics and wiring.

It presents the definitions of electrical power, power factor, its causes, methods of measurement and correction.

The electrical devices employed in induction motors starting methods are treated intensely with the description of their characteristics, through technical concepts and with the helping of figures, making it possible to understand and size the wiring diagrams most used in electric drives.

It deals with electronic starters: soft-starter and variable frequency drives, introducing its principle of operation, characteristics, forms of connection in addition to concepts of installation and parameterization.

It provides the main electrical diagrams used in practice and the description of the symbology adopted by international standards National Electrical Manufacturers Association (NEMA) (USA) and International Electrotechnical Commission (IEC) (Europe).

It is recommended to technicians and engineers and people who work in the areas of automation, mechatronics and electrical, as well as professionals who wish to keep updated.

Acknowledgments

To the colleagues, of the do Núcleo de Pesquisa e Desenvolvimento em Engenharia Elétrica (NUPEDEE) and Colégio Técnico Industrial (CTISM) for the aid with equipment for the book figures.

To WEG that allows the use of figures that gave a practical aspect to the book.

Special thanks to colleague Professor Felix Alberto Farret for his encouragement and the help me in contact with the Publisher which was essential to this book becoming a reality.

To Professor Rafael Concatto Beltrame for the valuable help in editing the images of the book.

The CRC Press/Taylor & Francis publisher for always believing in this project.

To Vanessa Garrett for patience, professionalism and attention, fundamental for the composition of this work.

To the other unnamed friends and colleagues who contributed directly or indirectly to this work.

To God for giving me health and intellectual conditions to complete this task.

Author

Claiton Moro Franchi received a bachelor's degree in electrical engineering from the Federal University of Santa Maria (UFSM) and master and doctoral degrees in Chemical Engineering in Universidade Estadual de Maringá (UEM), Maringá, Brazil.

He is currently a professor at UFSM. Previously, he worked as an engineer and technical consultant in industrial automation and maintenance in different companies in Brazil.

He has experience in electrical and chemical engineering, with emphasis on process control, wind power generation and industrial automation. Acting mainly on the following topics: supervision systems, programmable logic controllers, industrial networks, and industrial process control, with several book publications and experience in research and professional.

1

Electric Motors

The electric motor is a machine that transforms electrical energy into mechanical, kinetic energy, (i.e., in a motor), using electric energy, whether continuous or alternating, ensuring movement on an axis, which can be exploited in different ways, depending on the motor application.

The drive of machines and mechanical equipment by electric motors is a subject of great economic importance. It is estimated that the world market for electric motors of all types is around billions of dollars a year. In the field of industrial drives, it is estimated that from 70% to 80% of the electric energy consumed by all industries is transformed into mechanical energy by electric motors. Figure 1.1 shows a typical induction motor used in the industry.

According to the power supply, the electric motors can be divided into motors of direct current (DC) and alternating current (AC).

The following are the types of electric motors most used in the industry.

1.1 Electric Motors Types

There are many types of motors, Figure 1.2 presents the motors most used in practice.

This book will be focused on the motors in alternating current, since most of the electric motors used in the industry belong to this category. The following will present some basic concepts of the main types of DC motors.

1.1.1 Direct Current Motors

The substantial evolution of power electronics has made possible the construction of static converters employing reliable, low-cost and simple maintenance thyristors. With this, DC motors, despite their high cost, have become

FIGURE 1.1
Three-phase induction electric motor.

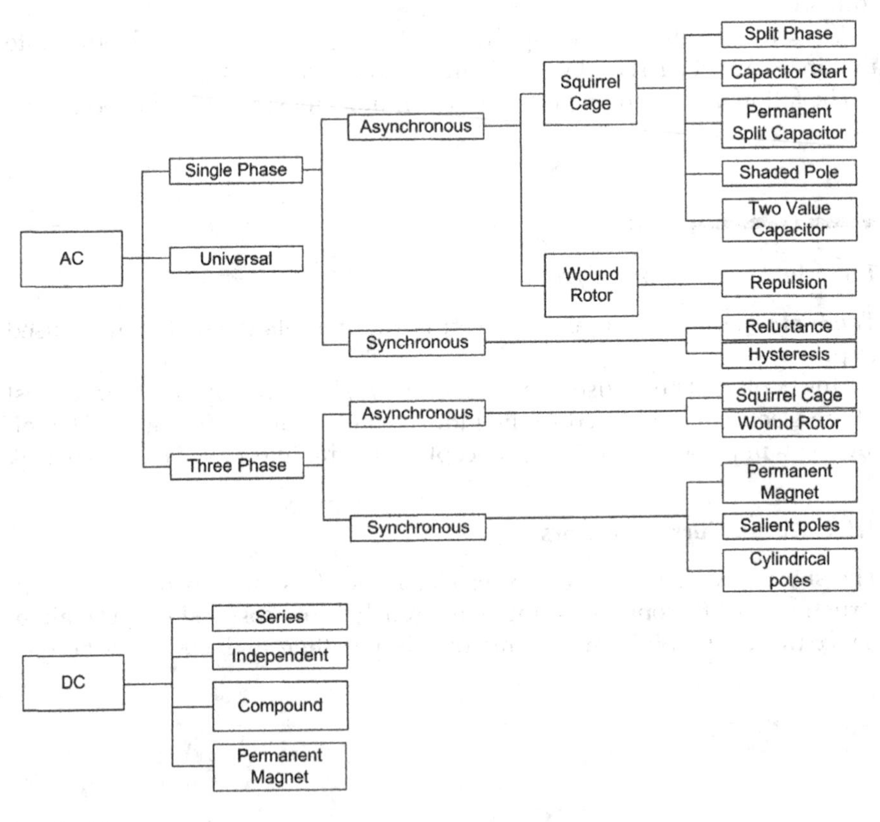

FIGURE 1.2
Electric motor types.

FIGURE 1.3
DC motor. (Courtesy of Weg.)

an alternative in a series of applications that need this fine tuning of speed and high torque. Figure 1.3 shows a typical DC motor.

The DC motors can be divided into two main magnetic structures:

Stator: Composed of a ferromagnetic structure with salient poles where there are coils placed that form the magnetic field. Typically, two windings are placed in the stator: the series field, consisting of a small number of turns with larger cross section coil wire, and the shunt field, made by a large number of coils with smaller cross section, as shown in Figure 1.4.

The magnetic field can also be produced by permanent magnets with this type of machine application restricted to applications of small power, such as toys, hot blowers, computer disc drives, etc.

Rotor: Is an electromagnet composed of an iron core with windings connected to a mechanical system called a commutator. The commutator is along with the rotor shaft and has a cylindrical surface with blades that are connected to the rotor windings and with brushes that are pressed together with these blades and connected to the power supply. The commutator has the function of transforming DC into AC in a suitable way for motor torque development.

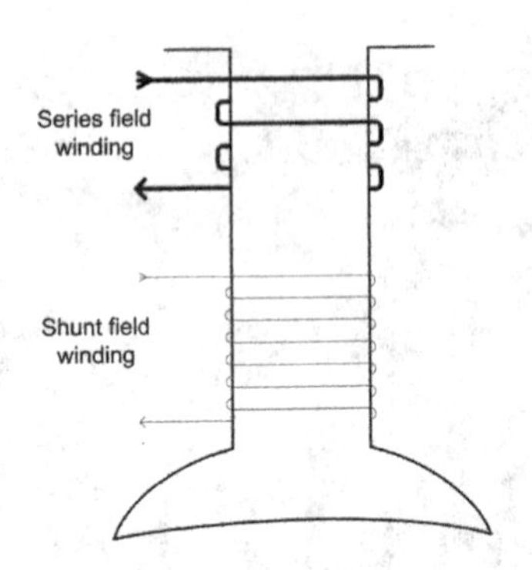

FIGURE 1.4
Series and shunt field in a stator.

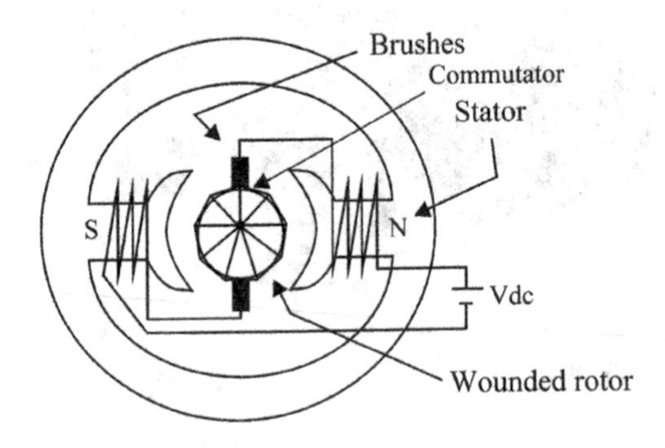

FIGURE 1.5
Stator and rotor integrated in a 2 poles DC motor.

Figure 1.5 shows the stator and rotor integrated in a two-pole DC motor.

The DC motor can be divided into two distinct circuits: armature (rotor) and field (stator). In this analysis, it will be considered a motor powered by two voltage supplies: one for the armature circuit (U_a) and one for the field circuit (U_f), as shown in Figure 1.6.

The DC motor operation is based on the produced force from the interaction between the magnetic field and the armature current in the rotor making the rotor moves.

Applying Kirchhoff's law in the armature circuit we have:

$$U_a = I_a.R_a + E \tag{1.1}$$

where:

U_a is the armature voltage
R_a is the armature resistance
I_a is the armature current
E is the induced electromotive force

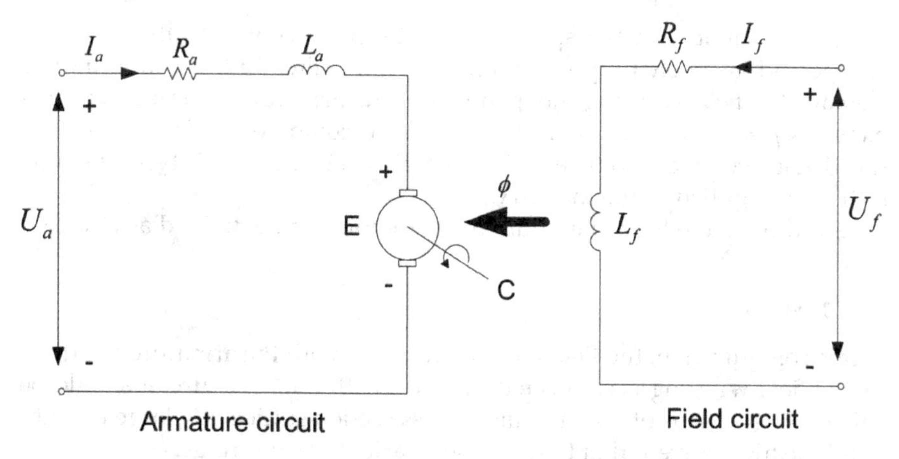

FIGURE 1.6
Armature and field circuit in a DC motor.

The air gap flux (ø) is proportional to the field current (I_f):

$$ø = k_2.I_f \tag{1.2}$$

where k_2 is the field constant.

The motor torque is given by:

$$C = k_3.ø.I_a \tag{1.3}$$

where k_3 is the torque constant.

Considering Faraday's Law, the induced electromotive force (E) is proportional to the magnetic air gap flux (ø) and the rotation (n), thus:

$$E = k_1.ø.n \tag{1.4}$$

where:

n is the speed

k_1 is the constant considering rotor dimensions, number and pole connection

Arranging Equations (1.1) and (1.4), we have the motor speed expressed by:

$$n = k_1 \frac{U_a - I_a.R_a}{ø} \tag{1.5}$$

In practice, the armature resistance is very small, causing a small voltage drop in the armature ($R_a.I_a \cong 0$), so we will have the following equation to for speed:

$$n = k_1 \frac{U_a}{ø} \tag{1.6}$$

We can conclude that the speed is directly proportional to the armature voltage and inversely proportional to the air gap flux. In this configuration, we have the field winding independent of the armature winding, and it is named *separately excited DC motor,* which we could see in Figure 1.6. The speed adjusted by the voltage variation in this field, being this type of motor, is the most applied in the industry.

According to the type of excitation, the motors can be divided as follows.

1.1.1.1 Series

In this configuration, the field coils are in series with the armature winding. As the field winding is connected in series with the armature, it should be built with few turns of wire having a cross section sufficiently large to withstand the high current that flows through armature windings.

In the series, excited motor, the magnetic air gap flux (ø) per pole depends on the armature current (I_a), which depends on the load applied to the motor. This causes the machine to have a high torque at low speeds and the speed can be very high when the motor is in an unload condition. This occurred due to the low field current that causes low magnetic air gap flux (ø), which is in the denominator of Equation (1.5), resulting in high velocities for small field values. It is recommended to always keep load on the shaft, since the speed can increase to very high values, resulting in high centrifugal forces that can damage the machine. Figure 1.7 shows a series excited motor configuration.

1.1.1.2 Shunt Excited

In this type of machine, the field coils are in parallel with the armature windings considered a type of self-excited motor. As the windings are connected

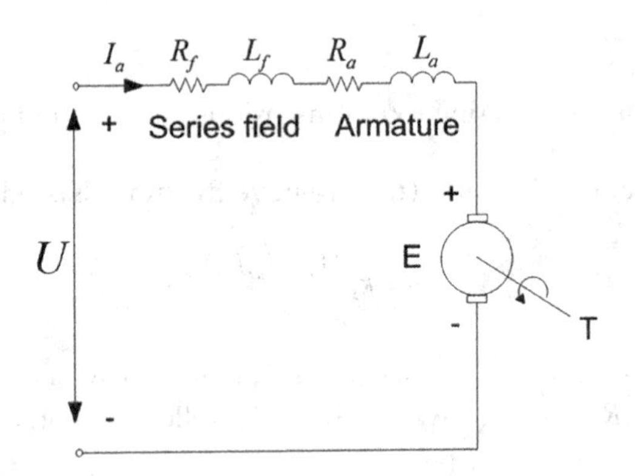

FIGURE 1.7
Series excited motor configuration.

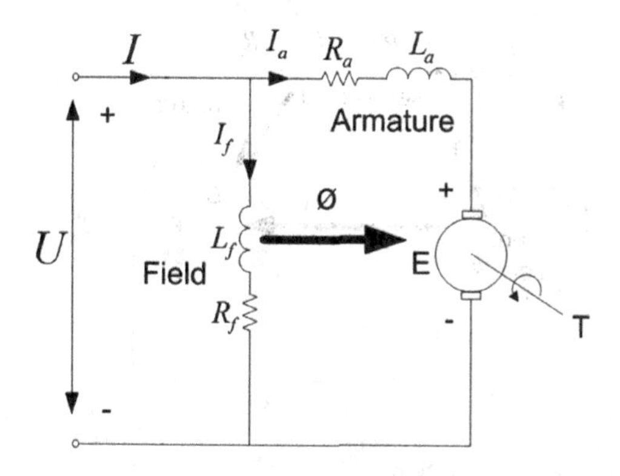

FIGURE 1.8
Shunt excited motor configuration.

in parallel (subjected to the same voltage supply) the field winding coils are made with a large number of turns of small cross sections.

In this configuration, there are separate branches: one for the armature current (I_a) and the other one for the field current (I_f), which we can see in Figure 1.8.

The overall current is divided in two parts: armature (I_a) and field (I_f), and could be obtained by the following equations:

$$I_f = \frac{U}{R_f} \tag{1.7}$$

$$I = I_a + I_f \tag{1.8}$$

So when this motor is in the running condition, the supply voltage is constant and the current shunt is obtained by Equation (1.7), making the field current and, consequently, the flow constant, so this motor is considered constant flux or constant speed. This characteristic is desired in industrial applications.

1.1.1.3 Compound Excited

This motor has the constructive characteristics of the series and the shunt excited motors. Figure 1.9 shows the wiring diagram for this type of motor.

The magnetic field created by the shunt winding is always greater than that generated by the series winding. When the motor operates at low load, the current in the armature I_a that flows in the series winding is small, causing the magnetic field to be negligible. However, the shunt winding can be completely energized, keeping the operating characteristics of the machine.

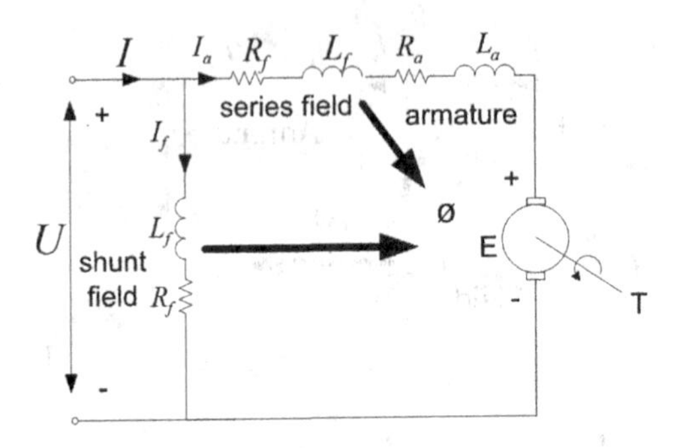

FIGURE 1.9
Compound excited motor configuration.

When the load is applied on the shaft, the current in the series field increases and, consequently, the magnetic field increases, but the magnetic field in the shunt winding remains constant. This causes the speed (which depends on the flow ø) to fall from no-load to full-load conditions between 10% and 30%.

There are two connection types for this cumulative compound motor where the magnetic field of the shunt fields and series are summed and the differential compound where the series field is connected in order to oppose the magnetic field is generated by the shunt winding.

NOTE: The constant development of power electronics should lead to a progressive reduction in the use of DC motors. This is because variable frequency drives developed for induction motors, especially the squirrel cage, are already becoming more attractive options in terms of speed control due to association with the low cost and maintenance of this type of motor.

1.1.2 Alternating Current Induction Motors

Most applications have their most economical configuration with the use of squirrel cage induction motors; it is estimated that this type of motor has around 90% (in units) of the motors manufactured in the world.

The induction motor has become the most used type in the industry because most systems of electric power distribution are AC. Compared with the DC motor, the induction motor has the advantage of simplicity, which translates into low cost and maximum efficiency with minimal maintenance. The efficiency is high for medium and maximum loads, with an acceptable power factor when a correct selection is made.

We will start our study by presenting, in the following, the main single-phase induction motors.

1.1.2.1 Single-Phase Induction Motors

Single-phase motors are so called because their field windings are connected directly to a single-phase supply. Among the various types of single-phase electric motors, cage rotor motors stand out for their manufacture simplicity and, mainly, robustness, reliability and reduced maintenance.

Because these motors have only one supply phase, they do not have a rotating field like polyphase motors, but have a pulsating magnetic field. This prevents them from having starting torque, taking into account that magnetic fields are induced in the rotor in line with the stator field.

To solve the starting problem, auxiliary windings are used, which are sized and positioned so as to create a second dummy phase, allowing the rotating field creation required to start the motor. Thus, it will have a two-part armature winding. The first part is a main winding, which is connected directly to the energy supply. The other part is the secondary winding, which can be connected in series with a capacitor, and this circuit is connected in parallel with the main circuit. In this way, the electric current flowing through the auxiliary winding is leading in relation to the main winding current.

Single-phase induction motors are the natural alternative to multiphase induction motors where three-phase power supply is not available, and are often used in homes, offices, workshops and in rural areas in applications such as water pumps, fans and drives for small machines. It is not recommended to use single-phase motors larger than 3 horse power (HP), since they are connected only to one supply phase, causing a considerable load imbalance in the grid.

The use of single-phase motors is justified for what was said earlier, however, there are some drawbacks of this type of motor in comparison with three-phase motors:

- Taking into account the cost, the single-phase motor has a higher cost than a three-phase motor of the same power.
- The single-phase motor suffers mechanical wear of centrifugal switch required at the motor start.
- The single-phase motor reaches only 60%–70% of the power compared with a three-phase motor of the same size.
- The single-phase motor features lower power factor and efficiency.
- Due to the fact of having a single phase, they drain more current than similar three-phase motors.
- Are an imbalanced load to the electrical distribution system.

1.1.2.2 Types of Single-Phase Induction Motors

The main single-phase induction motors can be divided into the categories listed below:

1.1.2.2.1 Shaded Pole

The shaded pole motor, thanks to its starting process, is the simplest, most reliable and economical of single-phase induction motors, as shown in Figure 1.10.

FIGURE 1.10
Commercial shaded pole motor. (Courtesy of Weg.)

Constructively, there are several types, and one of the most common forms is the salient poles. Each pole has a part (usually 25%–35%) with a copper ring or bar surrounding a portion of each pole. This single-turn winding is named the shading coil, as shown in Figure 1.11.

The current induced in this single-turn winding makes the magnetic field through it suffer a delay in relation to the magnetic field of the part not

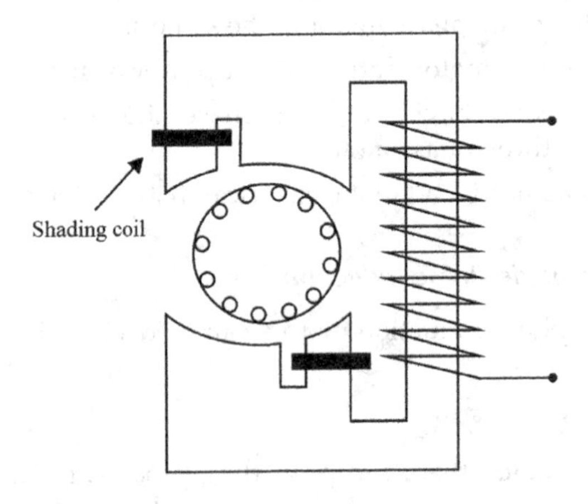

FIGURE 1.11
Shading coil in a shaded pole motor.

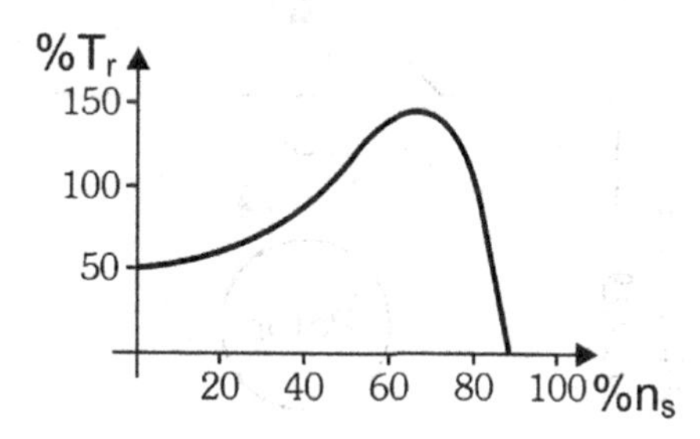

FIGURE 1.12
Torque by speed curve for shaded pole motor.

embraced by it. The result is similar to a rotating field that moves from the direction of the not embraced to the embraced part of the pole. This produces the torque that causes the motor to start and reach the rated speed.

The direction of rotation, therefore, depends on the side where the part of the pole is located. Consequently, the distorted field motor has a single direction of rotation. This can generally be reversed by changing the position of the rotor shaft tip in relation to the stator. There are other methods to obtain the change of direction of rotation, but they are much more costly.

When considering performance, these types of motors have low starting torque (15%–50% of rated), low efficiency and power factor. Figure 1.12 shows a typical curve rated torque ($\%T_r$) by synchronous speed ($\%\ n_s$) of this type of motor. Thus, they are usually manufactured for small powers, ranging from a few thousandths of HP to 1/4 HP. In general, speed control in this type of motor is achieved by reducing the applied voltage.

Due to their simplicity, robustness and low cost, they are ideal in the following applications: fans, exhaust fans, room cleaners, refrigeration units, clothes and hair dryers, small pumps and compressors, and household applications.

1.1.2.2.2 Split Phase

This motor has a main winding and an auxiliary (dummy) winding for starting, both lagged 90°, as shown in Figure 1.13. The auxiliary winding creates a phase shift that produces the torque required for initial rotation and acceleration. When the motor reaches a specific speed, the auxiliary winding is disconnected from the power supply by a centrifugal switch that normally acts by a centrifugal force or, in specific cases, by current relay, manual switch or other special device. As the auxiliary winding is designed to operate only at start-up, if it is not switched off immediately after starting, it is damaged.

The centrifugal switch maintains a contact block with the auxiliary winding by means of springs, so that the circuit is closed at the start. As the motor speed increases, weights are shifted outward, exceed the spring tension and move the contact block away, opening the auxiliary winding circuit, which remains open while the motor is running.

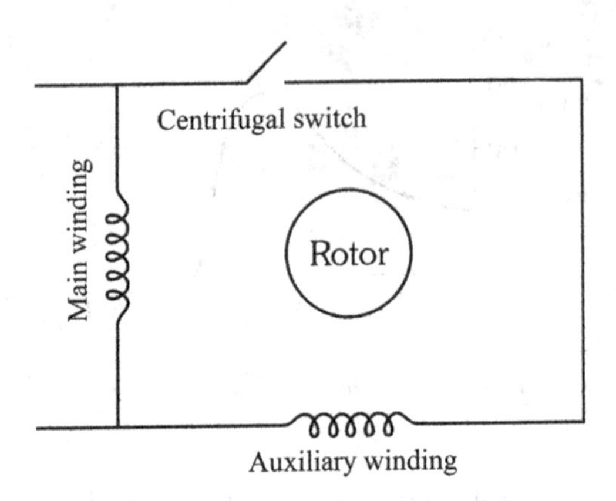

FIGURE 1.13
Split-phase motor wiring diagram.

The lag angle that can be reached between the main winding and auxiliary winding currents is small, so the starting torque is proportional to the sine of the angle between the currents in the main and auxiliary windings at the time of starting. It has starting torque equal to or slightly above rated.

For this type of motor, the auxiliary winding is disconnected from the energy supply through a centrifugal switch when the rotation is between 75% and 80% of the synchronous speed, because in this speed range, the torque produced by the pulsating magnetic field of the main winding exceeds the one developed by the two combined windings. Figure 1.14 shows the torque behaviour with the motor speed variation, and Figure 1.15 shows a commercial split phase motor.

The locked rotor current changes from five to seven times the rated current, but is not a problem. Since the rotors of this type of motor are of reduced size, exhibiting low inertia even when connected to the load, the relatively high starting current drops almost instantly.

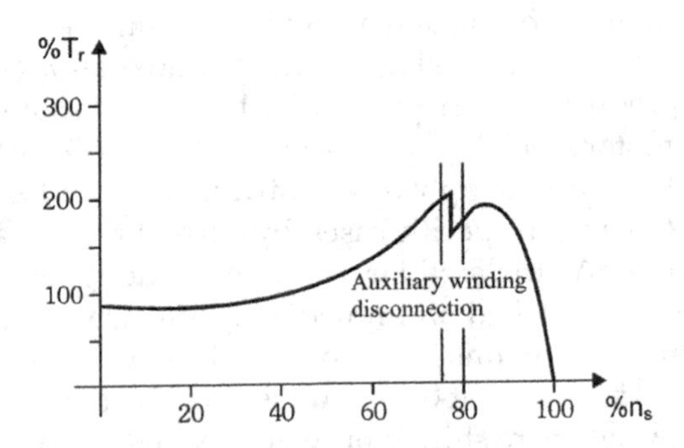

FIGURE 1.14
Torque by speed curve for the split-phase motor.

FIGURE 1.15
Commercial split-phase motor.

In order to reverse the direction of rotation of the split-phase motor, it is necessary to reverse the polarity of energy supply connection terminals in relation to one of the main or auxiliary windings. The reversal of the direction of rotation can never be made under running conditions.

Speed control in split-phase motors must be performed in a very limited range, which is above the centrifugal switch operating speed and below the synchronous speed. Its speed control is very difficult, since its synchronous speed is determined by the grid frequency and by the number of poles in the main winding.

This type of motor has application to fractional powers and to loads that require small starting torque, such as office machines, fans and exhaust fans, small polishers, hermetic compressors, centrifugal pumps, etc. They are usually built in fractional powers that do not exceed 3/4 of HP.

1.1.2.2.3 Capacitor Start

This motor is similar to a split phase. The main difference lies in the inclusion of an electrolytic capacitor in series with the starting auxiliary winding. The capacitor allows a greater lag angle between the main and auxiliary winding currents, providing high starting torques. Figures 1.16 and 1.17 show a commercial starting capacitor motor and wiring diagram, respectively.

As in the split-phase motor, the auxiliary circuit is switched off when the motor reaches between 75% and 80% of the synchronous speed.

In this range of speeds, the main winding alone develops almost the same torque as the combined windings. For larger speeds, between 80% and 90% of the synchronous speed, the torque curve with the combined windings crosses the torque curve of the main winding, as shown in Figure 1.18. In this way, for speeds above this point, the motor develops less torque for any slip.

FIGURE 1.16
Commercial capacitor start motor.

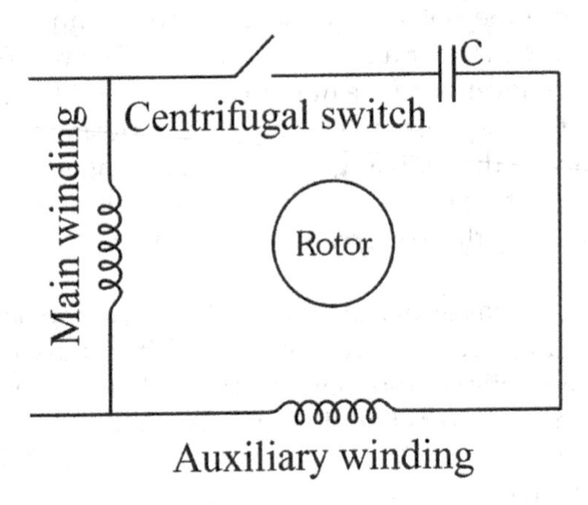

FIGURE 1.17
Capacitor start wiring diagram.

Because the curves do not always intersect at the same point, and the centrifugal switch does not always open at exactly the same speed, it is common practice to make them open just before the curves intersect.

Because the sizing of the auxiliary winding and the starting capacitor is based only on its motor start operation, trouble in the centrifugal switch can cause damage not only to the motor windings but also to the capacitor.

It is the same way for split-phase motors. In order to reverse the direction of rotation, it is necessary to reverse the polarity of the supply connection

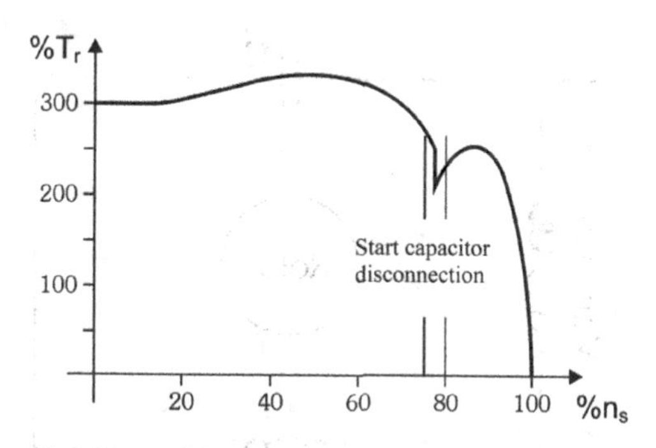

FIGURE 1.18
Torque by speed curve for starting capacitor motor.

terminals in relation to one of the windings. This makes it possible to reverse the direction of rotation while the motor is running.

With its high starting torque (between 200% and 350% of rated torque), the start capacitor motor can be used in a wide variety of applications and manufactured for power from 1/4 to 15 HP.

1.1.2.2.4 Permanent Split Capacitor

In this motor, the auxiliary winding and the capacitor are permanently connected, being an electrostatic type capacitor. Figures 1.19 and 1.20 shows this type of motor and its electric diagram.

FIGURE 1.19
Commercial permanent split capacitor. (Courtesy of Weg.)

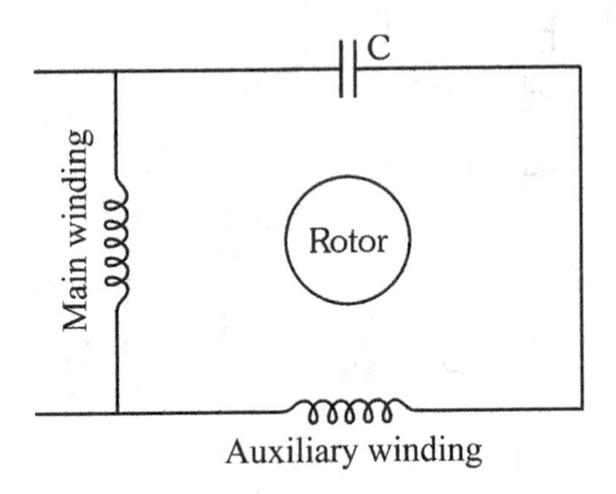

FIGURE 1.20
Permanent split capacitor wiring diagram.

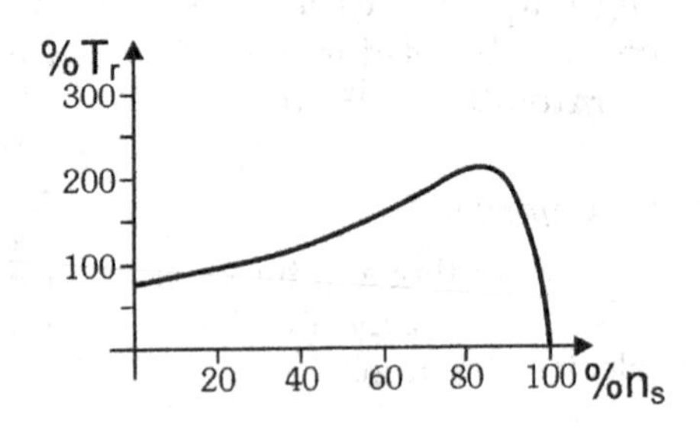

FIGURE 1.21
Torque by speed curve for the permanent capacitor motor.

The purpose of this capacitor is to create magnetic field conditions very similar to those found in multiphase motors, thereby increasing maximum torque, efficiency and power factor, and significantly reducing noise. Figure 1.21 shows the torque behaviour with the speed change.

Constructively, they are smaller and maintenance-free because they do not use contacts and moving parts, as in previously presented motors. However, its starting torque is lower than split-phase motors (50%–100% of the rated torque), which limits its application to equipment that does not require high starting torque, such as office machines, fans, exhaust fans, blowers, centrifugal pumps, grinders, small saws, drills, air conditioners, sprayers, etc. They are usually manufactured for power from 1/50 to 1.5 HP.

1.1.2.2.5 Two Value Capacitor

It is a type of motor that uses the advantages of the two previous ones: starting as the start capacitor motor and operating in the same way as the permanent capacitor motor. Figures 1.22 and 1.23 show a commercial two-value

FIGURE 1.22
Commercial two-value capacitor.

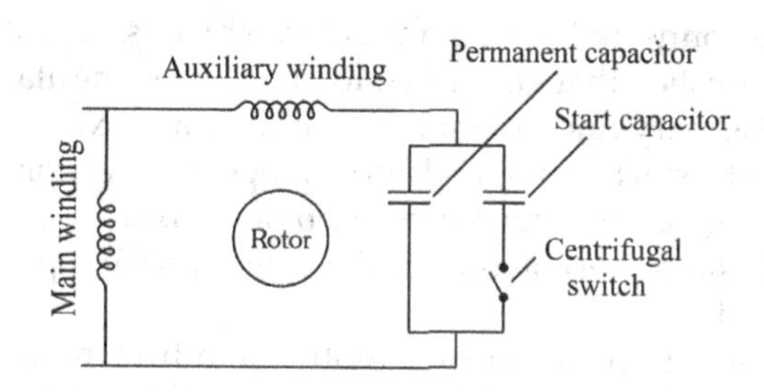

FIGURE 1.23
Two-value capacitor wiring diagram.

capacitor motor diagram. Due to its high cost, it is usually manufactured only for powers over 1 HP.

In this motor, two capacitors are used during the starting period. One is a starter capacitor with a reasonably high capacitance, about 10–15 times the value of the permanent capacitor, which is disconnected from the circuit by a centrifugal switch when the motor speed reaches 75%–80% of the synchronous speed, as shown in Figure 1.24.

It is possible to reverse its direction of rotation, because when in operation, if the polarity of the connection supply terminals is inverted in relation to one of the windings, its direction of rotation is also reversed. Frequent inversions reduce the centrifugal switch lifespan. Thus, when frequent reversals are required, preference should be given to the use of a permanent capacitor motor.

1.1.2.2.6 Repulsion

This type of motor has a stator and winding rotor with no connection between them. The current in the rotor is generated exclusively by induction,

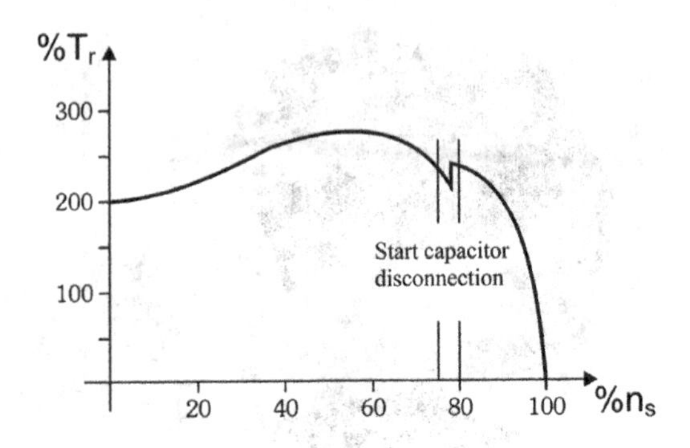

FIGURE 1.24
Torque by speed curve for the two-value capacitors motor.

with its winding connected to a commutator, which is in contact with a pair of short-circuit brushes that can move to change its angular position with respect to an imaginary line drawn through the rotor axis.

This motor can be started, stopped and change its direction of rotation and speed by the change in the angular position of the brushes.

In Figure 1.25, the representation of the single-phase repulsion induction motor is presented.

The connection of the repulsion stator winding motor to a single-phase circuit generates a magnetic field due to current flow in the stator windings.

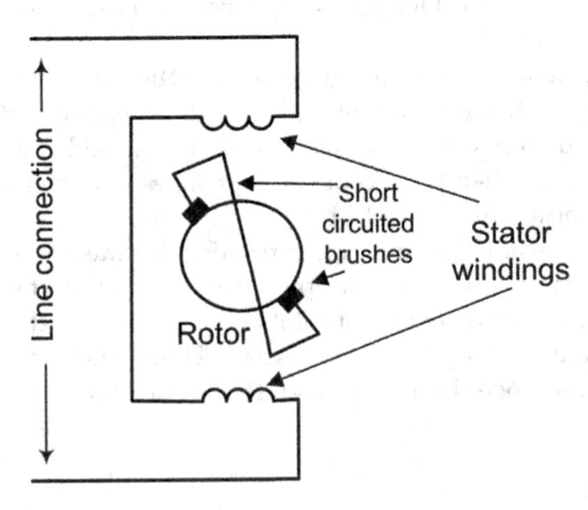

FIGURE 1.25
Repulsion wound rotor.

The stator field induces a voltage and a resulting current in the rotor windings. If the brushes are placed in a defined position of the commutator, the current in the armature windings will create magnetic poles in the armature. These poles have a relationship with the displacement stator field poles of approximately 15 electrical degrees. The rotor poles polarity are the same as the adjacent poles of the stator, which cause a repulsion torque and the motor armature rotation.

This type of motor is mainly used for applications that require constant torque, such as fans, blowers, etc. This motor has the disadvantages of sparking in the brushes, low power factor at reduced speeds and, in case of unloaded operation, will produce high speeds, which can seriously damage the motor.

1.1.2.2.7 Universal

Several home appliances, especially kitchen appliances, and portable tools use this type of single-phase motor, called universal, whose operating principle is completely different from the induction motor. The designation of a universal motor derives from the fact that it can operate under both AC and DC power supply. Strictly speaking, this is a series DC motor. For AC operation, the stator and rotor must be made of laminated plates to avoid losses due to hysteresis and stray currents.

Typically, the stator is a set of salient poles with coils wound on them. The rotor consists of a winding distributed in grooves and connected in series with the coils of the stator (armature). The terminals of the rotor coils are welded to a collector ring integral with the shaft, and the connection with the power supply is made by a set of graphite brushes. In Figure 1.26 we have the representation of this motor.

It is a variable speed motor, with low speeds for large torque and high speeds for small loads. The starting torque is also high. Because of this, they are commonly used in small appliances, such as electric drills and sanders,

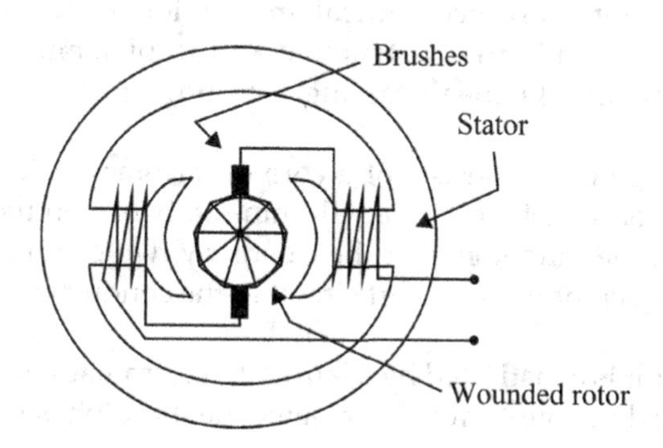

FIGURE 1.26
Universal motor.

which require high torque, and in blenders, vacuum cleaners and centrifugal pumps, which require high speed.

They are usually manufactured for fractional powers of up to 3/4 HP. For powers above a few HP, they work in an unsatisfactory way on alternating current. It presents great sparking on the brushes, and the efficiency and power factor decrease.

1.1.2.3 Three-Phase Electric Motor

The three-phase induction motor is the most widely used in both the industry and home environment because current systems in electric power distribution are usually three-phase alternating current. The use of three-phase induction motors is advisable from 3 HP. For lower powers, the single phase is justified.

The three-phase induction motor has a relative advantage over the single-phase, since it has easier starting, the noise is smaller, and it is cheaper.

In the following are presented the most three-phase motors used.

1.1.2.3.1 Squirrel Cage Induction Motor

The squirrel cage rotor is the toughest type of motor. It does not require the use of brushes or commutator, which avoids many problems related to wear and maintenance.

The simplest form of the squirrel cage rotor motor has a relatively low starting torque and the peak current in the starting reaches up to ten times the rated motor current. These aspects can be partially improved by the construction of the rotor itself. In particular, the bars forming the cage influence these characteristics. Better performance motors are equipped with high bar cage rotors, wedge-shaped or double bars.

This is the most widely used motor in the industry today. It has the advantage of being more economical in relation to the other types of three-phase motors in both constructive and implementation aspects. In addition, by choosing the ideal starting method, it has a fairly large range of applications.

The squirrel cage rotor consists of a core of ferromagnetic plates, isolated from each other, on which are placed aluminium bars (conductors) arranged parallel to each other and joined at their ends by two conducting rings, also in aluminium, which provide a short circuit in the conductors as illustrated in Figure 1.27.

The motor stator is constituted by a laminated ferromagnetic core in whose cavities are placed the windings that receive the three-phase alternating current from the power supply.

The conductor bars of the cage are generally placed with a certain slope to avoid the trepidations and noise by the electromagnetic action between the stator and rotor cavities.

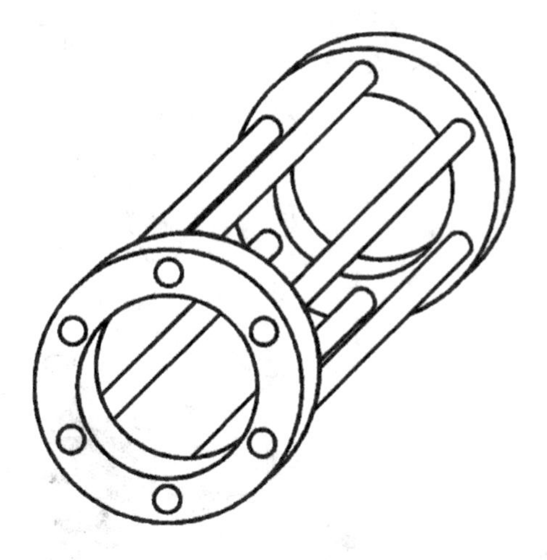

FIGURE 1.27
Squirrel cage rotor.

The advantage of this rotor over the wound rotor is that it results in a faster, more practical and cheaper armature construction.

It is a robust, inexpensive, fast-producing motor that requires no commutator (a sensitive and expensive device), providing a fast and easy connection to the grid.

The main drawback is the low starting torque in relation to the wound induction rotor. Figure 1.28 shows a three-phase induction motor with a squirrel cage rotor.

1.1.2.3.2 Wound Rotor Motor

The wound rotor has a winding consisting of three coils, similar to the motor stator. These coils are normally connected in series, with the three free terminals connected to brushes on the rotor shaft. These brushes allow the connection of rheostats (variable resistors) in the rotor coil circuit to change the starting characteristics, such as improving the starting torque and reducing the starting peak current.

It differs from the squirrel cage rotor motor only as regards the rotor, consisting of a laminated ferromagnetic core on which are housed the coils to the three phases. Generally, the winding are connected in star. The three free terminals of the three-phase winding are connected to the brushes placed on the rotor shaft. To start the motor, these three rings are connected externally to a starting rheostat consisting of three variable resistors, also connected in star. In this way, the rotor windings are also in closed circuit.

The starting rheostat task, connected to the rotor windings, is to reduce the high starting currents in the case of high power motors.

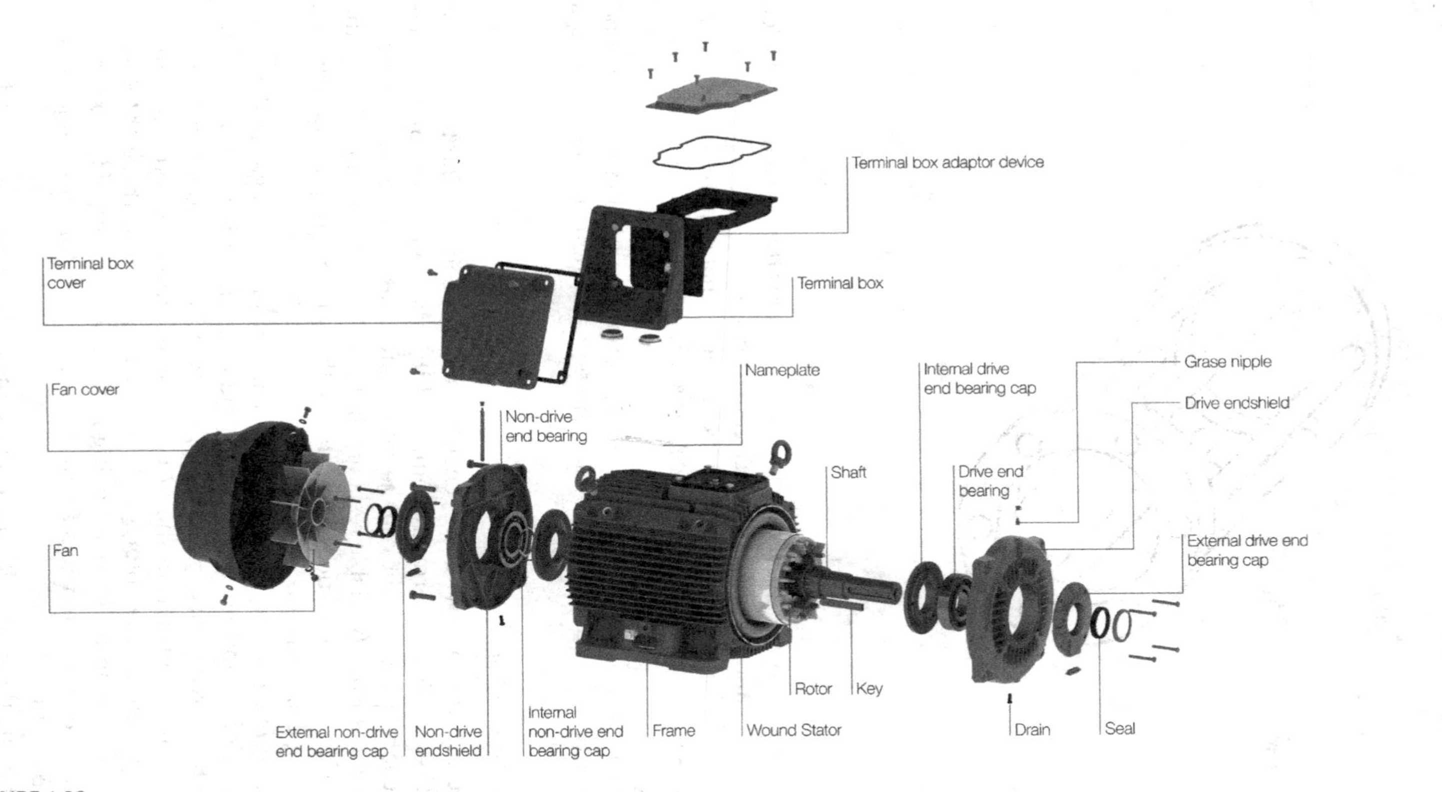

FIGURE 1.28
Three-phase squirrel cage induction motor. (Courtesy of Weg.)

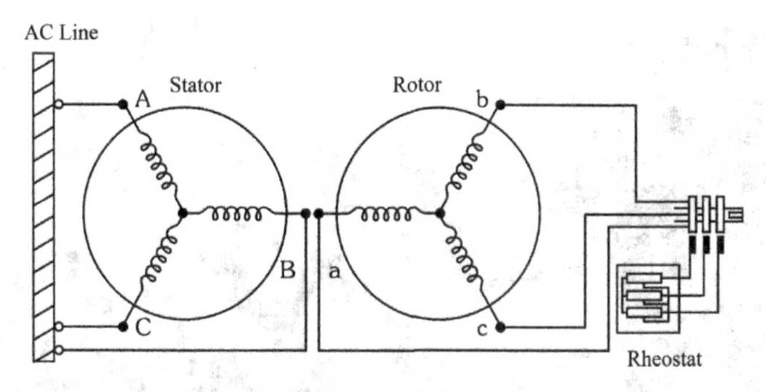

FIGURE 1.29
Wiring diagram in a wound rotor motor.

As the motor increases the speed, the resistors are progressively withdrawn from the circuit until short-circuited, when the motor starts operating at its rated speed, as shown in Figure 1.29.

The wound rotor motor also works with short-circuit rotor elements (such as the squirrel cage rotor motor) when it reaches its rated speed.

The wound rotor motor replaces the squirrel cage rotor at very high power due to the lower starting current allowed by the rotor configuration. The wound rotors are very useful when starting the full armature voltage, with a high starting torque and moderate current at start.

The rheostat causes the motor to work with much greater slip than the conventional (>5%), resulting in a larger starting torque. Figures 1.30 and 1.31 show,

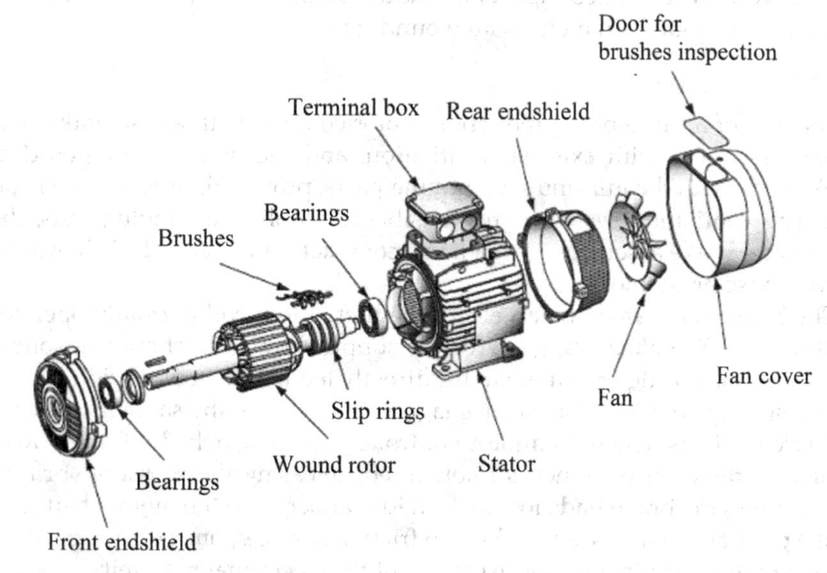

FIGURE 1.30
Exploded view of a wound motor.

FIGURE 1.31
Commercial wound rotor induction motor. (Courtesy of Weg.)

respectively, the exploded view of an induction motor with wound rotor and a commercial induction motor with wound rotor.

1.1.2.3.3 Brake Motor

It consists of a three-phase induction motor coupled with a disc brake. The motor is closed, with external ventilation, and the brake is composed of brake pads with the minimum of moving parts, providing little friction heating. The ventilation system is responsible for cooling the motor, thus, the motor and brake assembly form a fairly compact unit. Figure 1.32 shows the three-phase brake motor.

The brake is activated by an electromagnet whose coil normally operates within a ± 10% voltage range, which is supplied by a direct current source consisting of a bridge rectifier circuit directly fed by the power grid.

The supply circuit of the electromagnet is driven by the same motor control circuit. Thus, when the motor control circuit is switched off, the power source of the electromagnet is interrupted, releasing the pressure springs, which press the brake pads in a disk rigidly attached to the motor shaft. The brake pads are compressed by the two friction surfaces, one being formed by the cover and the other by the armature of the electromagnet itself.

FIGURE 1.32
Three-phase brake electric motor. (Courtesy of Weg.)

In order for the armature of the electromagnet to be displaced by the action of the spring, it is necessary that the electromagnetic force is less than the force exerted by the spring, which occurs when the motor is disconnected from the energy supply. In the same way, when the motor is started, the electromagnet is energized, attracting its armature in the opposite direction to the force of the spring, causing the braking disk to spin free, without friction.

The brake motor application is restricted to industrial activities, when there is a need for rapid stops for safety requirements, as well as precision in the positioning of the machines, such as cranes, elevators, rotors, conveyor belts, winders, etc.

It is not advisable to apply the brake motor in activities that could lead to the penetration of abrasive particles, as well as water, oil, and others, in order to reduce the efficiency of the braking system or even to damage it. The heat generated by the friction during the braking operation must be removed by the motor cooling system.

In general, the brake motors can be divided into three different categories:

1.1.2.3.3.1 Slow Braking In this type of brake, the rectifier bridge is connected directly from the motor terminals with the current sent to the electromagnet brake coil. Figure 1.33 shows the wiring diagram to slow braking motor.

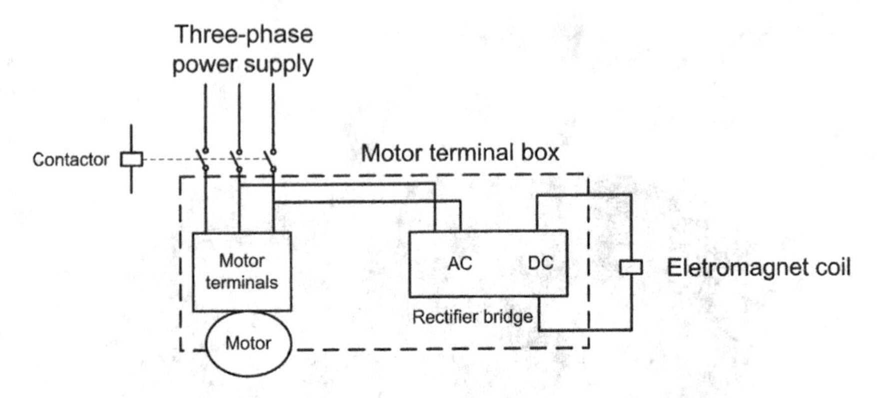

FIGURE 1.33
Wiring diagram of slow braking motor.

1.1.2.3.3.2 Medium Braking In this case, the rectifier bridge is connected to the source using alternating current. This circuit must be connected to an auxiliary contact of the motor control contactor so that the brake is switched on or off together with the motor, avoiding enabling the protection elements in the starting. Figure 1.34 shows the medium braking wiring diagram.

1.1.2.3.3.3 Rapid Braking The rectifier bridge is powered from the AC source. This time, the direct current supply circuit of the rectifier bridge is interconnected to a normally open auxiliary contact of the motor control contactor as presented in Figure 1.35. In the following Table 1.1, the typical characteristics of three-phase brake motors is presented.

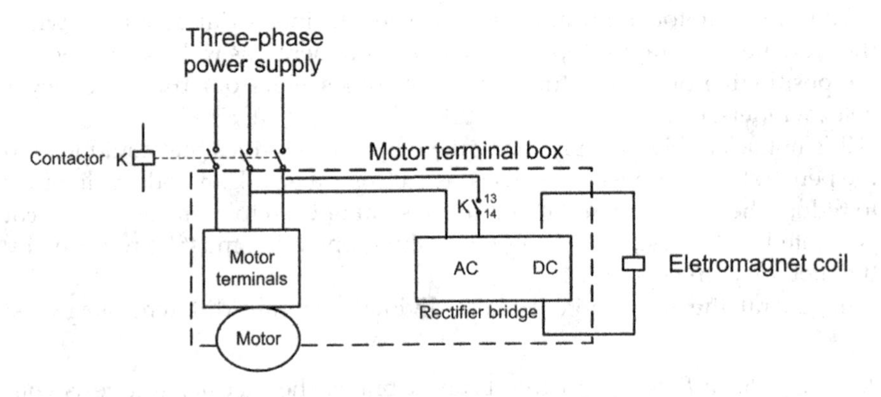

FIGURE 1.34
Wiring diagram of medium braking motor.

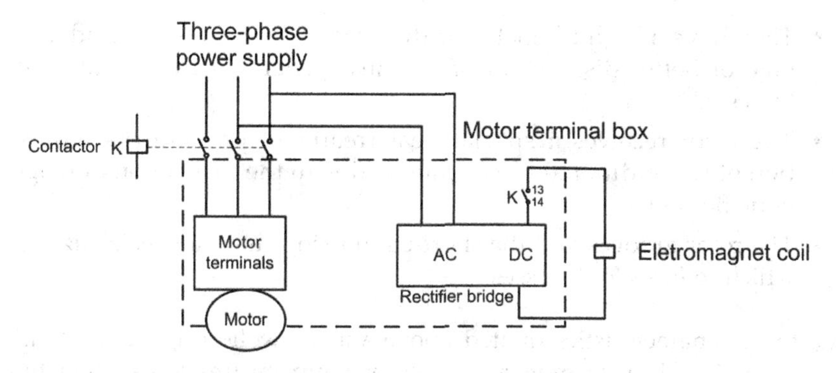

FIGURE 1.35
Wiring diagram of rapid braking motor.

TABLE 1.1

Typical Characteristics of Three-Phase Brake Motors

	Brake Time (ms)					
Frame	**Slow Brake**	**Medium Brake**	**Rapid Brake**	**Brake Torque (N.m)**	**Brake Power (W)**	**Brake Current (A)**
90S/L	650	300	170	25	40	**0.2**
100L	700	350	220	40	50	**0.25**
112M	800	450	250	70	60	**0.3**
132S/M	1000	600	300	80	100	**0.5**
160M/L	1200	800	370	160	120	**0.55**

1.1.2.3.4 High Efficiency Motors

These motors use better quality materials in both the stator and the rotor and, for the same power in the shaft, consume less energy during the same conditions of operation.

The high efficiency motors have some special features:

- Employ higher quality silicon steel plates, which provide a reduction in the magnetizing current and, consequently, increases the efficiency of the motor.
- A greater amount of copper is used in the windings, which cause reduction in losses due to heating as a function of the electric current (Joule effect).

- They have a high fill factor of the grooves, which has the advantage of better dissipation of the heat generated by the internal losses.
- The rotor receives a special heat treatment providing a reduction of the additional losses caused due to the dispersion of magnetic flow.
- The rotor grooves and the short-circuit rings have a special sizing, which reduces Joule losses.

Due to the characteristics quoted above and its well-designed ventilation system, high efficiency motors operate at temperatures lower than those of conventional motors, which makes possible a greater overload capacity, allowing a service factor generally higher than 1.10.

Figure 1.36 shows a commercial three-phase high efficiency motor.

1.1.2.3.5 Dahlander

The Dahlander winding is an option in an application that requires a two-speed motor. The number of rotations at a slower speed is always half the number at a higher speed. The motor efficiency at a higher speed is better than at a lower speed. The motor power at a higher speed is 1.5–1.8 times higher than at lower speed. The Dahlander winding consists of six coils, which can be combined in two ways. The motor has six terminals, such as the motor for one speed, but cannot be adapted for two voltages.

FIGURE 1.36
Three-phase high efficiency motor.

NOTE: There are three-speed motors in which a Dahlander winding is attached to a separate winding. To obtain four speeds, it joins two separate Dahlander windings on a single motor.

1.1.2.3.6 Synchronous Motors

They are named synchronous motors because the speed is synchronized with the rotating field that is established in the stator. The speed of the synchronous motor is determined by the equation:

$$Ns = \frac{120.f}{p} \tag{1.9}$$

where:

Ns is the synchronous speed in rpm
f is the frequency in Hertz
p is the number of poles

As the grid frequency that feeds the motor is constant, as well as its number of poles, we can consider the synchronous motor AC a constant speed machine.

In the same way, the operation of synchronous motors requires the application of an alternating voltage in the motor stator. The excitation of the rotor field is done with direct current source, directly from an external source, or an exciter connected to the motor shaft. A small portion of the motor torque will be used to generate the direct current for field excitation.

Figures 1.37 and 1.38 show the excitation field and a commercial synchronous motor, respectively.

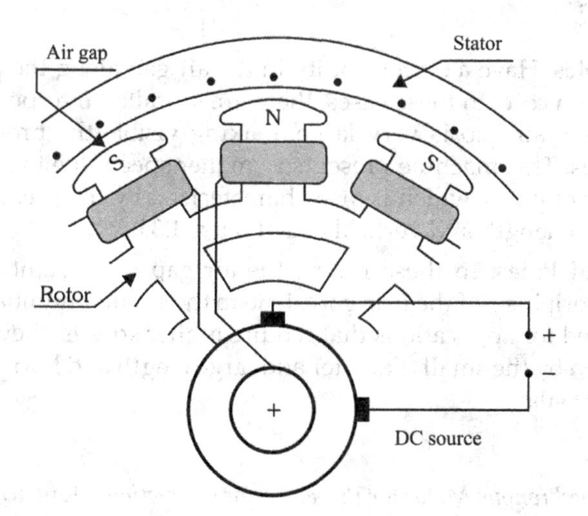

FIGURE 1.37
Excitation field in a synchronous motor.

FIGURE 1.38
Commercial synchronous motor. (Courtesy of Weg.)

According to the type of rotor, we can divide the synchronous motors into two categories:

Salient Poles: Have a discontinuity in the air gap along the periphery of the iron core. In these cases, there are so-called interpole regions, where the air gap is very large, making visible the protrusion of the poles. This machine presents a smaller speed due to the greater number of poles, which is also characterized by the great diameter and small length and vertical axis (Figure 1.39).

Cylindrical Poles: In these rotors, the air gap is constant along the entire periphery of the iron core. Due to the smaller number of poles employed for applications that require higher speeds, they are characterized by the small diameter and large length and horizontal axis (Figure 1.40).

1.1.2.3.6.1 Synchronous Motor for Power Factor Correction Due to the possibility of field excitation control, the synchronous motor has a characteristic that no other motor has, which is to allow the change of the field excitation and, consequently, a correction of the power factor.

FIGURE 1.39
Salient poles synchronous motor. (Courtesy of Weg.)

FIGURE 1.40
Cylindrical poles synchronous motor. (Courtesy of Weg.)

In a synchronous motor, when a load is applied, there is a phase angle shift of the rotor relative to the field. Although the motor speed remains synchronous, there is a lagging power factor under these conditions. In order for the motor to re-operate under unit power factor conditions, the DC excitation must be increased. If the current increase is maintained, the power factor is leading. It is possible to state that for a given load, the power factor is directly dependent on the DC excitation current.

It occurs because when the excitation current is reduced, the electromotive force induced in the stator is small, which causes the stator to absorb from the power supply a sufficient reactive power for the magnetic field creation, causing a low power factor. If the excitation current is increased for the same load, there will be an increase in the electromotive force on the stator, which causes the stator current, which was previously lagged, to be in phase with the power supply voltage, characterizing a unit power factor. If the increase of excitation current continues, then we have a lead stator current, which characterizes a leading power factor.

This behaviour can be seen in Figure 1.41 which shows that the power factor of a motor submitted to a given load, represented by the different curves, depends on its excitation current (I_F), where I_A represents the stator current.

Thus, the synchronous motor can be an alternative for the correction of power factor, compared to the traditional method that employs capacitors. However, some care must be taken regarding the type of load used, because variations in torque produce variations in the factor of power. It is also not advisable to use synchronous motors for power factor correction with powers below 50 HP.

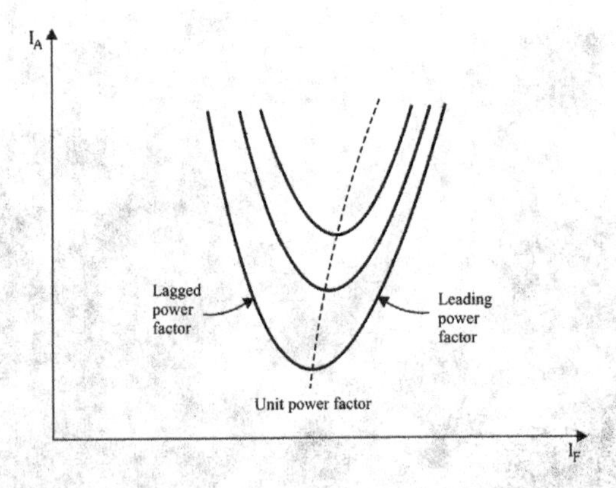

FIGURE 1.41
Power factor change as a function of excitation current change.

1.1.2.3.6.2 Advantages of Synchronous Motors in Relation to Induction Motors The following are some synchronous motors advantages in relation to induction motors:

- Besides being used to provide mechanical power, they have the characteristic of correcting the power factor.
- They have higher efficiency than equivalent induction motors when working with unit power factor.
- The rotors of the synchronous motors allow the use of larger air gaps, allowing smaller tolerances.

1.1.2.3.6.3 Disadvantages of Synchronous Motors in Relation to Induction Motors Due to its peculiarities, the synchronous motors application is very restricted, because as previously mentioned, they need DC excitation source, as well as constant maintenance.

A further disadvantage of the synchronous motor with respect to the induction motor is that it is not able to start only with the application of an alternating current in the stator, since it is necessary that the motor is brought to a sufficient speed close to the synchronous speed so that it synchronizes with the rotating field.

In order to start the motor, some techniques must be used. The use of a DC motor coupled to the motor shaft and the use of compensating windings (damper windings) are the most common.

When compensating windings, which may be of the squirrel cage or wound rotor type, are used, the DC field winding is short-circuited by applying supply voltage to the stator terminals, bringing the motor from no load to synchronism, as if it were an induction motor. After starting, the short-circuit connection is undone, with direct current being applied to the excitation circuit.

The motor can also be driven by means of a small induction motor or DC motor coupled to its shaft, so when the rotor approaches the synchronous speed, it is energized by a DC source, and, hence, the rotor goes on to follow the rotating field and is able to operate.

1.1.2.3.7 Reluctance Motor

The reluctance motor is a type of motor that induces non-permanent magnetic poles in a ferromagnetic rotor. The rotor of this motor has no windings and the torque is generated through magnetic reluctance.

Considering that the reluctance has the role of resistance to the circulation of a magnetic flux.

The operating principle of this type of motor is based on the principle that a metal part will move to complete the magnetic flux path with minimal

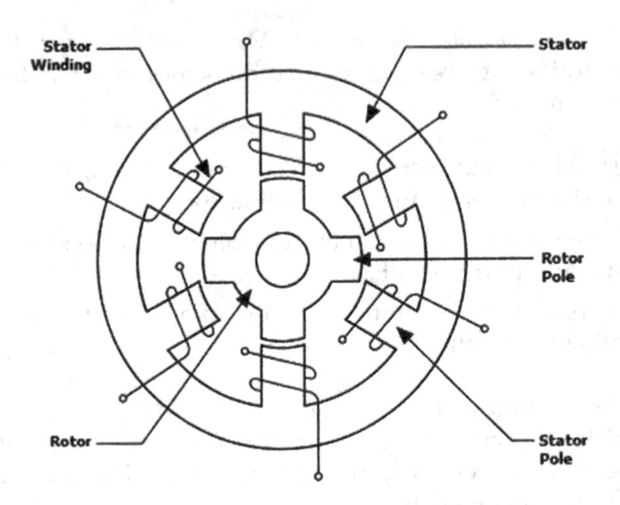

FIGURE 1.42
Reluctance motor.

reluctance and the reluctance is a function of the rotor position as shown in Figure 1.42.

There are several ways of controlling the synchronize reluctance events to rotor position. This machine requires a sophisticated control to avoid oscillations and acoustic noise. This machine also produces a relationship between torque and highly non-linear current and low power factor.

Main advantages are simple brushless construction, no commutator, a permanent magnet and rotor windings. It also features high starting torque, and lower cost for higher power density, typically recommended for higher power motors.

1.1.2.3.8 Permanent Magnet Motors

In this type of motor, permanent magnets are used to generate the magnetic field rather than windings in the rotor. The permanent magnet motors are developed with NdFeB magnets, a neodymium magnet the most widely used type of rare-earth magnet, is a permanent magnet made from an alloy of neodymium, iron and boron to form the $Nd_2Fe_{14}B$ structure. The permanent magnets are inserted inside the rotor, instead of conventional aluminium alloy. Figure 1.43 shows a typical permanent magnet motor.

As the Joule losses in the induction motor rotor are responsible for a significant portion of total losses in the motor, replacing squirrel cage induction motors with a permanent magnet motor ensures high efficiency levels when compared with a conventional induction motor. The use of a neodymium magnet to avoid circulating current through the rotor winding, making the motor operate cooler, results in a increase of its life time.

Table 1.2 shows the comparison of a 150 kW permanent magnet motor with a conventional squirrel cage induction motor in a compressor system.

FIGURE 1.43
Typical Permanent magnet motor.

TABLE 1.2
Test Values Comparison of 150 kW Permanent Magnet and Induction Motor in a Compressor System

Parameter	Permanent Magnet	Squirrel Cage Motor
Efficiency (%)	97.3	94.5
Weight (kg)	443	850
Frame	250S/M	280S/M
Current (A)	40	50

TABLE 1.3
Frame Comparison between Permanent Magnet and Squirrel Cage Induction Motors

Power (kW)	Permanent Magnet	Squirrel Cage Motor
22	160L	132M
30	200L	180M
37	200L	180L
75	280S/M	225S/M
110	315S/M	250S/M

AC induction motors are the most common solution found in motor-driven systems, and are larger and less efficient when compared with permanent magnet motors. Permanent magnet motors have a higher initial cost, however, they are offered in a smaller volume (approximately 43% smaller) and weight reduced up to 35%. Table 1.3 shows a comparison of typical permanent magnet and squirrel cage frames.

Permanent magnet motors are built at least one frame size smaller compared with squirrel cage induction motors and, in some cases, two frame sizes smaller.

NOTE: The reduction of motor size also allows a reduction in cooling fan size, ensuring a small noise level.

In this type of motor, the magnetic field is constant. These motors require a drive, like variable frequency drives, cannot be started directly from the grid, and are designed to operate in a wide speed range with constant torque and efficiency with no need of forced ventilation unit.

Despite all the advantages presented by the permanent magnet motor, cost is considerably higher than the squirrel cage induction motor, making, in most cases, its application prohibitive.

1.2 Motor Selection

For motor selection, several factors are crucial. The importance of these factors depends on the use to which the motor is subject and the available financial resources. The following are the main factors that should be taken into account in the process of selecting a motor:

- **Power supply**: Type (AC/DC), amplitude, voltage, frequency.
- **Environmental conditions**: Aggressiveness, dangerousness, altitude, temperature, etc.
- **Load requirements and service conditions**: Power, duty cycle, mechanical stresses, torque requirements, reliability, etc.
- **Consumption and maintenance**: Standard or high efficiency motors.
- **Controllability**: Position, torque, speed, starting current (according to load requirements).

Exercises

1. Despite its high cost, in which situations can DC motors be a good solution?
2. Describe the characteristics of the stator and rotor of a DC motor.
3. How is it possible to control the speed of a DC motor with independent excitation? Show with the help of equations and diagrams.
4. Why in the DC motor series configuration is it recommended that there is always load on the shaft?

5. The DC shunt excited motor is considered constant flux or constant speed. Why?
6. How is it possible to solve the low start torque in single-phase AC motors?
7. List the main characteristics torque characteristics and auxiliary windings of different types of single-phase induction motors.
8. What is the operating principle of a repulsion motor type?
9. Describe the main characteristics of a universal motor.
10. Why is the squirrel cage induction motor actually the most used in the industry?
11. When is the use of wound motor recommended?
12. Where is the brake motor applied? Describe, with the help of diagrams, the braking types of this motor.
13. What are the characteristics of high efficiency motors?
14. What is a synchronous motor? Describe the main characteristics and types.
15. How is possible to correct power factor using synchronous motors?
16. List the advantages and disadvantages of synchronous motors.
17. Describe the operating principle of a reluctance motor.
18. How is the magnetic field generated in the rotor in a permanent magnet motor? Present its advantages in relation to squirrel cage motors.
19. What factors should be considered when selecting a motor?

2

Three-Phase Motors

2.1 Introduction

The three-phase system is the most common method of power generation, transmission and distribution employed to transport electrical energy. This way, three-phase motors are the natural choice for use in industrial applications allied to the characteristics of cost, robustness and reliability of this type of motor. The main features of this type of motor will be presented below.

2.2 Three-Phase Induction Motor Construction

The squirrel cage induction motor consists, basically, of the following elements:

- **A static magnetic circuit**: Composed of ferromagnetic sheets stacked and isolated from each other, called a stator, where the motor frame is located; the structure also has a support function. It has a robust construction in cast iron, steel or is aluminium-injected, is corrosion resistant and has fins for cooling.
- **Coils**: The number of groups characterize the single-phase or polyphase motor. They are located in open cavities in the stator and are powered by the alternating current supply.
- **Rotor**: Composed of a ferromagnetic core, also laminated, on which is a winding, or a set, of parallel conductors, in which currents are induced by the alternating current of the stator coils.

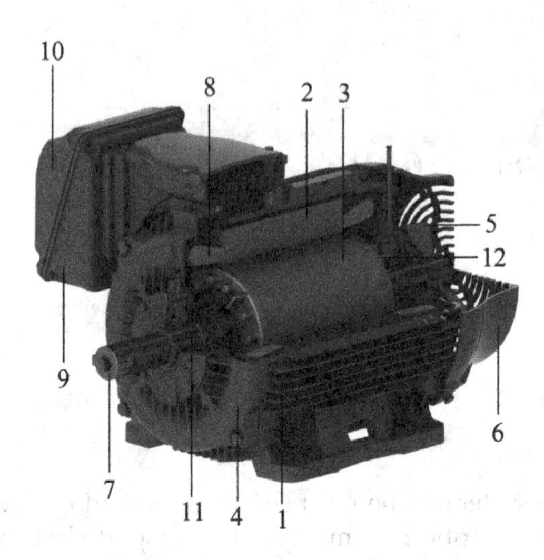

FIGURE 2.1
Squirrel cage induction motor. (Courtesy of Weg.)

The rotor is supported in a cavity which transmits the mechanical energy to the load. The air gap (distance between the rotor and the stator) is greatly reduced in order to decrease the current when there is no load in the motor. Figure 2.1 shows the main elements in an asynchronous motor with a squirrel cage rotor.

From Figure 2.1, it is possible to divide the motor into two main parts:

1. Stator
    - Frame (1)
    - Laminated metal plates (2)
    - Three-phase winding (8)
2. Rotor
    - Shaft (7)
    - Laminated metal plates (3)
    - End ring (12)
3. Other parts
    - Endshield (4)
    - Fan (5)
    - Fan cover (6)
    - Terminal Box (9)
    - Terminals (10)
    - Bearing (11)

When the motor is energised, it acts as a short circuit secondary transformer, so it requires a much larger current than the rated current from the power supply, which can reach about seven times the rated current value.

As the rotating field drags the rotor, increasing its speed, the current decreases until it reaches the rated current, at that time, the rotation reaches its rated value.

The fixed part of the motor is called a stator. In the motor frame, there is a core consisting of thin iron blades of about 0.5 mm, with slots that house the three-phase winding.

The phase windings and the stator core are responsible for generating the magnetic field. The motor speed is determined by the number of poles. When the motor is switched on at its rated frequency, the magnetic field speed is called the synchronous motor speed.

In order to better understand the three-phase induction motor operation, it is fundamental to know some electromagnetism concepts. In the following, the laws of Faraday and Lenz, essential for understanding the motor operation, will be presented.

2.3 Faraday's Law

To begin the Faraday's Law study, it is important to review some electromagnetism concepts.

A conductor carrying electric current (I) generates around it a magnetic field, as shown in Figure 2.2.

The magnetic field is proportional to the electric current, which the conductor is carrying, and the direction of the magnetic field is shown by the right-hand thumb rule, in which the thumb points towards the electric current direction (I), with the fingers wrapping around the conductor, indicating the direction of the magnetic field (B), as seen in Figure 2.3.

Thus, considering a solenoid, the magnetic field is shown Figure 2.4.

As illustrated earlier, a conductor carrying electric current always produces a magnetic field. The inverse situation also occurs, i.e., a magnetic field can also produce electric current. In 1831, Michael Faraday, in England, and Joseph Henry, in the United States, working independently, conducted the following experiment with a loop of electrically conductive material connected to a galvanometer. In this situation, no indication can be expected in the instrument, since there is no current source in the circuit, as shown in Figure 2.5.

Faraday's first experiment was the arrangement presented in Figure 2.6. If a magnet is moved towards a spiral conductor, the galvanometer indicates

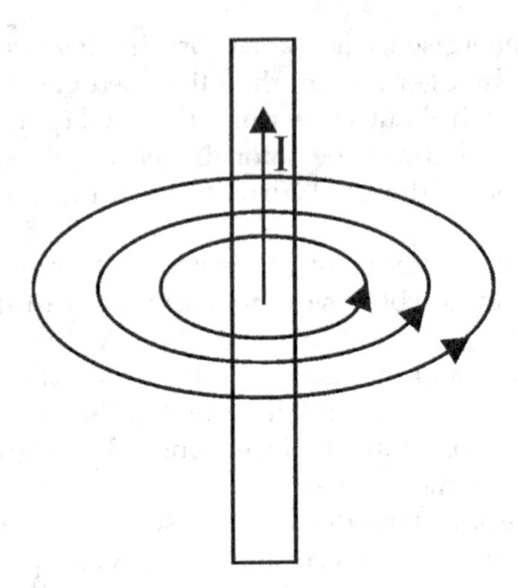

FIGURE 2.2
Magnetic field created in a conductor carrying electric current.

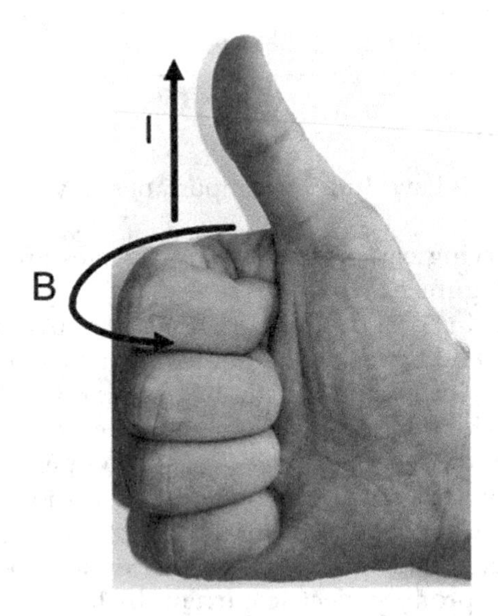

FIGURE 2.3
Right-hand thumb rule.

a current. If it is moved away, it also indicates current, but in the opposite direction, as Figure 2.6 illustrates.

So, it is possible to conclude that, with the magnet at rest, there is no indication—that is, there would only be indication if there was magnetic

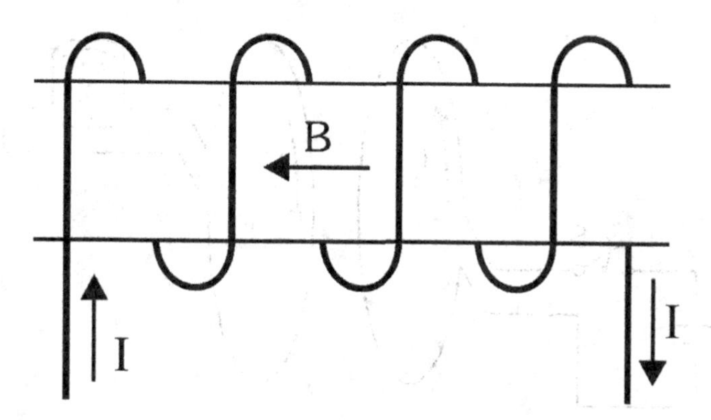

FIGURE 2.4
Magnetic field in a solenoid carrying current.

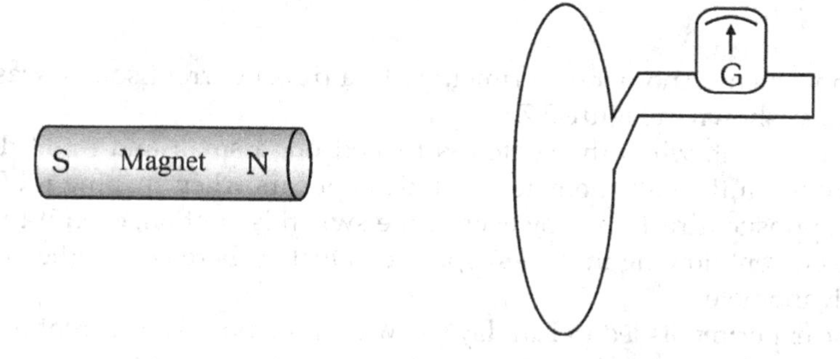

FIGURE 2.5
Galvanometer indication equal to zero when there is no variation in the magnetic field.

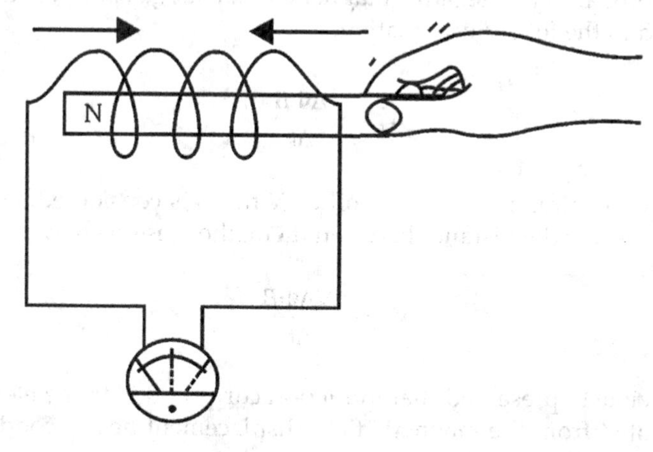

FIGURE 2.6
Current display in the galvanometer submitted to a magnetic field change.

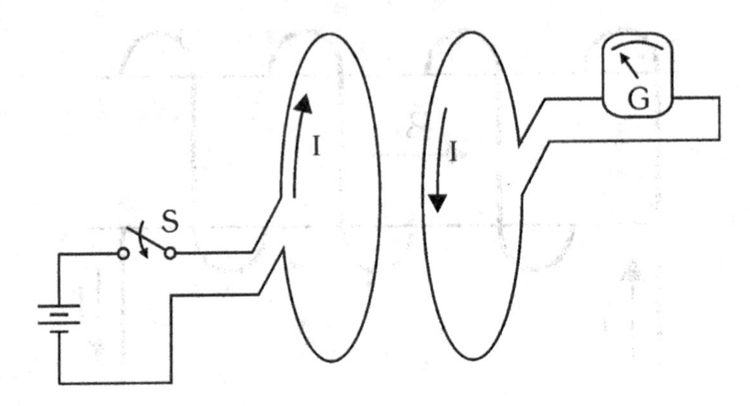

FIGURE 2.7
Direct current source experiment.

field movement. Another experiment with a direct current source was performed, as shown in Figure 2.7.

In this circuit, when the switch is turned on, a small and rapid deflection occurs in the galvanometer, and also appears when turning it off, but in the opposite direction. However, if the switch is kept on, even if there is a large current flowing in the left spiral conductor, there is no indication on the galvanometer.

These experiments led to Faraday's Law of Induction. The current flowing through the loop with the galvanometer is named induced current, which is produced by an electromotive force induced (V_e). Faraday concluded that it is proportional to the negative magnetic flux change (ΦB) over time, as is represented in the following equation.

$$V_e = -\frac{\Delta \Phi B}{\Delta t} \tag{2.1}$$

If, instead of a spiral conductor, a coil of N turns is considered sufficiently compact to negate the distance between them, the V_e is given by:

$$V_e = -\frac{N \Delta \Phi B}{\Delta t} \tag{2.2}$$

It was previously presented that induction corresponds to the electric current generated from the magnetic field displacement near a conductor, or vice versa.

When an electric current is flowing in a coil, a magnetic field is generated. If the electric current is variable, the magnetic field will also be variable. Thus, movement of the magnetic field is relative to the conductor.

If next to this coil (first coil or inductor coil) there is a second coil, it will also be cut by the lines of force. As a consequence, a voltage, (Figures 2.8 and 2.9) known as induced voltage, appears in the second coil, and its magnitude depends on:

- Voltage applied in the inductive coil.
- Number of turns of the primary winding.
- Number of turns of the secondary winding.

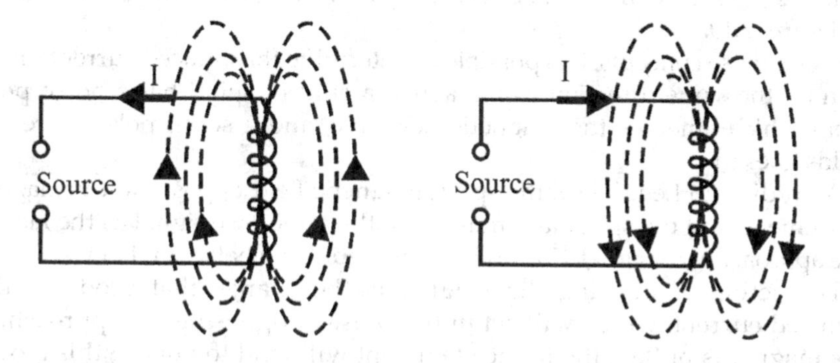

FIGURE 2.8
Magnetic field produced by flowing electric current

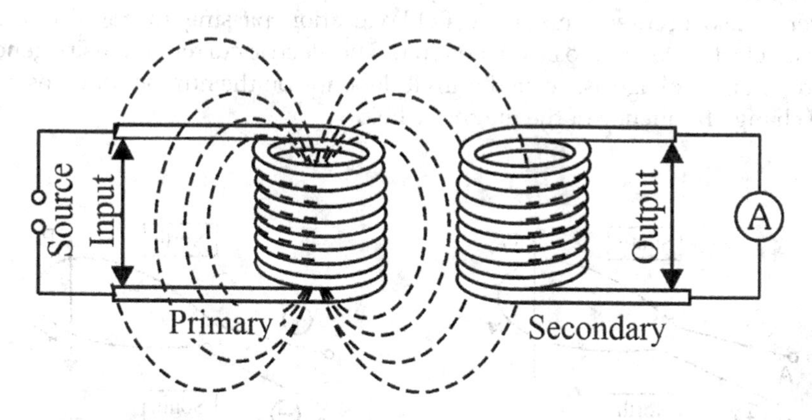

FIGURE 2.9
Voltage induced on the secondary winding near to the primary winding.

2.4 Lenz's Law

This law states that the "electric current induced by a source such as a changing magnetic field always creates a counterforce opposing the force inducing it." In this law, the sense of induced current is clearly defined.

As was previously said, a conductor carrying electric current produces a magnetic field around it, as illustrated in Figure 2.2. The magnetic field direction can be determined by the "right hand thumb rule": holding the wire with the right hand at a position where the thumb points to the current that results in a magnetic field (*B*) with the direction coinciding with the position of the fingers around the wire. Thus, if we have an electromagnetic field, its polarity will alternate according to the current direction, as shown in Figure 2.10.

Using these concepts, it is possible to determine the induced current direction by the same direction as a magnet: a current spiral has a north pole, from which emerges the magnetic field line, and a south pole where the fields goes in.

According to Lenz's law, the spiral magnetic field opposes to the magnet movement, and the spiral face must have the same pole signal as the face of the approaching magnet. So, it will create repulsion between them.

The action of pushing the magnet is the change that produces the induced current, which will act in the sense of opposing the approach. If the magnet is pulled, the induced current will tend to oppose this movement, creating a south pole to attract the magnet. Figures 2.11 and 2.12 shows how it happens.

We saw that a conductor carrying current produces a magnetic field. The inverse also occurs: a magnetic field variation passing through a spiral causes electric current to be created in it. The electric current intensity generated and the voltage associated with it depends on the number of turns and the change frequency of the magnetic field.

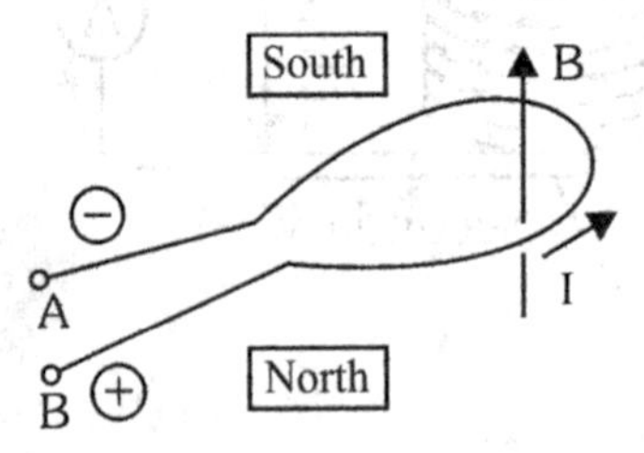

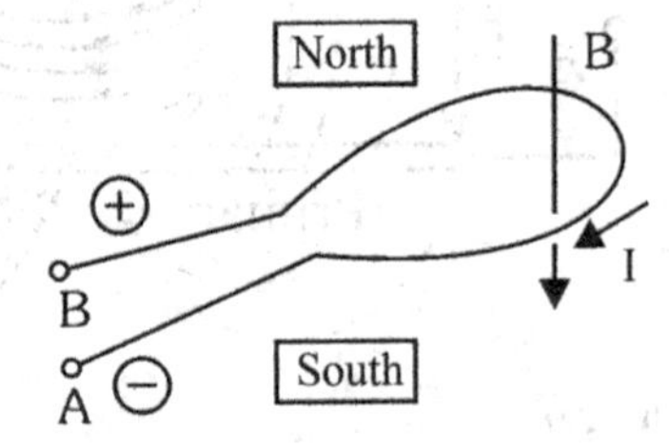

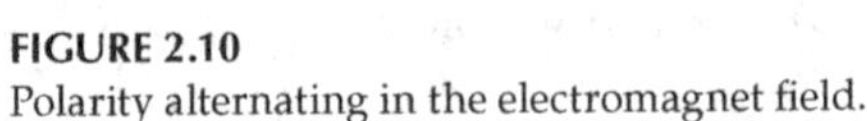
FIGURE 2.10
Polarity alternating in the electromagnet field.

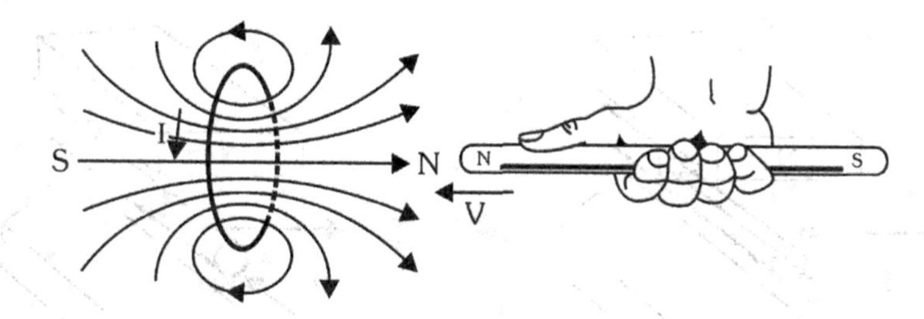

FIGURE 2.11
Magnet field direction induced with the magnet approach.

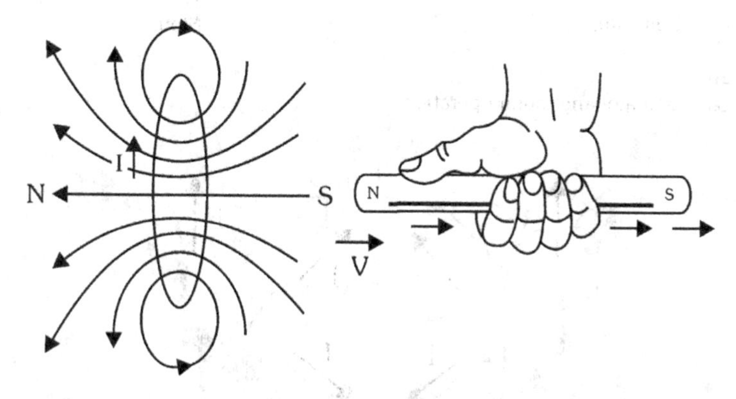

FIGURE 2.12
Magnetic field direction induced with magnet spacing.

2.5 Operation Principle of a Three-Phase Induction Motor

The principle of an induction motor operation is based on Faraday's law of electromagnetic induction, which states that if a conductor is moved through a magnetic field (B), a voltage is induced. If the conductor is a closed circuit, a current (I) will circulate. When the conductor is moved, a force (F), which is perpendicular to the magnetic field, will act on the conductor. This is the electric generator principle.

In the motor, the induction principle acts in an inverted way—that is, a conductor carrying current is positioned within a magnetic field. The conductor is then influenced by a force (F) that moves it out of the magnetic field. The following Figure 2.13 illustrates these principles.

The current direction can be determined by the left-hand rule for the generator and the right-hand rule for the motor, as shown in Figure 2.14.

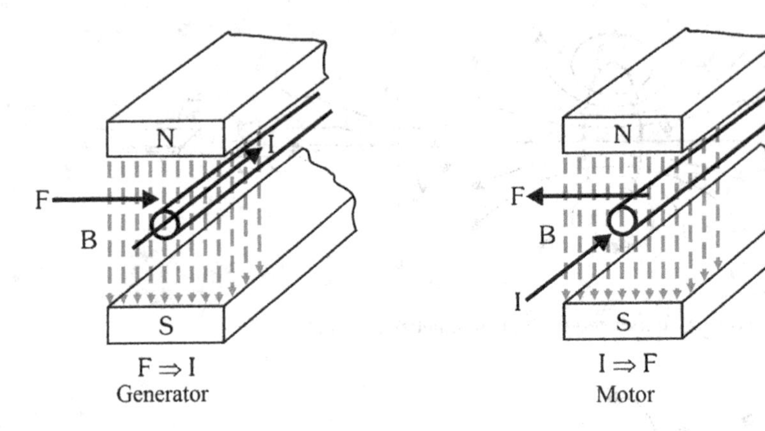

FIGURE 2.13
Principles of generator and motor operation.

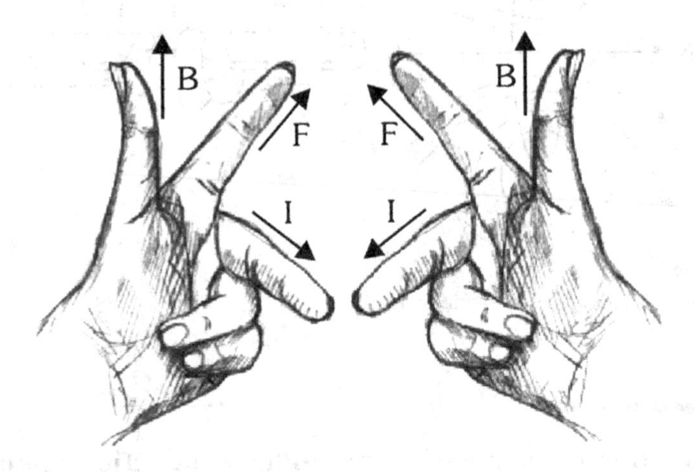

FIGURE 2.14
Left-hand and right-hand rules used to determine the direction of current in the generator and motor, respectively.

The magnetic field is generated in the stationary part (stator), and the conductors are influenced by the electromagnetic forces that are in the rotating part (rotor), as shown in Figure 2.15.

Figure 2.15a shows a single-phase winding carrying a current I_1, and the field H is created by it. The winding consists of a pair of poles (north and south) whose effects add up to establish field H. The magnetic flux crosses the rotor between the two poles and closes through the stator core.

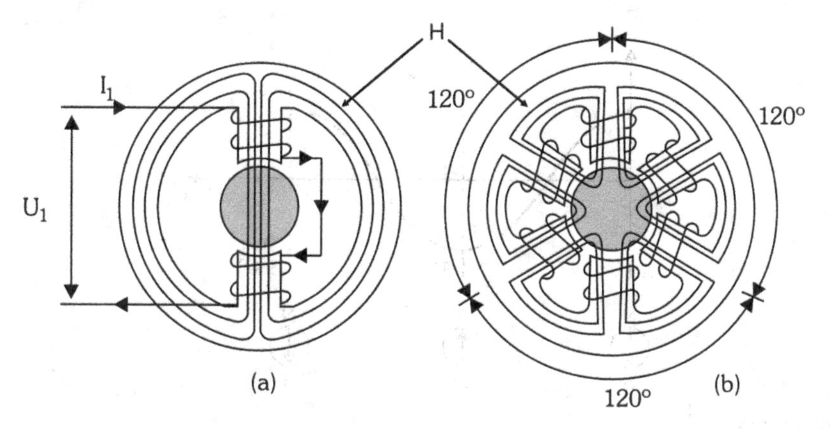

FIGURE 2.15
Single-phase motor winding (a) and three-phase motor winding (b).

If the current I_1 is alternating, the field H is also; its value, at each instant, is represented by a graph similar to Figure 2.16, including inverting the direction in every half cycle.

The H field becomes alternating because its intensity changes proportionally to the current, always in the north-south direction, as shown in Figure 2.17.

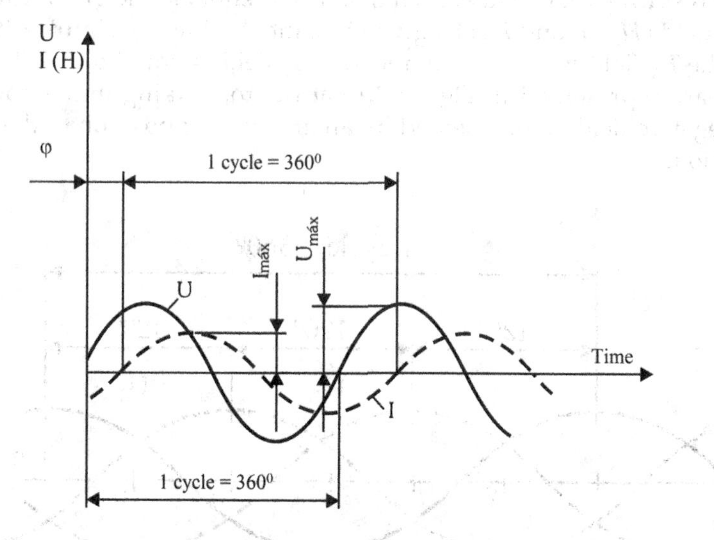

FIGURE 2.16
Alternating current and voltage in a single-phase circuit.

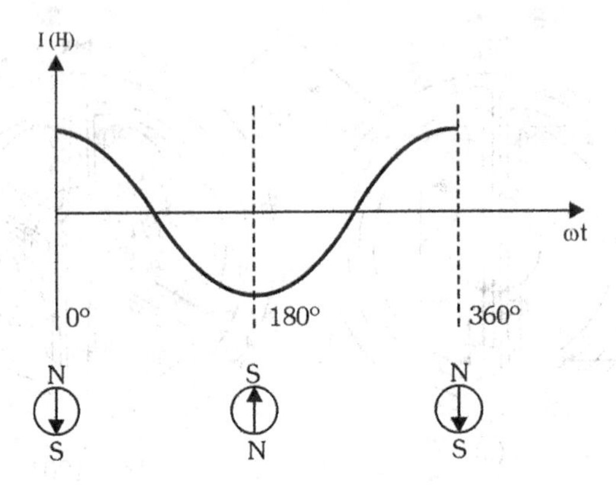

FIGURE 2.17
Alternating magnetic field.

Figure 2.15b shows a three-phase winding made with three single-phase windings out of phase spaced from each other at 120°. If this winding is connected in a three-phase system (see Figure 2.18), the currents I_1, I_2, I_3 will, likewise, create their own magnetic fields H_1, H_2, H_3, these fields being also spaced 120°. Moreover, as they are proportional to the respective currents, the magnetic fields become lagged at time 120° to each other as shown in Figure 2.19.

The resulting total field at each instance is equal to the graph sum of the three fields H_1, H_2 and H_3 at a given instant. At time (1), Figure 2.19 shows that the H_1 field is maximum and the H_2 field is equal to 0.5. The three fields are represented in Figure 2.19 at the top, taking into account that the negative field is represented by an arrow in opposition to the normal direction.

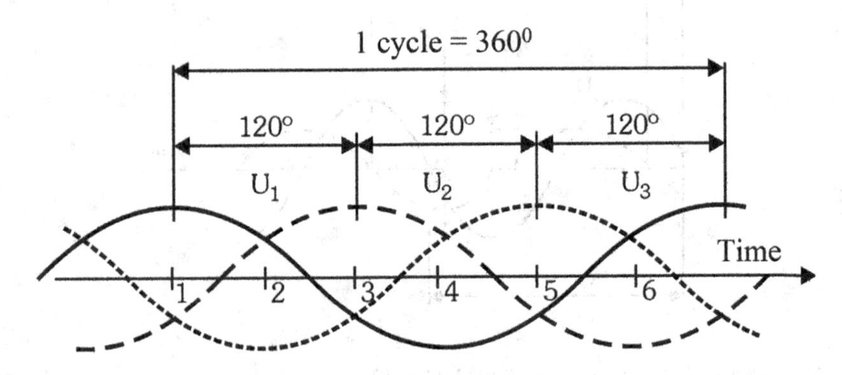

FIGURE 2.18
Current and voltage in a three-phase circuit.

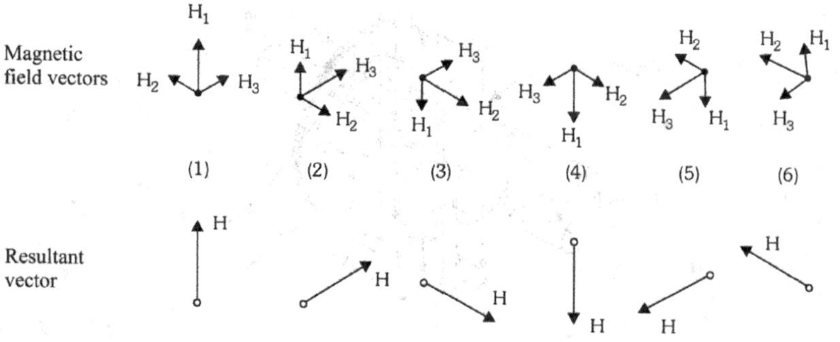

FIGURE 2.19
Components and resulting magnetic fields in a three-phase circuit.

It is possible to observe that the resulting field *H* has constant intensity, but its direction rotates, completing one turn at the end of the cycle.

Thus, when a three-phase winding is fed by three-phase currents, a rotating field is created as if there were a single pair of rotating poles with constant intensity. This rotating field, created by the three-phase winding of the stator, induces voltages in the rotor bars (flow lines cut rotor bars) which generates currents and, consequently, magnetic fields in the rotor with polarity opposite that of the rotating field. As a result, the opposing fields attract and the rotating stator field rotates: the rotor tends to follow the rotation of that field, developing a torque in the motor that causes the rotor to move.

In the three-phase induction motor, the stator and the rotor are made up of spaced ferrous silicon blades. As was presented in Chapter 1, there are two main types of rotor: wounded and short-circuited.

The rotors with blades in short-circuit are called squirrel cages, and are the most commonly used in the industry. This type of rotor has aluminium bars that pass through the slots with an aluminium ring placed at each end of the rotor to short circuit the bars.

In this way, when a rotor bar is placed in a rotating field, the magnetic field passes through the bar and induces a current (I_W) in the rotor that is affected only by force (F), as illustrated in Figure 2.20.

The resulting force is proportional to the flux density (B), the induced current (I_W), the rotor size (l) and the angle (θ) between the force and the flux density. In this way we will have:

$$F = B \cdot Iw \cdot l \cdot \text{sen}(\theta) \quad (2.3)$$

Considering $\theta = 90°$, we have:

$$F = B \cdot l \cdot Iw \quad (2.4)$$

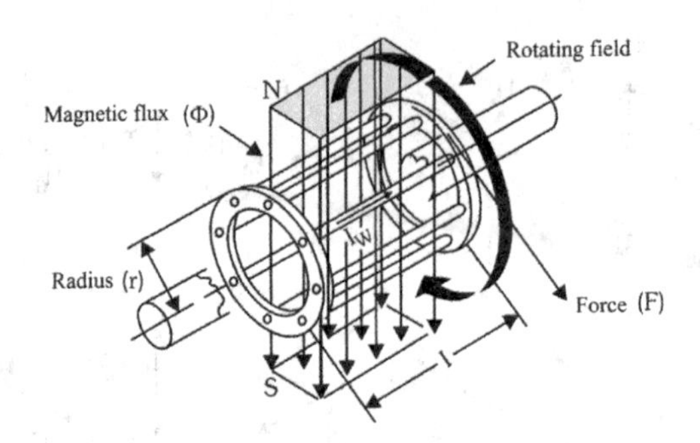

FIGURE 2.20
Rotating field and short-circuited rotor.

The next pole that will pass to the rotor has the polarity reversed. This induces a current in the opposite direction. Since the direction of the magnetic field has also changed, the force acts in the same direction as before (see Figure 2.21).

When the entire rotor is placed in the rotating field, the rotor speed will not reach the rotating field speed, since at the same speed, no current would be induced in the rotor.

To illustrate the induction motor operation principle, consider the device shown in Figure 2.22. In the figure, a permanent magnet is suspended by a wire on an aluminium disk in a bearing support on a fixed iron plate. The lines of force that constitute the magnetic field will be completed through the plate. For this experiment, we consider the pivot with a zero friction. As

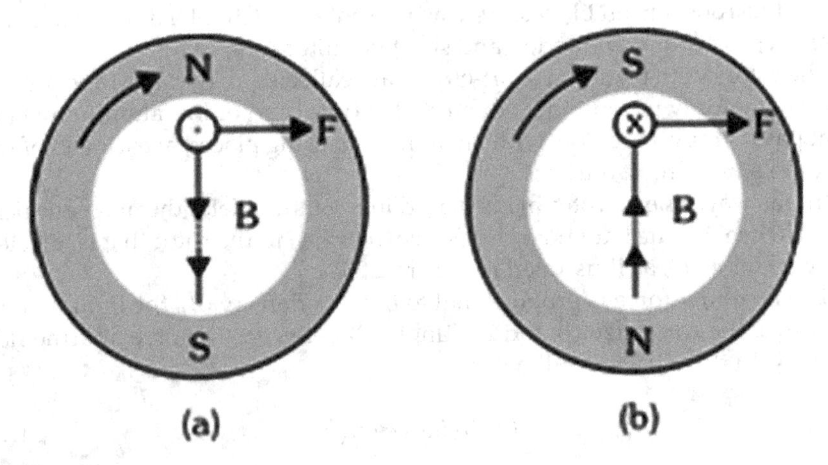

FIGURE 2.21
Reverse polarity in the rotor (right-hand rule). (a) Current direction "entering" in the conductor, (b) Current direction "coming out" in the conductor.

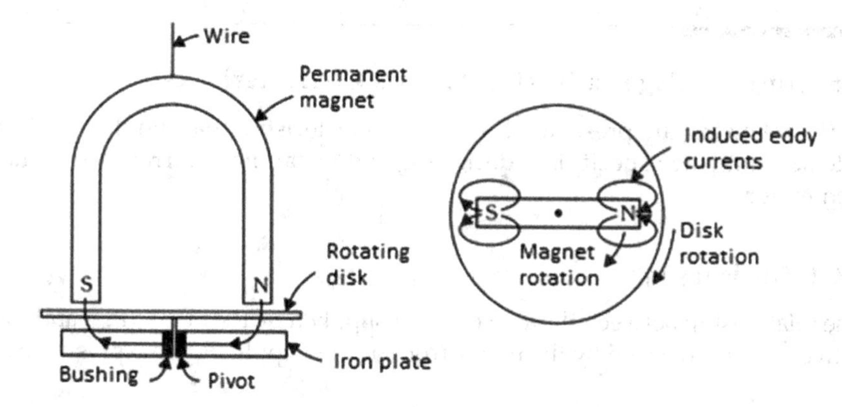

FIGURE 2.22
Experiment that illustrates the principle of operation of the induction motor.

the magnet rotates, the disk below it rotates following the magnet rotation direction.

In this way, the disk will follow the movement of the magnet due to the induced parasitic (eddy) currents that appear due to the relative movement of the disk relative to the magnetic field. According to Lenz's Law, the induced voltage direction and its parasitic currents produces a field that will tend to oppose the force (motion) that produced the induced voltage.

The eddy parasitic currents, which have been induced on the disk, will produce a pole S on the disk at a point situated under the rotating pole N of the magnet and a pole N on the disk under the rotating pole S of the magnet. Thus, whenever the magnet is in motion, it will continue to produce eddy currents and poles of polarity opposite the element (magnet) that created them. In this way, the disk will rotate in the same direction as the magnet, but should rotate at a slower speed than the magnet.

If the disk were to be driven at the same magnet speed, there would be no relative movement between the conductor and the magnetic field and no eddy currents would be produced on the disk. Thus, the rotational speed of the disc will never be equal to that of the magnet.

There is a speed difference between the magnet and the disk movement, called slip, which is essential for the movement and torque production in the motor. Every induction motor has different speeds produced between the synchronous speed of the rotating magnetic field and the rotor speed.

In order to better understand the induction motor, we will present the main characteristics of three-phase squirrel cage induction motors.

2.6 Squirrel Cage Induction Motors Characteristics

In the following are presented the main characteristics that should be taken into account when specifying, designing and installing squirrel cage induction motors.

2.6.1 Efficiency (η)

The relationship between the active power supplied by the motor (P_{am}) and the active power requested by the motor from power supply (P_{as}) is expressed by:

$$\eta = \frac{P_{am}}{P_{as}} \tag{2.5}$$

As load is applied to the motor, the efficiency increases and can reach 96% on the high-power motors. We must take into account two curves to analyse the motor efficiency.

2.6.1.1 Motor Efficiency Based on Rated Power

Due to characteristics relating to iron mass and air gap thickness, with low loads, the three-phase induction motor has a considerable armature current. Because the conductors have their own electrical resistances, this leads to considerable values for the copper losses of the armature winding.

When small loads are applied, the rotor has very low slip, causing the currents induced in the rotor winding to be small and, consequently, small localised losses are produced.

The rotor iron mass, although appreciable, when under small slips, leads to small losses. Therefore, it is possible to conclude that, when there is no load in the machine, the present losses are due only to the stator: iron core and winding.

If the machine-rated power is small, comparatively, it has high losses, leading to relatively smaller efficiency. In general, the efficiency (η) increases when the rated power (P_R) increases, as we can see in Figure 2.23.

Table 2.1 shows the motor efficiency for different rated power motors.

2.6.1.2 Motor Efficiency in Function on the Shaft Load

As load is applied to the motor shaft, an increase in its efficiency will occur. Thus, the closer to the rated load, the greater the motor efficiency. Figure 2.24 shows the efficiency x load (L(%)) on the shaft curve for a 5 HP and 3470 rpm three-phase induction motor.

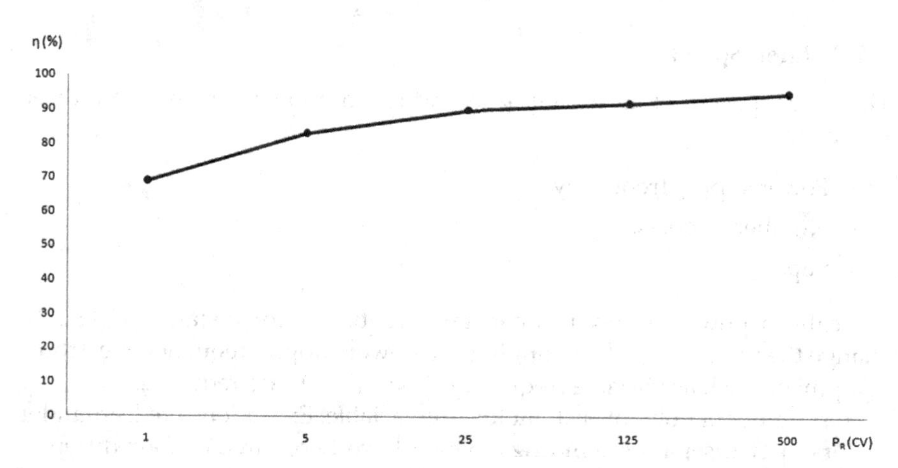

FIGURE 2.23
Motor efficiency based on rated power.

TABLE 2.1

Efficiency as a Function of Rated Power

Rated Power (HP)	Speed (rpm)	Efficiency (%)
1	1705	69
5	1730	83
25	1750	90
125	1770	92
500	1785	95

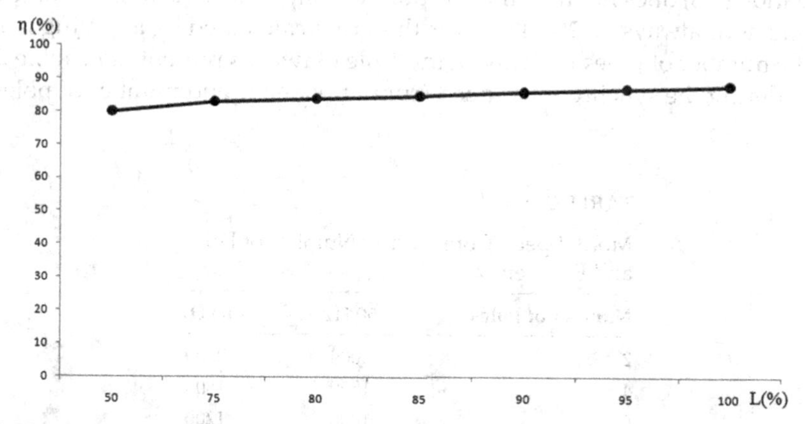

FIGURE 2.24
Efficiency as a function of the load applied to the motor.

2.6.2 Rated Speed

The rated speed of a three-phase induction motor depends on three parameters:

- Power supply frequency.
- Number of poles.
- Slip.

Typically, a power supply has constant frequency, for instance 60 Hz. To change the motor speed by changing the power supply frequency, electronic equipment, such as variable frequency drives (VFD), are required.

The number of poles of each motor is invariable. Exceptions are Dahlander motors or two separate windings, which have two, three or four different pole numbers. These types of motor can be connected through appropriate electrical circuits to run at different speeds.

The following equation shows the relationship between the supply frequency, the number of poles and the speed:

$$n = \frac{120.f}{P} \tag{2.6}$$

where:

n is the speed in rpm (revolutions per minute)

f is the power supply frequency in Hz

P is the number of motor poles

Equation (2.6) does not take into account the slip. Thus, the number of revolutions will always be 2%–4% lower than that calculated by applying load.

The number of poles is always a multiple of two, as presented in Table 2.2, considering the synchronous speed (rpm), frequency and number of poles.

TABLE 2.2

Motor Speed Considering Number of Poles and Frequency

Number of Poles	50 Hz	60 Hz
2	3000	3600
4	1500	1800
6	1000	1200
8	750	900
10	600	720

The use of motors with more than six poles is quite rare, and their acquisition must be made to specific order. Due to production of a very small number, the larger the number of poles of the motor, the greater its cost.

2.6.3 Slip

If the motor rotates at a speed that is different than the synchronous speed—i.e., is, different from the rotating field speed—the rotor windings cut off the lines of magnetic field and, by the laws of electromagnetism, flow induced currents.

The higher the load, the greater the torque needed to drive it. To obtain the torque, the speed difference must be greater so that the induced currents and the magnetic fields produced are larger. Therefore, as the load increases, the motor speed drops. Figure 2.25 shows synchronous and motor speed.

When the motor load is zero, the rotor rotates almost with synchronous rotation.

Another important feature that we should take into account is that the slip decreases as the rated power of the motor increases. For example, a 10 cv four pole motor has a slip of 2.78%, while a 500 cv motor with the same number of poles has a slip of 0.83%.

The difference between the motor speed (n) and the synchronous speed (n_s), slip (S), can be expressed in rpm, as a fraction of the synchronous speed or as a percentage of this.

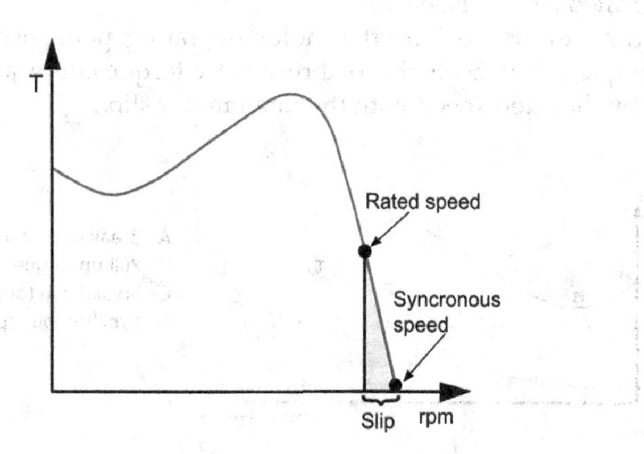

FIGURE 2.25
Synchronous and motor speed.

$$S(\text{rpm}) = n_s - n \tag{2.7}$$

$$S(\%) = \frac{100 * (n_s - n)}{n_s} \tag{2.8}$$

Example: What is the slip of a 60 Hz four-pole motor if its rated speed is 1730 rpm?

$$S(\%) = \frac{100 * (1800 - 1730)}{1800} = 3.88\%$$

2.6.4 Torque Speed Design

According to the mechanical load in the shaft, there is a torque requirements curve. In ventilation loads, for example, the resistant torque is proportional to the square of the speed, whereas, in cranes, hoists and overhead cranes, the resistant torque is practically constant.

The curve in Figure 2.26 shows how the motor output torque T_{Motor} changes when the motor runs from stop to rated speed under a constant supply voltage and frequency. The dashed line shows the mechanical load torque requirements.

Upon starting, the motor will run when the starting torque exceeds the load breakaway torque by applying a *breakaway torque*, which is defined as the torque required to start motor rotation from a stationary position. The motor will accelerate if the motor torque exceeds the load torque, and, as the speed increases, the motor torque will increase to a maximum located at point C called a *breakdown torque.*

According to Figure 2.26, the motor operating point occurs where the motor torque curve meets the load resistance torque curve, and this point is defined by the rated speed with the rated motor slip.

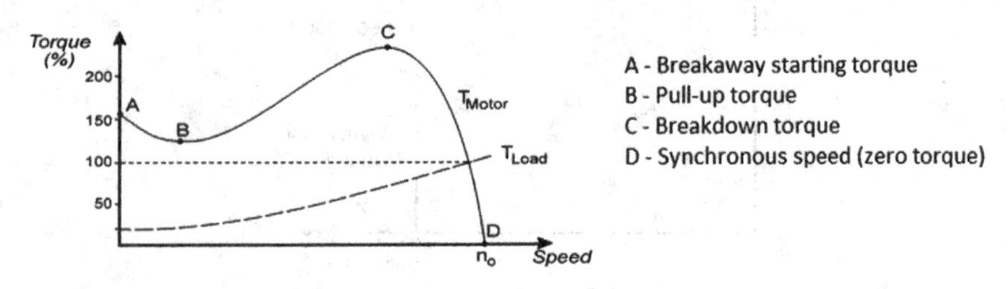

FIGURE 2.26
Typical Speed torque curve for a AC Three phase induction motor

When the load torque increases, the motor speed drops slightly, slip increases, stator current increases and the motor torque increases to reach the load torque requirements. Therefore, the range between C and D, on the torque–speed curve is considered stable operating range for the motor, as seen in Figure 2.26.

When the load torque increases to a point beyond breakdown torque, the motor will stall because it enters the unstable torque curve portion (ABC) and any increase in load torque requirements will promote a continuous reduction in drive speed and torque.

The *pull-up torque* is the minimum torque developed by the motor as it accelerates from standstill to the speed at which breakdown torque occurs.

In order to accelerate the motor, it must have enough torque to overcome the inertia of the load. The accelerating torque can be considered as the difference between the available motor torque and the load resistance torque curve. The load resistance torque curve may have different profiles that depend on each application.

NOTE: Even motors with the same nominal HP may have different characteristics, such as start current, torque curves, speeds and other parameters.

The National Electrical Manufacturers Association (IEC), a National Electrical Manufacturers Association (NEMA) standard, classifies squirrel cage motors into categories according to the characteristics of torque in relation to the speed and current and slip. The main categories are:

Design N: Also called B by the NEMA standard, has normal starting torque, normal starting current and low slip. Most of the motors found in the market fit into this category. It is used to drive normal loads with low starting torque, such as pumps, machine tools and fans.

Design H: Also called C by the NEMA standard, features high starting torque, normal starting current and small slip. It is used for loads requiring larger starting torque, such as loaded conveyors and mills.

Design D: Also called D by the NEMA standard, has high starting torque, normal starting current and high slip (>5%). It is used in presses and similar machines, where the load has periodic peaks, and in elevators, where the load needs high starting torque.

Figure 2.27 shows the characteristic curves of torque by speed for the three categories:

Each design will have a defined geometric shape, as shown in Figure 2.28.

Design N has a simple geometric shape with a circular design, while in design H, the geometric drawing is more complex, the cage can practically

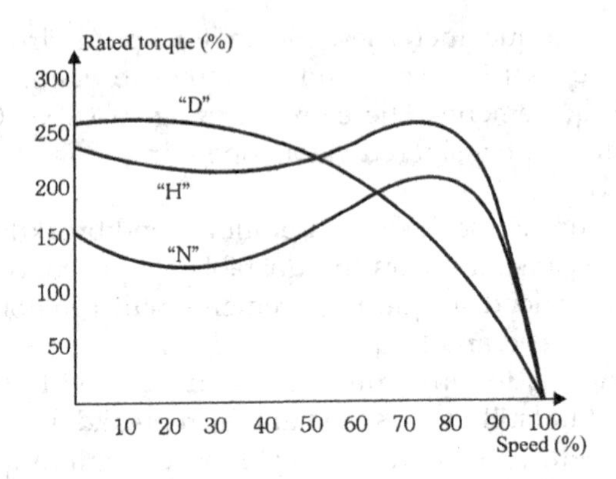

FIGURE 2.27
Torque speed designs according to NEMA and IEC standards.

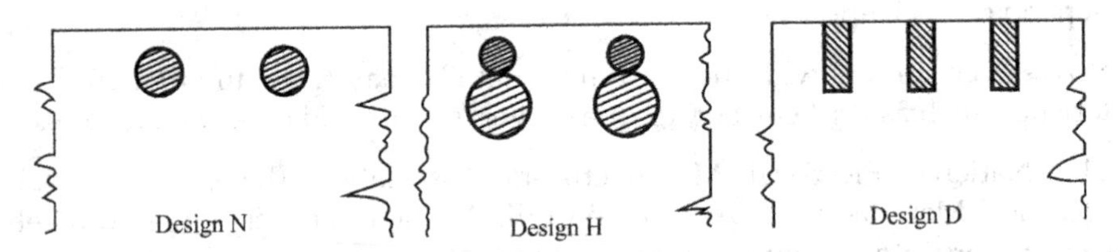

FIGURE 2.28
Geometric shapes of rods of a rotor squirrel cage.

be divided into two, with one of smaller sections more superficial and one of larger sections straight. Thus, the induced currents in the departure go to the surface cage, and since this cage has a smaller straight section, its electrical resistance is larger, so the starting torque is larger than the N category, because the eddy currents are limited by the high resistance of the outer cage. As the motor accelerates, the currents, under the effect of the dynamic characteristics of the circuit, distribute themselves on the total surface of the bar.

In design D motors, the squirrel cage bars are narrow and begin in the periphery of the rotor, deepening gradually. These cages have high electrical resistance. The large heat dissipation in the rotor rod resistors requires a larger ventilation capacity for the motor if it runs continuously.

In addition to these four classes, NEMA also has the design classes E and F, which are called soft-start induction motors, which present very low starting currents and are used in loads that require small starting torques in situations where starting currents are a problem. Currently, these design classes are considered obsolete.

2.6.5 Insulation Class

Every conductor carrying electric current dissipates energy in the form of heat by means of the Joule effect, known by the following equation.

$$P = R.I^2 \tag{2.9}$$

where:
P is the electric power (watts)
R is the electric resistance (ohms)
I is the electric current (amperes)

The stator and rotor windings, or their bars, in the case of a squirrel cage rotor, dissipate heat. The variable flux acting on the magnetic stator-rotor core also induces undesirable currents in the steel sheets, since they are also conductive material. These currents are called parasites, or eddy currents. They are also a source of heating by Joule effect.

The stator and rotor cores are laminated, i.e. made of very thin sheets, precisely to minimize these currents. Therefore, the blades are made in the flow direction to reduce the area subject to induction and increase the electrical resistance of the parasitic circuit. Magnetic hysteresis, due to alternating flux, is also a source of internal motor heat generation.

The main motor task is to provide mechanical work by converting electric power. Thus, heat is a form of energy that is not employed to produce mechanical work, being considered lost energy, and the higher the losses, the lower is the motor efficiency.

Typically, three-phase induction motors for normal application are fully enclosed, and the internally generated heat ends up promoting an internal temperature rise. Due to the temperature difference established between the interior of the motor and the outside environment, a heat transfer process occurs.

The heat generated internally is dissipated by means of the frame, which is finned to make easy the heat exchange with the environment, as shown in Figure 2.29.

Behind the rear cover of the frame, there is a fan driven by the motor shaft itself, which has the purpose of increasing air circulation by the cooling fins. A steel or cast iron fan cover serves to direct the airflow. This is the fully enclosed motor configuration with external ventilation. There are also motors that are open and with internal ventilation.

The winding constitutes the most critical part of the motor from the thermal point of view, where the winding wires are insulated. The wires are covered with synthetic enamel of an organic nature, and a reinforcement of the insulation by the impregnation of the varnish is placed on the winding.

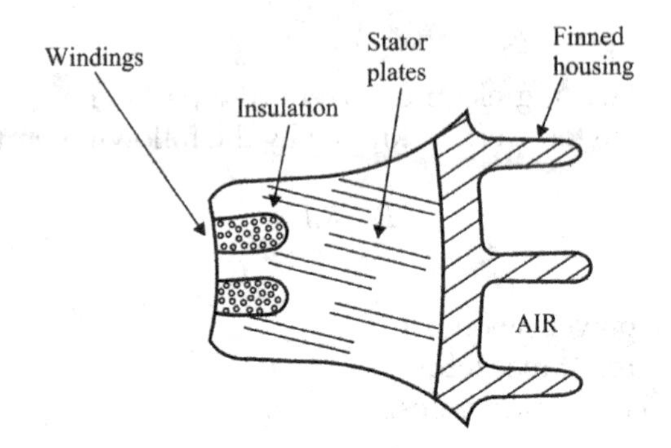

FIGURE 2.29
Heat exchange of motor windings.

If the windings temperature exceeds a certain value, the insulating coating material of the wires ends up burning, placing it in a short circuit condition.

If the motor operates within its temperature limits, the insulation must have an unlimited life. If the winding temperature rises, even if insufficient to cause the burn, it contributes to the degradation of the insulation until it is lost at some point.

The highest motor temperature is inside the slot where the windings are located. Varnish impregnation and compaction of the coils must be perfect, because the air is a bad heat conductor and the existence of empty spaces makes difficult the heat extraction and the temperature increases, damaging the winding life. A warmer point inside the winding may compromise the motor even though the average winding temperature as a whole remains below the maximum limit.

2.6.5.1 Temperature Rise

The temperature is a key point to be considered in motor selection, i.e. it must be ensured that the temperature in the windings is due to a load connection that does not exceed a critical temperature allowed by the windings, that the insulation stator and rotor can withstand without permanent damage.

In electrical machines, designers want to limit the rise in temperature in the winding, considering ambient temperature. This total difference (Δt) is called the *temperature rise* of the motor, as shown in Figure 2.30.

Temperature rise could be obtained by the equation:

$$\Delta t = (t_{\max} - t_{\text{ext}}) - (t_{\text{ext}} - t_a) \tag{2.10}$$

Where:

$t_{\max}$ = maximum temperature in windings
t_{ext} = temperature in the external surface
t_a = ambient temperature

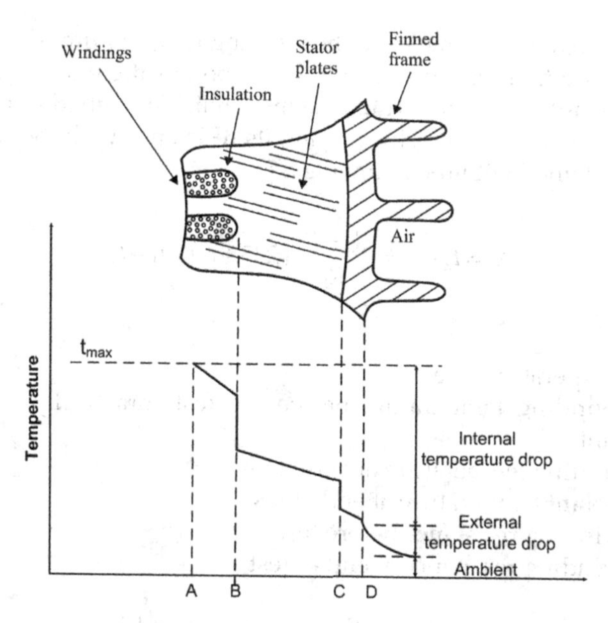

FIGURE 2.30
Temperature rise in a motor frame.

In order to increase the motor efficiency, it is important to reduce the internal temperature drop (heat transfer by conduction) to promote the greatest possible external heat transfer.

The internal temperature drop depends on the temperature at different points of the motor, described in Table 2.3, and previously presented in Figure 2.30.

TABLE 2.3
Temperature Points in an Electrical Motor

Point	Description
A	Hottest point of the winding located inside the slots, where the heat from the losses in the conductors is generated.
AB	Temperature drop resulting from heat transfer from the hottest point to the leading edge of the wire.
B	Insulation contact point of the slot with the conductors on one side and with the plates of the core on the other side.
BC	Temperature drop by conduction through the core plates.
C	Temperature drop between the core and the frame.
CD	Temperature drop by conduction through the frame thickness.

It is not an easy task to measure the winding temperature of a motor with sensors, because there is considerable variation at different points, and it is difficult to identify hot spots. Thus, the most reliable method of temperature measurement of a winding is by varying its resistance with the temperature that can be obtained with the following equation:

$$\Delta t = t_2 - t_a = \frac{R_2 - R_1}{R_1}(235 + t_1) + t_1 - t_a \tag{2.11}$$

where:

- Δt is the temperature rise
- t_1 is the winding temperature before the test (practically equal to the coolant)
- t_2 is the winding temperature at end of test
- t_a is the coolant temperature at end of test
- R_1 is the winding resistance before test
- R_2 is the winding resistance at end of test

The temperature rise limits for the materials used in the motors insulation are based on standards, and the insulation class is defined internationally by NEMA and IEC 34.1. These standards specify the maximum possible temperatures for the various insulation classes. A safe working temperature is the sum of the specified maximum ambient temperature and the maximum permissible temperature rise due to a load coupled to the motor.

The maximum critical temperatures and the maximum temperature rise for rotary electric machines for the main insulation class are shown in Table 2.4.

According to the standards, motors for normal application are installed in maximum ambient temperatures of 40°C. Above this temperature, working conditions are considered special. The maximum temperature in the slots is what is allowed by the class, subtracted at ambient temperature. In addition, the temperature is never uniform in the winding.

TABLE 2.4

Maximum Critical Temperatures and Maximum Temperature Rise for Electric Machines

Insulation Class	**A**	**E**	**B**	**F**	**H**
Maximum temperature (°C)	105	120	130	155	180
Maximum temperature rise (°C)	60	70	80	100	125

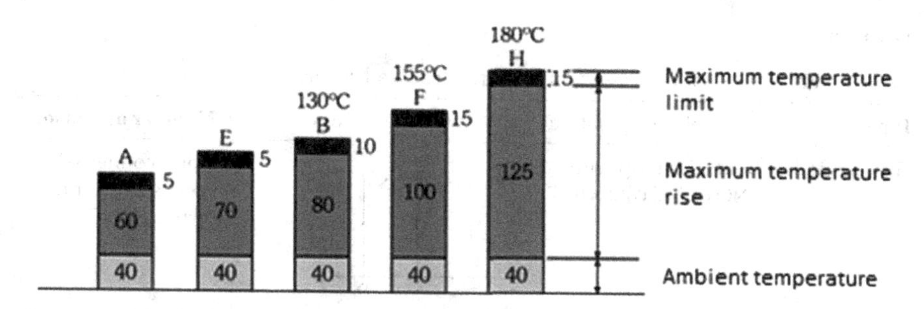

FIGURE 2.31
Maximum permissible temperatures for each class of insulation (IEC 34.1).

The standards consider a difference between the mean winding temperature and the maximum temperature point for each insulation class. The typical values for the isolation classes are: 5°C for classes A and E, 10°C for class B and 15°C for classes F and H. Figure 2.31 illustrates these characteristics.

From Table 2.4 and Figure 2.31, it is possible to conclude that rotary electric machines are designed for a total temperature rise, which is below the maximum specified for insulation materials.

For example, if we use the class of insulation F, we will have:

$$\text{Ambient temperature} + \text{max.temp.rise} = 40°\text{C} + 100°\text{C} = 140°\text{C}$$

In this case, we will have a temperature reserve of 15°C. The higher the thermal reserve, the longer the life expectancy of the insulation material. If the motor operates continuously at the maximum temperature of its insulation class, the life expectancy of the insulation material is, on average, 10 years. It is essential for the motor to work at temperatures below the maximum allowed, so that its useful life is around 20–25 years.

A common practice used by major manufacturers is to design the motor for class B insulation and use materials for class F insulation, making an extra thermal reserve of 20°C, making it possible to increase the life expectancy of the motor insulation.

Despite all the differences between the standards of IEC and NEMA, there is a similarity in the winding insulation, although there is no class E in the NEMA standard.

To protect the motor windings, thermal sensors are inserted in the heads of the windings. The temperature sensors are placed in "hot spots" of the motor windings. The most common temperature measurement techniques are given in Table 2.5.

TABLE 2.5

Typical Sensors Used to Measure Temperature in a Motor

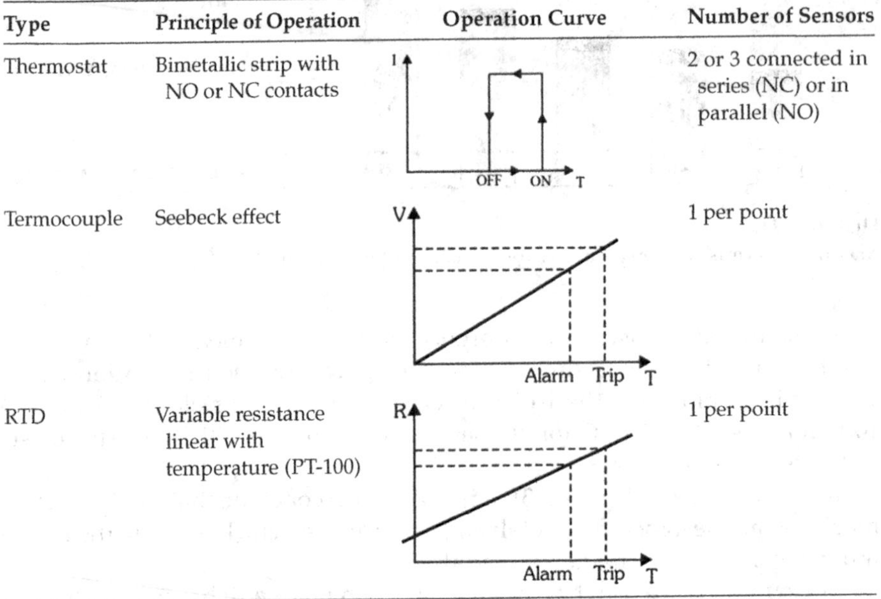

Type	Principle of Operation	Operation Curve	Number of Sensors
Thermostat	Bimetallic strip with NO or NC contacts		2 or 3 connected in series (NC) or in parallel (NO)
Termocouple	Seebeck effect		1 per point
RTD	Variable resistance linear with temperature (PT-100)		1 per point

2.6.6 Motor Cooling

The motor frame is intended to protect windings, bearings, mechanical and electrical parts from moisture, dust and mechanical damage. Also, it is responsible for the cooling process, by which heat is exchanged between the motor interior and the external environment.

There is a system that uses letters and numbers named IC (international cooling) to represent how the motor is cooled. The code covers all types of cooling from small fan-cooled motors to large liquid-cooled motors. This representation is shown in Figure 2.32.

Most of the applications are represented by a short code of two or three letter-numbers. According to IEC and NEMA standards, there are more than 20 designations in the market; in the following, the most used will be presented.

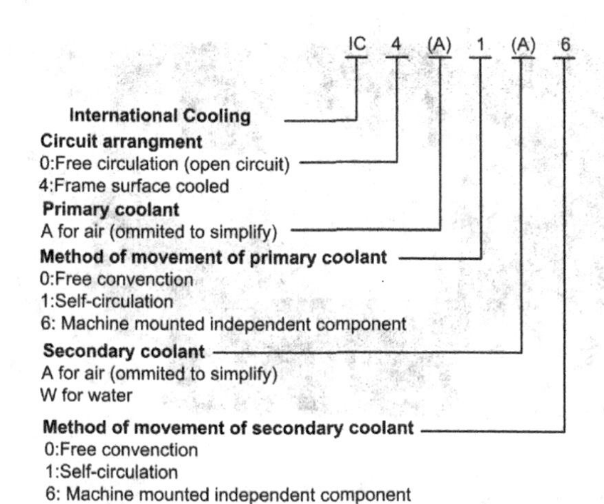

FIGURE 2.32
International cooling representation.

2.6.6.1 IC 01—Open Drip Proof (ODP)

The first digit (0) indicates that the airflow passes freely through the motor windings and the second digit (1), which is produced by an integral fan (self-circulation).

In this cooling type, the ambient air circulates inside the motor, removing heat from the heated parts of the machine. Figure 2.33 shows how the airflow flows through the motor.

The frame of these motors is smooth. They are manufactured in the powers of 1/3, 1/2, 3/4, 1, 1 ½, 2 and 3 HP. Typically, the most common degree of protection for these motors is IP 21, which avoids liquid from falling into the

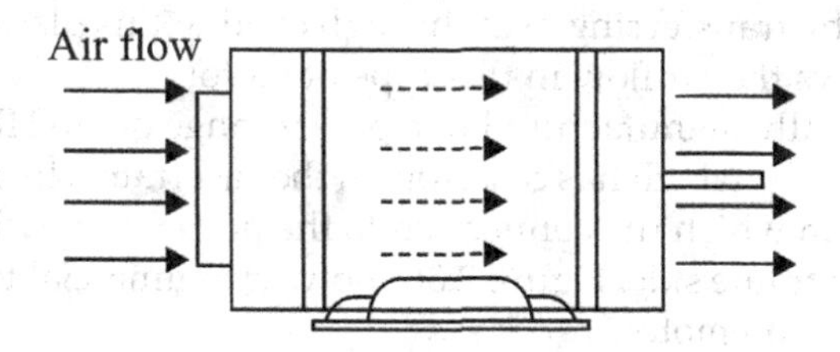

FIGURE 2.33
Air flux in an open motor.

FIGURE 2.34
Open drip proof motor.

motor within a 15 degree angle from vertical, and are typically employed indoors in relatively clean, dry places. Figure 2.34 shows a commercial type of this motor.

2.6.6.2 IC 41—Totally Enclosed Fan-Cooled (TEFC)

The first digit (4) indicates that the frame surface is cooling, and the second (1) indicates that the airflow is generated by a fan integral. In the totally enclosed fan-cooled (TEFC), there is no heat exchange between the internal motor and the exterior. In the motor, there are gaps in the gaskets, which allow the internal cooling medium to escape when it starts operating, warming up. These gaps also allow penetration of the external coolant when it is turned off and begins its cooling process. The heat exchange of these motors is done by transferring heat through cooling fins placed on its frame. Figure 2.35 shows the air flow in this type of motor.

They are typically manufactured in a power range of 1/6 HP up to 500 HP. In this motor family, which has cast iron, in the cast frame, there is a terminal connection box in which the connection to the power supply is made, which is located on the frame side. Figure 2.36 shows a commercial totally enclosed fan-cooled induction motor.

2.6.6.3 IC 40—Totally Enclosed Non-Ventilated (TENV)

The first digit (4) indicates that the frame surface (external) is cooled, and the second digit (0) represents that the cooling is provided without the use of fan.

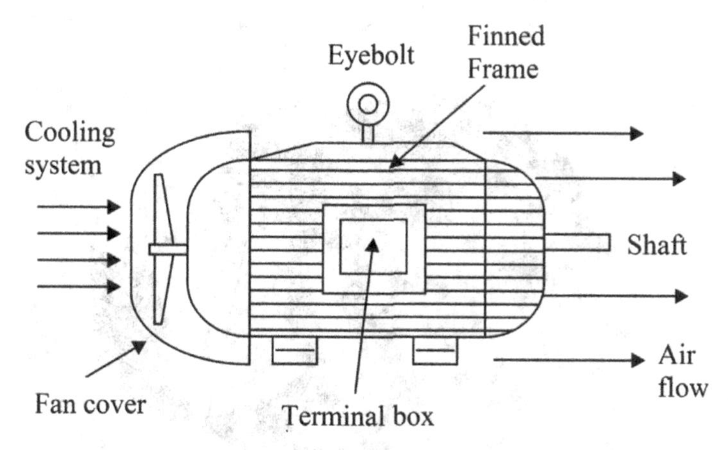

FIGURE 2.35
Totally enclosed fan cooled motor airflow representation.

FIGURE 2.36
Totally enclosed fan cooled motor.

This motor is similar to the TEFC; the difference is that it has no fan for cooling and openings are fully enclosed to prevent air entrance. They are used for environments subject to dust and moisture, are not recommended for explosive locations. They are also recommended for VFD applications with constant torque for all speed ranges. Figure 2.37 shows a typical TENV enclosure motor.

FIGURE 2.37
Totally enclosed non-ventilated (TENV) motor. (Courtesy of Weg.)

2.6.6.4 IC 418—Totally Enclosed Air Over (TEAO)

The first digit (4) indicates that the external frame surface is cooled, the second digit (1) represents that the cooling is provided without the use of fan and the third (8) indicates that the motor is in the air stream of a driven fan or blower.

It is designed in a totally enclosed frame to operate in environments with humidity and corrosion, with typical applications being in chillers or cooling towers with special sealing to avoid particles entrance. It uses cast iron with epoxy paint to provide robustness for humid and corrosive environments (see Figure 2.38).

2.6.6.5 Special Applications Frames

The following are some special motors applications that require an appropriate frame.

2.6.6.5.1 Washdown Duty

Designed to operate in environments with water presence. It meets the requirements of food, pharmaceutical and local areas where the motor needs to be constantly washed. It has antibacterial paint shaft and screws with stainless steel, caps and connection boxes with polycarbonate resin to prevent water entrance in the connection box, as well as internal anticorrosive paint. It is available in TEFC and TENV frames. Figure 2.39 shows this motor and its application.

FIGURE 2.38
Totally enclosed air over (TEAO). (Courtesy of Weg.)

2.6.6.5.2 Explosion-Proof Enclosures (EXPL)

Recommended to be used in classified areas where there exists high-risk flammable materials, such as fuels, or places where there is danger of explosion due to the presence of highly flammable gases, such as warehouse tunnels. These motors must withstand mechanical stress, as any damage to the insulation can cause major accidents.

It is designed with a highly robust frame and structure, with all the elements: screws, joints, etc., compatible with the applied forces. The seals should be flawless, since a small spark can start an explosion. This construction is

FIGURE 2.39
Wash duty motor and application. (Courtesy of Weg.) *(Continued)*

FIGURE 2.39 (Continued)
Wash duty motor and application. (Courtesy of Weg.)

closed with external ventilation, with reinforced covers and connection box. Its production control and quality is very strict.

Explosion-proof motors are designed, manufactured and must be tested under the rigid requirements of Certification Laboratories. Figure 2.40 shows a commercial motor with an explosion-proof enclosure.

FIGURE 2.40
Explosion-proof enclosures motor. (Courtesy of Weg.)

2.6.7 Duty Cycle

The duty cycle is defined as the regularity of load to which the motor is subjected. The main limiting factor of the developed power is the maximum temperature that the motor reaches. The temperature rise is not immediate. It happens according to an exponential function. In the same way, when the motor is switched off, the temperature decreases exponentially. If the motor operating regime is intermittent—that is, variable—it is important to know the maximum temperature, since it defines the rated motor power to be specified.

There are situations where the motor is effectively switched off during its duty cycle. An example of this is a motor that drives a water pump from a supply system, usually connected for a long period, under constant load. A motor that drives the vehicle lift system for oil change at a gas station has a very short running time compared to the total rest time. In practice, there are infinite duty cycles to which the motor can be submitted.

The motor is not a piece of equipment that can be exposed to constant starting and stopping, such as turning on/off a flasher. It is very common that the required mechanical load of the shaft changes from an "unloaded" situation to an overloaded situation. Therefore, it is recommended that there is a small power gap, over than that established by the duty cycle, to overcome small overloads.

The NEMA standard defines duty cycles in three categories:

Continuous: Is a duty cycle defined as an essentially constant load for an indefinitely long time. This is the most common classification used in motors (approximately 90%).

Intermittent: The load is driven at load and no-load or load and rest intervals with defined intervals between each start.

Varying duty: Requires load operation at different time intervals that may be subject to wide variations. This calculation must take into account the required power peak and also the rms (root-mean-square) value, which will indicate the ideal motor from the heating point of view.

This method consists of multiplying the square of the required power for each part of the cycle by its duration in seconds and in the sequence dividing the sum of this result by the total time to complete each cycle, extracting the square root of this result, as in the following equation.

$$P_{rms}{}^2 = \frac{1}{T}\sum_{t=0}^{t} P(t)^2\,\Delta t \qquad (2.12)$$

where:

P_{rms} is the equivalent rms power required from the motor
$P(t)$ is the power required from the motor variable with time
T is the total cycle time

Figure 2.41 shows an example of an application of this method in a motor.

The value of the power (P_{RMS}) is obtained by applying the power values from Equation (2.12) in the periods of time presented in the figure, resulting in:

$$P_{RMS} = \sqrt{\frac{P_1^2.t_1 + P_2^2.t_2 + P_3^2.t_3 + P_4^2.t_4 + P_5^2.t_5 + P_6^2.t_6}{t_1 + t_2 + t_3 + t_4 + t_5 + t_6}} \tag{2.13}$$

If the motor is in the rest, only 1/3 of this period should be used for open motors, and ½ for closed motors. This is due to the effect of cooling reduction when the motor is at rest.

Example: Consider the machine operation in which an open motor operates at a load of 6 HP in 180 seconds, 5 HP in 30 seconds, 20 HP in 60 seconds and remains at rest for 240 seconds. Using Equation (2.13), we have:

$$P_{RMS} = \sqrt{\frac{6^2.180 + 5^2.30 + 20^2.60}{180 + 30 + 60 + \dfrac{240}{3}}} = 9.44\,\text{HP}$$

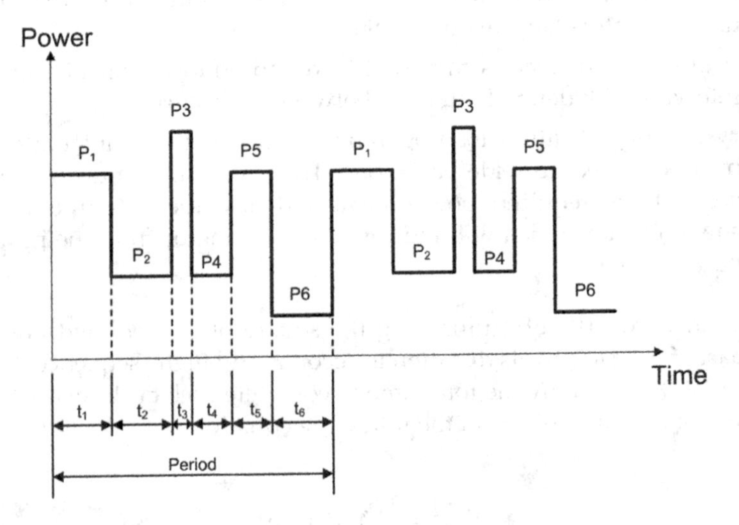

FIGURE 2.41
Application of equivalent rms power in a motor.

So, for this application, we can use a 10 HP motor.

The IEC 60034.1 standardizes ten different duty cycle types. Evidently, they do not show all the actual situations found in practice. Therefore, a real situation must be approximated to one of the standard situations that is more severe than the real one. Usually, the motors are designed for a continuous regime—i.e., constant load acting for an indefinite time—equal to the rated motor power. This duty cycle is named as continuous running (S1).

The ten duty cycles are described below:

S1 (Continuous running duty): The operation takes place with a constant mechanical load for a period until the thermal equilibrium is reached, as shown in Figure 2.42.

where:

N is the operation at constant load

T_{max} is the maximum temperature reached

This is the default duty cycle if there is no indication in the nameplate.

S2 (Short-time duty): Constant-load operation occurs in a shorter period of time and is necessary to reach equilibrium, followed by a stop and a de-energizing period. The 10, 30, 60 or 90 minutes are recommended for the duration of this duty cycle. Figure 2.43 represents this duty cycle.

S3 (Intermittent periodic duty): In this duty cycle, there is a sequence comprising an operating period with a constant load and a rest period. The cycle period is too short for the thermal equilibrium to be reached. Figure 2.44 shows this duty cycle.

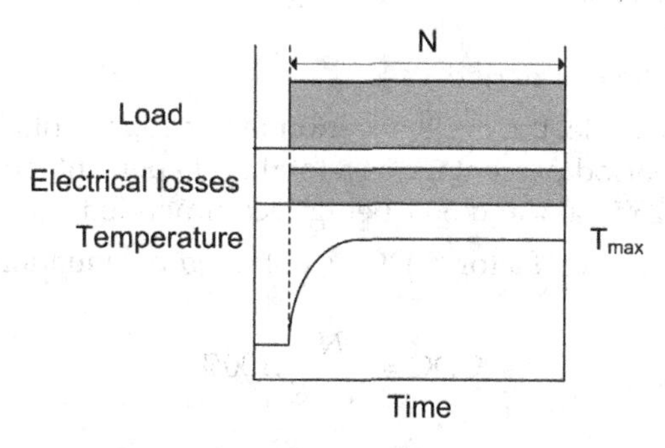

FIGURE 2.42
S1 (Continuous running duty).

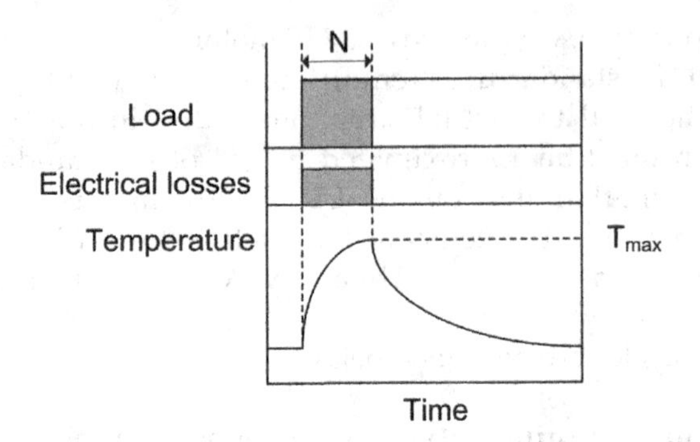

FIGURE 2.43
S2 (Short-time duty).

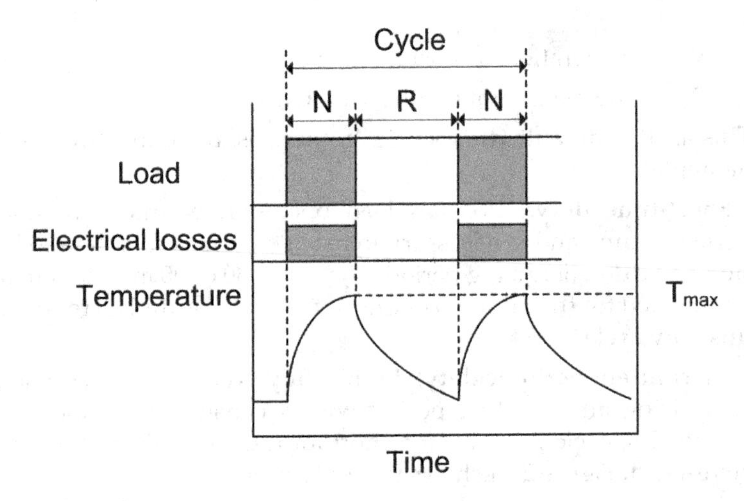

FIGURE 2.44
S3 (Intermittent periodic duty).

where R is the rest period.

In this duty cycle, the cycle duration factor represents the load activation period percentage as a total cycle percentage, with factors of 15%, 25%, 40% and 60% being recommended.

The cycle duration factor (CDC) is obtained by equation:

$$\mathrm{CDC} = \frac{N}{N+R}.100\% \tag{2.14}$$

S4 (Intermittent periodic duty with starting): A sequence of identical duty cycles is considered, each comprising: a period considering the starting current, a period with load and a stop period. The duty cycle period is too short to reach thermal equilibrium; the motor will be stopped due to load inertia or by mechanical braking (see Figure 2.45).

where D is the starting.

In this duty cycle, the duration factor (CDC) is obtained by equation:

$$\text{CDC} = \frac{D+N}{N+R+D}.100\% \tag{2.15}$$

S5 (Intermittent periodic duty with electric braking): A sequence of identical duty cycles, each comprising a starting moment, a period of operation with constant load, a period of rapid electrical braking and a period of motor stopping. Again, the period is too short for thermal equilibrium to be achieved (see Figure 2.46).

In this duty cycle, the duration factor (CDC) is obtained by equation:

$$\text{CDC} = \frac{D+N+F}{N+R+D+F}.100\% \tag{2.16}$$

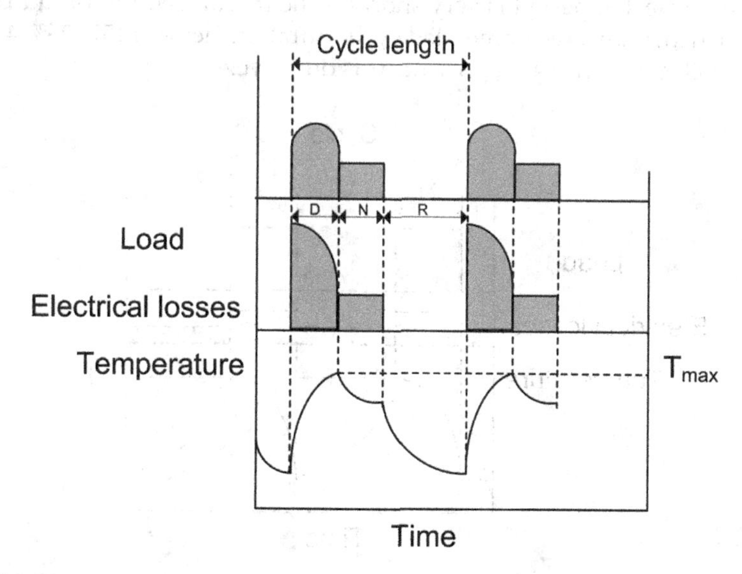

FIGURE 2.45
S4 (Intermittent periodic duty with starting).

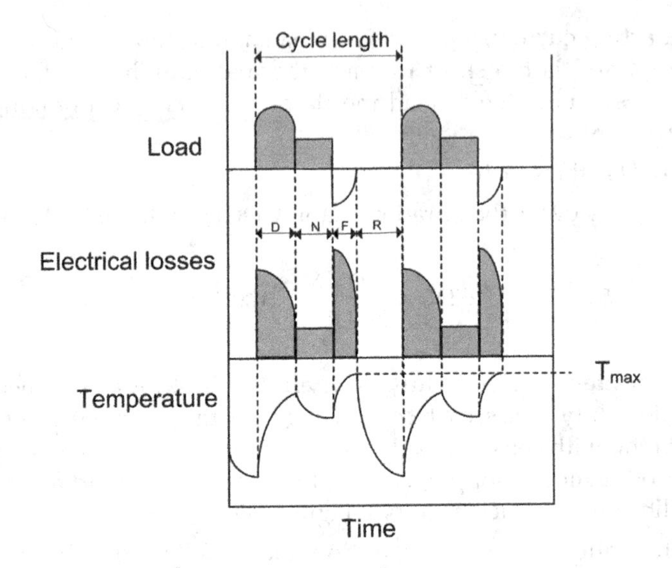

FIGURE 2.46
S5 (Intermittent periodic duty with electric braking).

S6 (Continuous operation periodic duty): A sequence of duty cycles where each cycle consists of a period of constant load and one of operation without load with the de-energisation of the motor. In this cycle, the period is very short for the thermal equilibrium to be reached, with a recommended cyclic duration factor of 15%, 25%, 40% and 60%. Figure 2.47 represents this duty cycle.

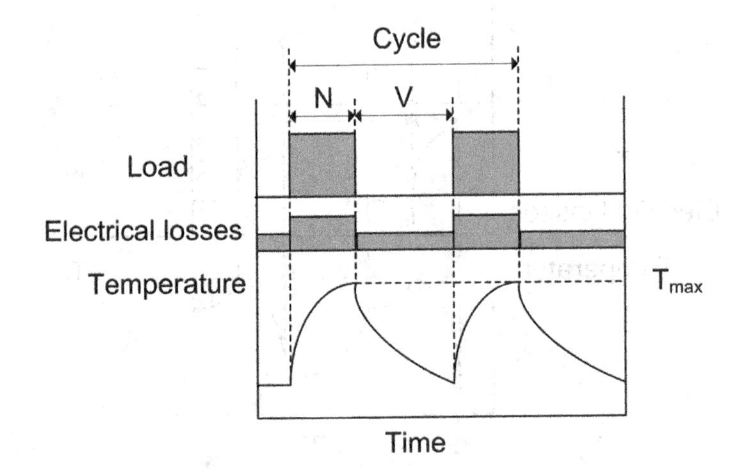

FIGURE 2.47
S6 (Continuous operation periodic duty).

where V is the operation at no-load.

In this duty cycle, duration factor (CDC) is obtained by equation:

$$\text{CDC} = \frac{N}{N+V}.100\% \tag{2.17}$$

S7 (Continuous operation periodic duty with electric braking): This duty cycle is a sequence of identical duty cycles, with each cycle consisting of a starting time, a time of operation at constant load, a time of electric braking without a time of de-energisation and at rest. In this duty cycle, the cycle duration factor is unitary, (see Figure 2.48).

S8 (Continuous operation periodic duty with related load/speed changes): A sequence of identical duty cycles, each consisting of a period of operation with constant load at a certain speed, followed by one or more periods of operation with other constant loads with different speeds. This type of duty cycle is used for pole changing motors, and there is a different cycle duration factor for each speed of rotation.

S9 (Duty with non-periodic load and speed variations): In this duty cycle, the load and speed does not change periodically. The motor is submitted to variable loads at variable speeds non-periodically within the operating range. This duty includes frequently applied overloads within the permissible temperature rise limits.

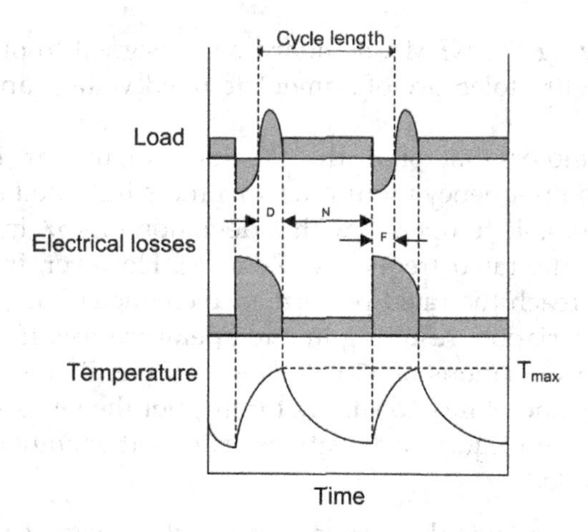

FIGURE 2.48
S7 (Continuous operation periodic duty with electric braking).

S10 (Duty with discrete constant loads and speeds): This duty cycle consists of a number of discrete load values not exceeding four per cycle, and each load cycle is executed in a way that allows the machine to achieve thermal equilibrium, and the load cycles can be different.

NOTE: Duty cycles S2 to S10 must be ordered directly from the manufacturers and are not considered for normal application.

2.6.8 Service Factor (SF)

The service factor (SF) is the factor which, when applied to rated power, indicates the allowable overload that can be continuously applied to the motor under specified conditions.

Example: SF = 1.15; The motor continuously withstands 15% overload over its rated power.

The service factor is a continuous overload capability, i.e., a power reserve that gives the motor operating conditions in unfavourable situations.

NOTE: Service factor is not to be confused with momentary overload capability.

2.6.9 Multiple Nominal Voltage and Frequency

Most motors are supplied with winding terminals, so that they can be connected to the power supply of at least two different voltages, for example, 220 V/380 V.

Motors following the NEMA standard are designed to operate within a range of up to ±10% tolerance of nameplate rated voltage and ±5% of rated frequency.

In the case of motors that follow the IEC standard, they are built to operate at the voltage and frequency within the tolerances indicated in Figure 2.49.

The motors are able to operate with a deviation of ±5% in the rated voltage and ±2% of the rated frequency (Zone A). However, in this zone, the motor does not reach the rated performance characteristics, due to voltage and frequency variation, resulting in a temperature rise. If the voltage and frequency suffer deviations of ±10% and ±3% (Zone B), the motor will run with restrictions and ability to supply torque, but the temperature will rise due the voltage and frequency deviations; extended operation in this region is not recommended.

NOTE: Operating outside these specifications can severely affect the motor life expectancy.

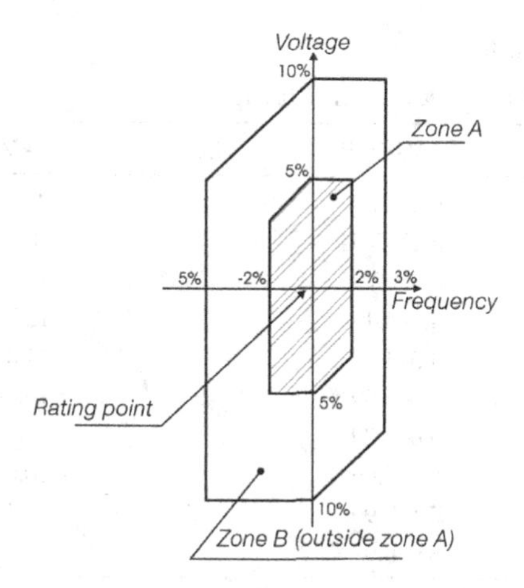

FIGURE 2.49
Operational areas to voltage and frequency according IEC.

2.6.10 Locked Rotor Current kVA Code Letter (CODE)

The locked rotor current (LRA) occurs in a short period of time when the motor starts; and the current is many times higher than the value of a full load running motor current (FLA). This intensity is obtained using the design code letter of the motor. The code letter represents a multiplier based on the design characteristics of the motor that will determine the maximum apparent power in kVA per HP (active power) with the motor in locked rotor conditions.

Table 2.6 shows the locked rotor kVA/HP multiplier for each code letter.

To better understand how the locked rotor current (LRA) is obtained, consider the following motor with: 220 V, 30 HP, 75,5A FLA and code letter E.

Analysing the table, we see that code letter E has a range of 4.5–4.99, so we will consider the half part 4.7.

TABLE 2.6

Locked Rotor kVA/HP Multiplier

Code	kVA/HP	Code	kVA/HP
A	0–3.14	L	9.0–9.99
B	3.15–3.54	M	10.0–11.19
C	3.55–3.99	N	11.2–12.49
D	4.0–4.49	P	12.5–13.99
E	4.5–4.99	R	14.0–15.99
F	5.0–5.59	S	16.0–17.99
G	5.6–6.29	T	18.0–19.99
H	6.3–7.09	U	20.0–22.39
J	7.1–7.99	V	22.4 and up
K	8.0–8.99		

The LRA is obtained by the following equation:

$$LRA = \left(\frac{CLM * 1000 * P}{V * \sqrt{3}}\right) \tag{2.18}$$

where:

P is the active power, HP
V is the motor voltage, V
CLM is the code letter multiplier in kVA/HP

So:

$$LRA = \left(\frac{4.7 * 1000 * 30}{220 * \sqrt{3}}\right) = 370.02\ \text{A}$$

LRA = 370.02 A, this is, 4.9 times motor FLA.

2.6.11 Speed Direction

To change the three-phase motor's direction of rotation is quite easy. Simply reverse two phases, as shown in Figure 2.50; it does not matter which phases will be changed.

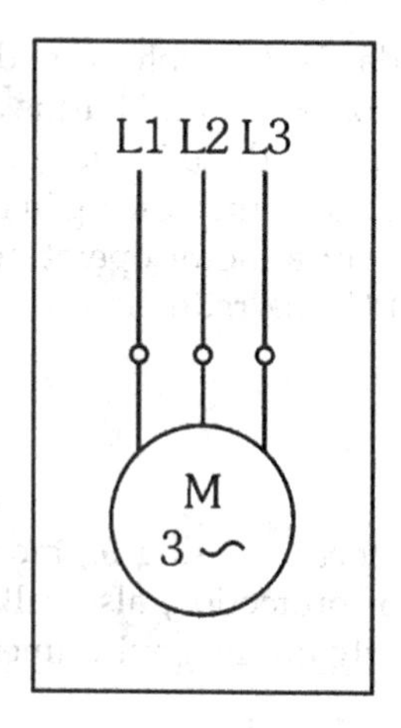

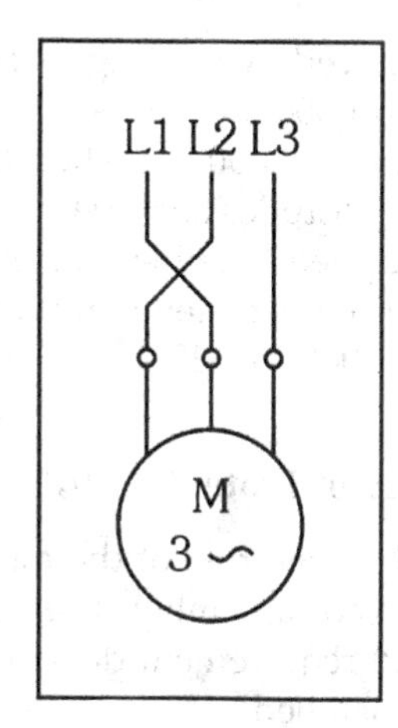

FIGURE 2.50
Change direction of rotation in a three-phase induction motor.

2.6.12 Environmental Conditions

AC induction motors are designed to operate under certain environmental conditions considering ambient temperature and altitude. When the ambient temperature and altitude are too high, current de-rating tables are provided by the manufacturers. Table 2.7 shows an example of a reduction for temperature and altitude.

At high altitudes, where atmospheric pressure is reduced, the cooling of electrical equipment is reduced by the decrease in the air's ability to remove heat from the motor. When the atmospheric pressure drops with increasing altitude, the air density drops and, as a consequence, the thermal exchange

TABLE 2.7
Typical Altitude and Temperature Reduction Factors

Ambient Temperature (°C)	Temperature Reduction Factor	Altitude above Sea Level (m)	Altitude Reduction Factor
30	1.07	1000	1
40	1	1500	0.96
45	0.96	2000	0.92
50	0.92	2500	0.88
55	0.87	3000	0.84
60	0.82	3500	0.84
70	0.65	4000	0.76

capacity is reduced. According to the standards, AC motors are designed for altitudes up to 1,000 m above sea level. Motor power and torque are reduced to altitudes above 1,000 m. When the motor needs to be designed for temperatures and altitudes above standards, the factors reducing in the Table 2.7 must be multiplied together. For example, for a motor operating at 2500 m altitude at an ambient temperature of 50°C, the reduction factors will be: (0.92 × 0.88) × 100%, or 81%.

2.6.13 Degree of Protection (IP)

The frame plays the role of the motor protection casing or, more precisely, of the stator-rotor assembly. The degree of protection, also called index of protection (IP) requirement depends directly on the environment in which the motor is installed.

A motor installed in an environment exposed to sun and rain, should require a higher degree of protection than a motor installed inside a clean and dry room. Environments considered to be aggressive to motors are those with the presence of dust, fibres, particulates as well as environments that are wet or subject to water jet.

The frame, as an enclosure, should provide effective protection to the motor in the environment in which it operates. The IEC 34.5 and Australian Standard (AS) 1359.20 standard establishes various degrees of protection for electrical enclosures. In general, the degree of protection of electric motors is usually expressed in two digits. The first one indicates protection against solid bodies, the second, protection against water. Tables 2.8 and 2.9 show the first and second digit protections respectively.

Three-phase induction motors for normal application are generally manufactured with the following degrees of protection:

- **IP54:** Complete protection against touch and against harmful dust accumulation. Protection against splashes from all directions. They are used in very dusty environments.
- **IP55:** Complete protection against touch and harmful dust accumulation. Protection against water jets in all directions. They are used in cases where the equipment is washed periodically with hoses.
- **IP(W)55:** Similar to the IP55, but are protected against bad weather, rain and sea. They are used outdoors. Also called marine motors.

Three-phase induction motors open for normal application are almost always IP21 rated. They are protected against finger touching and against solid foreign bodies over 12 mm (first digit 2). They also have protection against dripping vertically (second digit 1).

TABLE 2.8

First Digit of the Motor Degree of Protection

Digit	Test	Indication
0		No protection
1	50 mm	Protected against solid objects of over 50 mm (e.g., accidental hand contact)
2	12 mm	Protected against solid objects of over 12 mm (e.g., finger)
3	2,5 mm	Protected against solid objects of over 2.5 mm (e.g., tools, wire)
4	1,0 mm	Protected against solid objects of over 1 mm (e.g., small tools, thin wire)
5		Protected against dust (no deposits of harmful material)
6		Totally protected against dust

TABLE 2.9

Second Digit of the Motor Degree of Protection

Digit	Test	Indication
0		No protection
1		Protected against vertically dripping water (condensation)
2	15°	Protected against water dripping up to 15° from the vertical

(Continued)

TABLE 2.9 (*Continued*)

Second Digit of the Motor Degree of Protection

Digit	Test	Indication
3	60°	Protected against rain falling up to 60° from the vertical (rain)
4		Protected against water splashes from all directions
5		Protected against jets of water from all directions
6		Protected against jets of water comparable to heavy seas
7	0,15 m 1 m	Protected against the effects of immersion to depths of between 0.15 and 1 m
8		Protected against the effects of prolonged immersion at depth

There are also additional letters that are used to classify the level of protection to humans against access to hazardous parts. There also exists further letters that can be appended to provide additional information related to the protection of the device and person.

Table 2.10 presents these additional letters.

The description for the frame degree of protection is similar to NEMA's in relation to IEC (IP code). The NEMA standard is more descriptive and general, whereas the IEC is more precise and direct being given by a 2-digit code, as presented previously.

The NEMA standard takes into account other factors that are not addressed by the IEC, such as gasket aging, construction practices and corrosion resistance. In Table 2.11 are presented the NEMA-rated enclosures to non-hazardous locations.

TABLE 2.10

IP Ratings Additional Letters

Additional Letter	Description
A	Protected against access to hazardous parts with back of hand
B	Protected against hazardous parts with fingers
C	Protected against tools interfering with hazardous parts
D	Protection against wire from entering hazardous parts
Further Letter	**Description**
H	High-voltage device
M	Device moving during water test
S	Device remaining still during water test
W	Weather conditions

There are also classifications for hazardous location areas defined by NEC (National Electric Code) National Fire Protection Association (NFPA)-70 that are divided into three classes, as shown in Table 2.12.

As previously mentioned, NEMA presents some features that do not present in the IEC (IP). Thus it is not possible to find a relation between the IP code and NEMA, due to the fact that the IP code does not contemplate these additional items.

Table 2.13 shows the minimum NEMA specification that satisfies an IP code, being possible only in this sense, without having a direct correlation of IP to NEMA.

TABLE 2.11

NEMA-Rated Enclosures to Non-Hazardous Locations

Type	Description
1	Enclosures are intended for indoor use, primarily to provide a degree of protection against contact with the enclosed equipment.
2	Enclosures are intended for indoor use, primarily to provide a degree of protection against limited amounts of falling water and dirt.
3	Enclosures are intended for outdoor use, primarily to provide a degree of protection against windblown dust, rain, sleet and external ice formation.
3R	Enclosures are intended for outdoor use, primarily to provide a degree of protection against falling rain, sleet and external ice formation.
3S	Enclosures are intended for outdoor use, primarily to provide a degree of protection against windblown dust, rain and sleet, and to provide for operation of external mechanisms when ice laden.
4	Enclosures are intended for indoor or outdoor use, primarily to provide a degree of protection against windblown dust and rain, splashing water and hose-directed water.
4X	Enclosures are intended for indoor or outdoor use, primarily to provide a degree of protection against corrosion, windblown dust and rain, splashing water and hose-directed water.
5	Enclosures are intended for indoor use, primarily to provide a degree of protection against settling airborne dust, falling dirt and dripping non-corrosive liquids.
6	Enclosures are intended for indoor or outdoor use, primarily to provide a degree of protection against the entry of water during occasional temporary submersion at a limited depth.
6P	Enclosures are intended for indoor or outdoor use, primarily to provide a degree of protection against the entry of water during prolonged submersion at a limited depth.
7	Enclosures are for use indoors in locations classified as Class I, Groups A, B, C or D, as defined in the *National Electrical Code*.
8	Enclosures are for indoor or outdoor use in locations classified as Class I, Groups A, B, C, or D, as defined in the *National Electrical Code*.
9	Enclosures are for use in indoor locations classified as Class II, Groups E, F, or G, as defined in the *National Electrical Code*.
10	Enclosures are constructed to meet the applicable requirements of the Mine Safety and Health Administration.
11	Enclosures are intended for indoor use, primarily to provide, by oil immersion, a degree of protection to enclosed equipment against the corrosive effects of liquids and gases.
12	Enclosures are intended for indoor use, primarily to provide a degree of protection against dust, falling dirt and dripping non-corrosive liquids.
12K	Enclosures with knockouts are intended for indoor use, primarily to provide a degree of protection against dust, falling dirt and dripping non-corrosive liquids other than at knockouts.
13	Enclosures are intended for indoor use, primarily to provide a degree of protection against dust, spraying of water, oil and non-corrosive coolant.

TABLE 2.12

Hazardous Location Areas according to NEC

	Class I
Group A	Acetylene
Group B	Butadiene, ethylene oxide, hydrogen, propylene oxide, manufactured gases containing more than 30ydrogen by volume.
Group C	Acetaldehyde, cyclopropane, diethyl ether, ethylene.
Group D	Acetone, acrylonitrile, ammonia, benzene, butane, ethanol, ethylene dichloride, gasoline, hexane, isoprene, methane (*natural gas*), methanol, naphtha, propane, propylene, styrene, toluene, vinyl acetate, vinyl chloride, xylene.
	Class II
Group E	Aluminium, magnesium, and other metal dusts with similar characteristics.
Group F	Carbon black, coke or coal dust.
Group G	Flour, starch or grain dust.
	Class III
• Easily ignitable fibers, such as rayon, cotton, sisal, hemp, cocoa fiber, oakum, excelsior and other materials of similar nature.	

TABLE 2.13

Comparison between NEMA and IEC Degrees of Protection

Enclosure Type Number (NEMA)	Enclosure Designation (IEC)
1	IP10
2	IP11
3	IP54
3R	IP14
3S	IP54
4 and 4X	IP56
5	IP52
6 and 6P	IP67
12 and 12K	IP52
13	IP54

2.6.14 Frames

The dimensions standardised by the IEC and NEMA standards serve as a basis for the correct specification of the motor in relation to its size. These standards present great similarity differing mainly in the units used by IEC (mm) and NEMA (inches).

In both systems, frame size is given by the shaft height. In IEC, the frame size is the height of the shaft in millimetres, and the first two digits of the NEMA provide the height of the axis multiplied by 4 in inches.
Example:

Frame IEC 90S = Shaft height 90 mm

Frame NEMA 143 = Shaft height 14/4 = 3.5 in

According to NEMA definitions, the two-digit frames are intended for fractional motors although larger motors may display this encoding. Typically, three-digit frames are used for higher power motors. The third number does not directly represent a frame measure; it indicates the distance of mounting bolt holes in the base (dimension F), so the greater this number, the greater the distance, footless motors do not have this digit. So, a motor frame of 182 will have the F(57.2 mm) dimension smaller than the 184 (68.2 mm).

NOTE: If a frame by NEMA standards has only two numbers, they must be divided by 16.

Figure 2.51 presents an IEC frame example.

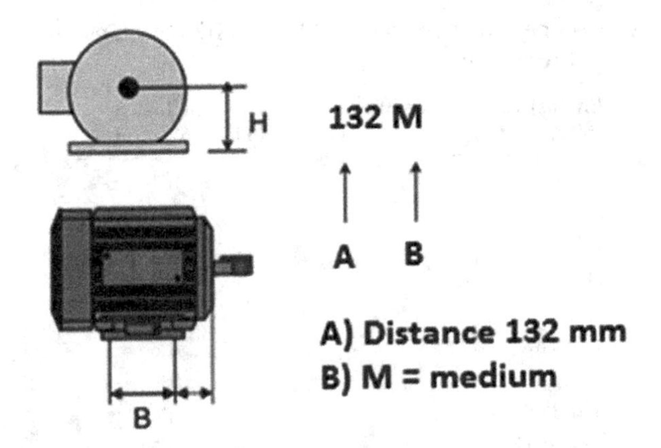

FIGURE 2.51
IEC Frame description.

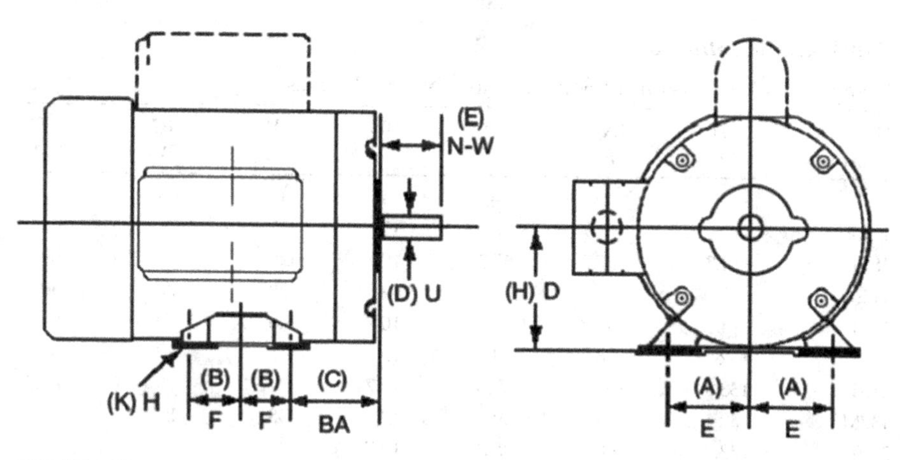

FIGURE 2.52
Dimensions according NEMA and IEC.

NOTE: In the IEC frame, there are additional letters (S = small, M = medium and L = large) that are related to dimension B.

Figure 2.52 shows a typical motor with dimensions according to NEMA and IEC standards.

Table 2.14 shows the comparative dimensions of some IEC and NEMA frames.

TABLE 2.14
Comparative Dimensions of Some IEC and NEMA Frames

IEC NEMA	(H) D	(A) E	(B) F	(K) H	(D) U	(C) BA	(E) N-W
56	56	45	35.5	5.5	9	38	20
N/A	–	–	–	–	–	–	–
63	63	60	40	7	11	40	23
42	66.7	44.5	21.4	7.1	9.5	52.4	28.8
71	71	56	45	7	14	45	30
48	76.2	54	34.9	8.7	12.7	63.5	38.1
80	80	62.5	50	10	19	50	40
56	88.9	61.9	38.1	8.7	15.9	69.9	47.6
90S	90	70	50	10	24	56	50
143T	88.9	69.8	50.8	8.7	22.2	57.2	57.2
90L	90	70	50	10	24	56	50
145T	88.9	69.8	63.5	8.7	22.2	57.2	57.2
100L	100	80	70	12	28	63	60

(*Continued*)

TABLE 2.14 (*Continued*)

Comparative Dimensions of Some IEC and NEMA Frames

IEC NEMA	(H) D	(A) E	(B) F	(K) H	(D) U	(C) BA	(E) N-W
N/A	–	–	–	–	–	–	–
112S	112	95	57	12	28	70	60
182T	114.3	95.2	57.2	10.7	28	70	69.9
112M	112	95	70	12	28	70	60
184T	114.3	95.2	68.2	10.7	28	70	69.9
132S	132	108	70	12	38	89	80
213T	133.4	108	69.8	10.7	34.9	89	85.7
132M	132	108	89	12	38	89	80
215T	133.4	108	88.8	10.7	34.9	89	85.7

It is essential to verify the frames in the manufacturers' manuals to confirm the frame dimension, because there may be variations in the dimension frames when considering different manufacturers.

In addition to the standard three-digit nomenclature, an alphabetical suffix could be added to assign modifications to the standard T frame design in NEMA motors. These additional suffixes are presented in Table 2.15.

TABLE 2.15

Additional Letter in NEMA Frame

Suffix	Definition
C	NEMA C face mounting.
D	NEMA D flange mounting.
H	Indicates a frame with rigid base having an F dimension larger than that one motor with the same frame with a suffix H.
J	NEMA C frame threaded shaft pump motor.
JM	Special pump shaft originally designed for a "mechanical seal" and also has a C face.
JP	Similar to the JM style of motor having a special shaft, the JP motor was originally designed for a "packing" type of seal and also has a C face.
S	The motor has a "short shaft" with shaft dimensions that are smaller than the shafts associated with the normal frame size. Used for belt-driven loads.
T	Indicates that the motor is of 1964 or later.
U	Indicates that the motor falls into the "U" frame size assignment (1952–1964).
Y	The motor has a special mounting configuration, which could have a special base, face or flange.
Z	Indicates a special shaft or having special features, such as threads, holes, etc.

2.6.15 Mounting

For the construction of a motor, one of the most important items is the way of fixing, which is made according to the mechanical design of the machine to be driven.

The IEC 34.7 standard defines the following forms, identified by the letters IM (Index of Mounting), in addition to a letter and one or two characteristic numbers.

Mounting positions must be specified so that mechanical elements, such as bearings, the frame, etc. are correctly located and dimensioned during assembly.

The system used to describe the arrangement types is shown below:

A prefix comprising the letters IM and also four numbers representing the type of construction and the mounting position.

A summary of the mounting types is shown below. Some standards used the letters B (horizontal mounting) and V (vertical mounting). IEC 34.7 does not use this system anymore. Figures 2.53 and 2.54 show the configuration types for the motors fixed from their base and flanged, respectively; the old system is shown in brackets.

Some mounting configurations according to the NEMA standard are shown in Figure 2.55.

There are significant differences in the approach of NEMA and IEC. Table 2.16 shows the matching of some mounting forms according to IEC and NEMA.

Foot-Mounted motors			
IM 1001 (IM B3) - Horizontal shaft - Feet on floor	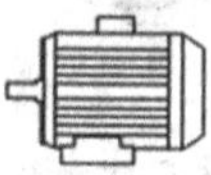	**IM 1071 (IM B8)** - Horizontal shaft - Feet on ceiling	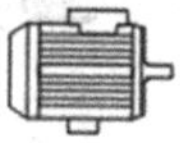
IM 1051 (IM B6) - Horizontal shaft - Feet wall mounted with feet on LHS when viewed from drive end	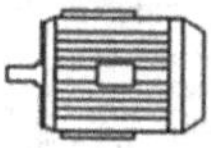	**IM 1011 (IM V5)** - Vertical shaft - Shaft facing down - Feet on wall	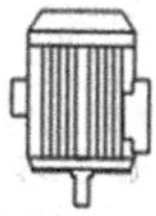
IM 1061 (IM B7) - Horizontal shaft - Feet wall mounted with feet on RHS when viewed from drive end	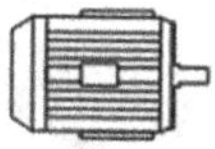	**IM 1031 (IM V6)** - Vertical shaft - Shaft facing up - Feet on wall	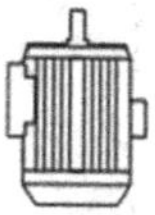

FIGURE 2.53
Foot mounted Motors (IEC 34.7).

Flange-Mounted Motors			
IM 3001 (IM B5) - Horizontal shaft		**IM 2001 (IM B35)** - Horizontal shaft - Feet on floor	
IM 3011 (IM V1) - Vertical shaft - Shaft facing down		**IM 2011 (IM V15)** - Vertical shaft - Shaft facing down - Feet on wall	
IM 3031 (IM V3) - Vertical shaft - Shaft facing up		**IM 2031 (IM V36)** - Vertical shaft - Shaft facing up - Feet on wall	

FIGURE 2.54
Flange mounted Motors (IEC 34.7).

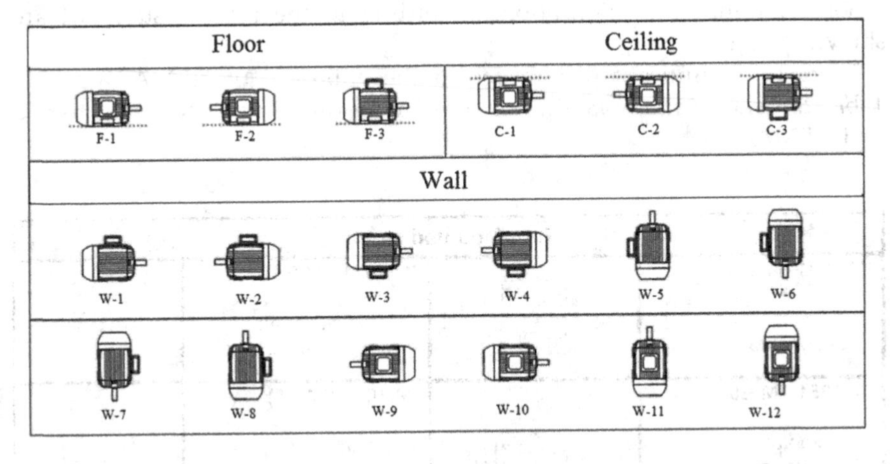

FIGURE 2.55
Standard mounting configurations according to NEMA.

TABLE 2.16
Equivalence between Mounting Forms according to IEC and NEMA

NEMA	IEC Code I	IEC Code II
F-2	IMB3	IM1001
W-1 to W-4	IMB7	IM1061
C-3	IMB8	IM1071
P Flange	IMB5	IM3001
D Flange	IMV1	IM3001

2.6.16 Nameplate

The nameplate details information relating to the construction and performance characteristics of the motor. It provides the key information to make clear to the end user about motors characteristics in accordance with the recommendations of NEMA MG-1 (Motors and Generators) and IEC 60034-1. Figures 2.56 and 2.57 show the nameplate of a typical motor and the representation of its characteristics in IEC and NEMA standards, respectively.

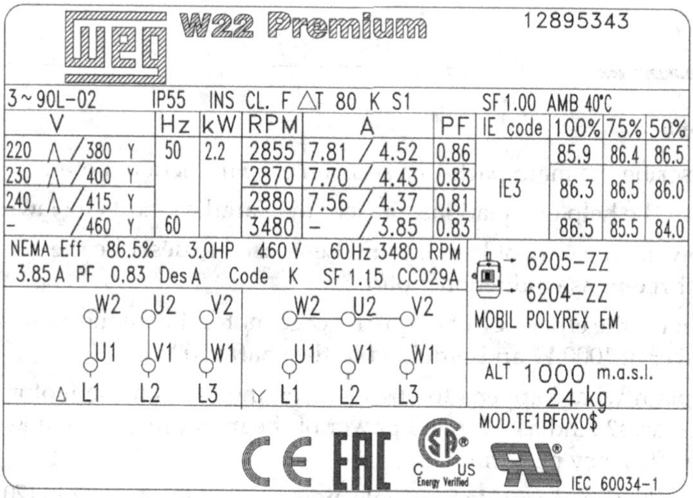

FIGURE 2.56
Typical motor nameplate (IEC). (Courtesy of Weg.)

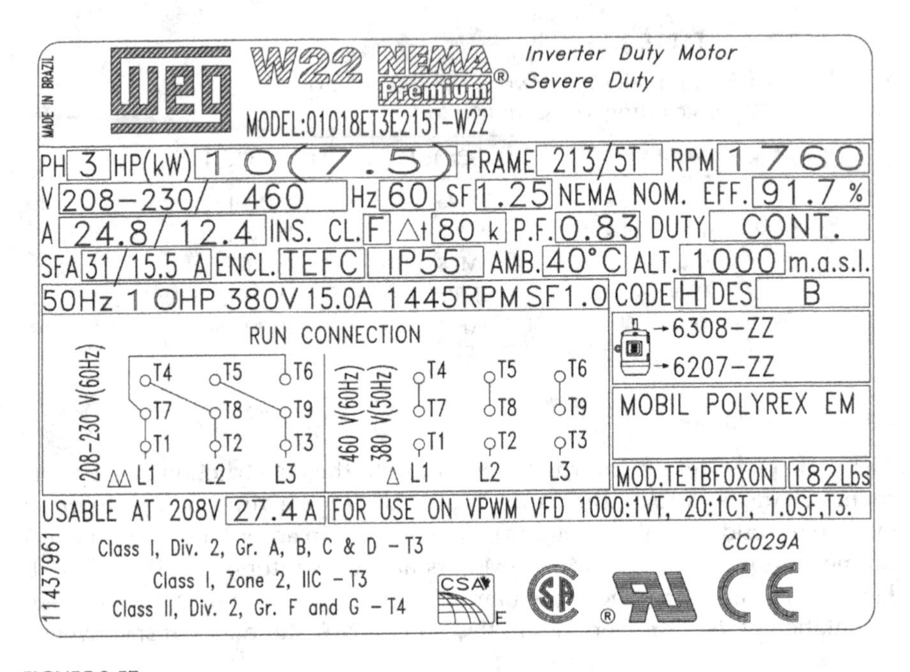

FIGURE 2.57
Typical motor nameplate (NEMA). (Courtesy of Weg.)

Exercises

1. Describe the main parts of a squirrel cage induction motor.
2. With the help of equations, present the Faraday and Lenz laws.
3. How are the laws of Faraday and Lenz used to describe the principle of three-phase induction motors?
4. What is the efficiency of a squirrel cage motor that requests an active power of 1000 W and provided in the shaft 950 W?
5. Explain what happens to efficiency as the load on the motor shaft increases? And as the rated power of the motor increases, how does the efficiency perform?
6. Find the slip of a 50 Hz four-pole motor with rated speed of 1470 rpm.
7. How does the motor torque change with speed when a load is applied in the shaft? With the help of a graph present the motor operating point and stable and unstable torque operation regions.

TABLE 2.17

Motor Nameplate

Ratings	Value
Frequency (Hz)	
Power (HP)	
Full-load current (A)	
Service Factor	
Enclosure type	
Duty	
Insulation Class	
Rotation speed (rpm)	
Design code letter	
Locked rotor current	

8. Present the motor torque speed design according to the IEC and NEMA standard.
9. Find the reserve temperature in a motor with insulation class A and an ambient temperature of 35°C.
10. List the temperature sensors and the operational principle used to measure temperature in a motor.
11. What is international cooling system (IC)? List the designations most used in the industry.
12. Define duty cycle and present the most important categories according to the IEC and NEMA.
13. Examine the motor nameplate (Figure 2.56) and find the following motor ratings in Table 2.17.

3

Electric Power

Whenever a load is connected to a given electric circuit in alternating current, there are three types of electric powers to be considered: real, reactive and apparent. In this chapter, the types of electric power, power factor and power factor correction methods will be presented.

3.1 Real Power

Also called true power, or active power, it can be defined as the electric energy transformed into any form of useful energy, such as, luminous, thermal and others, without the need for an intermediate transformation of energy.

Examples of real power consumer equipment:

- Resistors
- Heaters

The true power (P) in alternating current is given by the following equations:

Single-phase circuit:

$$P = V \cdot I \cdot \cos\varphi \tag{3.1}$$

where:

P is the true power in watts
V is the voltage in volts
I is the current in amperes
$\cos\varphi$ is the power factor

NOTE: In pure resistive AC the power factor ($\cos\varphi$) can be considered unitary.

Three-phase circuit:

$$P = \sqrt{3} \cdot V \cdot I \cdot \cos\varphi \tag{3.2}$$

Unit: Watts (W)

3.2 Reactive Power

It is the intermediate energy required for any equipment, such as, motors, transformers, reactors, capacitors and others. It is vital for these equipments to create their magnetic or electric field, making possible the use of the energy that actually performs the work, the real energy.

Reactive power consumers are transformers, reactors, induction motors and under-excited synchronous motors. They could be considered suppliers of reactive power—capacitors and super-excited synchronous motors—with the capacitor being the element most used for this purpose.

Examples of inductive reactive power devices:

- Induction motors
- Transformers
- Welding machines

The reactive power (Q) in alternating current is obtained by the following equations:

Single-phase circuit:

$$Q = V \cdot I \cdot \text{sen}\varphi \tag{3.3}$$

Three-phase circuit:

$$Q = \sqrt{3} \cdot V \cdot I \cdot \text{sen}\varphi \tag{3.4}$$

Unity: volt-ampere reactive (var)

3.3 Apparent Power

It is represented by the vector sum of the real power, with the reactive power being considered the total power consumed by the equipment. According to this apparent rated power, the equipment, for example, transformers, conductors, and others, are sized.

The apparent power (S) in alternating current is represented by the following equations:

Single-phase circuit:

$$S = V \cdot I \tag{3.5}$$

Three-phase circuit:

$$S = \sqrt{3} \cdot V \cdot I \tag{3.6}$$

Unit: volt-ampere (VA)

3.4 Power Triangle

The real, reactive and apparent powers can be represented by a right triangle called the power triangle. Recalling the fundamental concepts of trigonometry, we have the relationships between the sides and the angles of the right triangle as shown in Figure 3.1.

The Pythagorean theorem states that the hypotenuse (*H*) squared represents the square of the sum of the squares of the opposite leg (Ol) and the adjacent leg (Al), in accordance with the following equation:

$$H^2 = \text{Ol}^2 + \text{Al}^2 \tag{3.7}$$

The following trigonometric ratios, which relate to the angles of the right triangle, characterized by sine, cosine and tangent relations, are as follows:

Sine:

$$\sin\varphi = \frac{\text{Ol}}{H} \tag{3.8}$$

Cosine:

$$\cos\varphi = \frac{\text{Al}}{H} \tag{3.9}$$

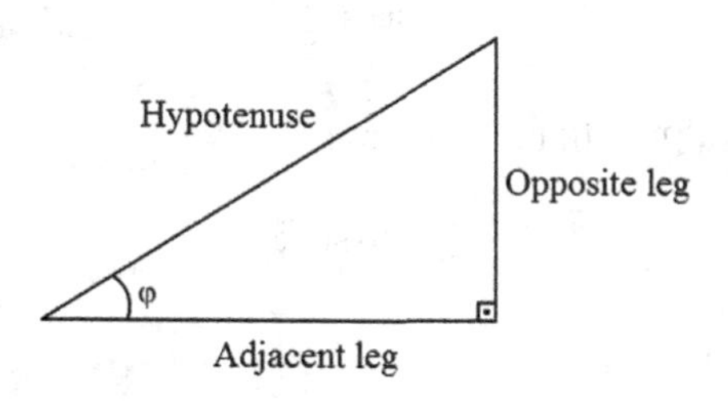

FIGURE 3.1
Right triangle.

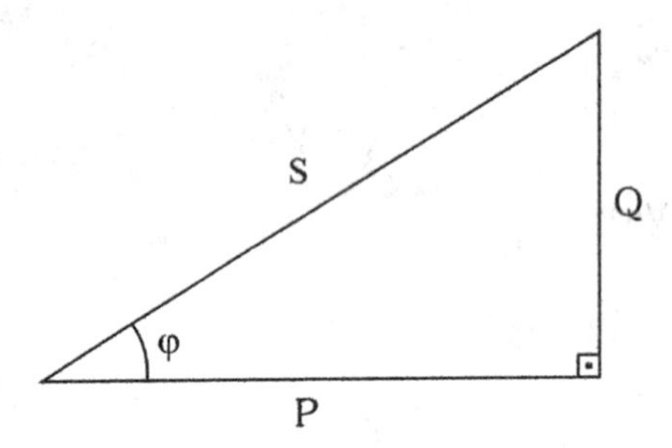

FIGURE 3.2
Electric power representation in a right triangle.

Tangent:

$$\tan\varphi = \frac{Ol}{Al} \tag{3.10}$$

The electric powers are represented in the right triangle in a graphical representation as shown in Figure 3.2.

The apparent power S is the geometric sum of the real power P with the reactive power Q. The powers are placed in the right triangle with Q as the opposite leg and P as the adjacent leg in relation to φ.

In this way, the relationship between the powers, given by the Pythagorean theorem, are:

$$S^2 = P^2 + Q^2 \tag{3.11}$$

It is possible to obtain the equations of the powers through the concepts of trigonometry, where:

Sine:

$$\sin\varphi = \frac{Q}{S} \tag{3.12}$$

The reactive power (Q), will be given by:

$$Q = S \cdot \sin\varphi \tag{3.13}$$

Cosine:

$$\cos\varphi = \frac{P}{S} \tag{3.14}$$

Thus, the active power (P) will be given by:

$$P = S \cdot \cos\varphi \tag{3.15}$$

The tangent function will give us the relation between reactive power (Q) and the real power (P).

Tangent:

$$\tan\varphi = \frac{Q}{P} \tag{3.16}$$

3.5 Power Factor

The power factor (cosφ) is the cosine of the phase shift between current and voltage. If the circuit is inductive, that is to say, a reactive energy consumer, when the voltage leads the current, the power factor is considered lagging.

In Figure 3.3, there is an example of a load applied to a purely inductive circuit, where the voltage leads currents by 90°.

If the circuit is capacitive, i.e., a reactive power supplier, the power factor is considered leading. In Figure 3.4, observe a purely capacitive circuit example, where the voltage lags current by 90°.

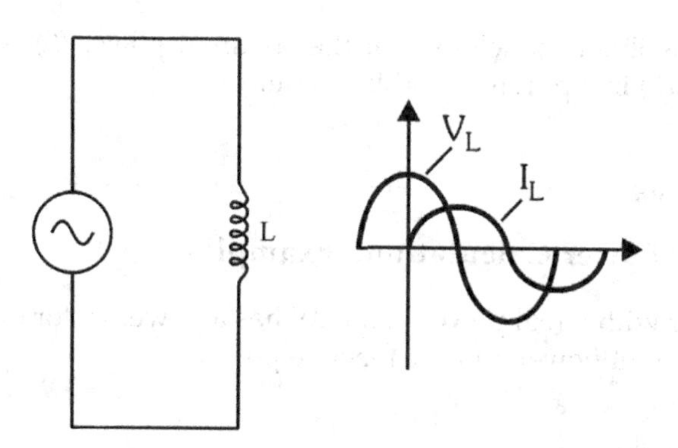

FIGURE 3.3
Purely inductive circuit.

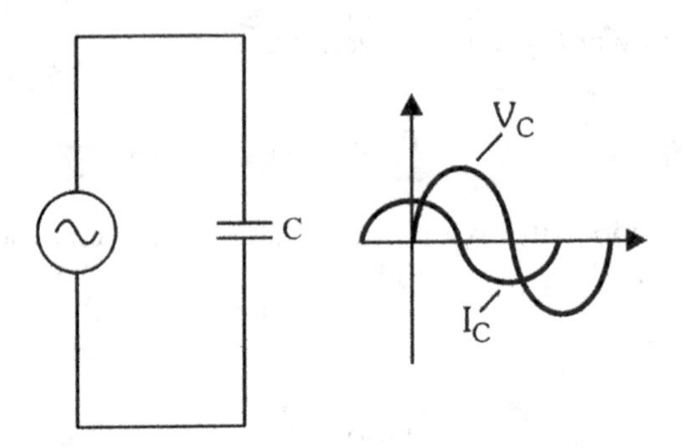

FIGURE 3.4
Purely capacitive circuit.

NOTE: In a purely resistive circuit, the voltage is in phase with the current, so there is no voltage lag or lead in relation to current. For this circuit, consider the following relationship:

$$\varphi = 0^0 \text{ and } \cos\varphi = 1$$

$$P = V \cdot I \cdot \cos\varphi$$

$$P = V \cdot I$$

Apparent power is:

$$S = V \cdot I$$

Thus:

$$S = P$$

Thus, it is possible to conclude that the apparent power (S) is equal to the active power (P) in a purely resistive circuit.

3.6 Electric Power Calculations Examples

1. A motor with a real power of 35 kW has a power factor of 0.85. Find the apparent power and the reactive power.

   Solution:

   Considering:

$$P = V \cdot I \cdot \cos\varphi$$

Apparent power:

$$S = \frac{P}{\cos\varphi} = \frac{35}{0.85} = 41.17\,\text{kVA}$$

Reactive power:

$$Q^2 = S^2 - P^2$$

$$Q = \sqrt{S^2 - P^2} = \sqrt{41.17^2 - 35^2} = 21.69\,\text{kvar}$$

2. A device has an apparent power of 100 kVA and a real power of 79 kW. Find the power factor and its reactive power.

   Solution:

   Considering:

$$P = V \cdot I \cdot \cos\varphi$$

Power Factor:

$$\cos\varphi = \frac{P}{S} = \frac{79}{100} = 0.79$$

Reactive Power:

$$Q = \sqrt{S^2 - P^2} = \sqrt{100^2 - 79^2} = 61.31\text{ kvar}$$

3.7 Power Factor in Industry

It is essential that a given industry make rational use of electricity. Through the installation power factor analysis, it is possible to know if electric energy is being used efficiently to achieve economic viability, avoiding waste and unnecessary expenses. A low power factor causes major problems with electrical installation, such as overloading the cables and transformers, increasing the voltage drop and increasing the electricity bill.

According to the established regulatory agencies of each country, it is always recommended to maintain the power factor as close as possible to the unit. It is also recommended that a reference limit be established for the inductive and capacitive power factor, as well as a form of evaluation and reactive energy billing criterion that determines that exceeding this defined limit adds penalties and, consequently, results in charges of the reactive power in electricity billing.

3.8 Causes of Low Power Factor

The main causes of low power factor are listed below:

- Oversized or small load motor
- Lamps with low power factor ballasts
- Air conditioning installations
- Welding machines
- Electronic equipment
- Over-sized transformers

3.9 Advantages of Power Factor Correction

The following are the main advantages of performing power factor correction in an electrical installation.

Voltage improvement: The current relative to the reactive power appears only in the inductive reactance. As this current is reduced by the capacitors, the total voltage drop is reduced to a value equal to the capacitor current multiplied by the reactance. Therefore, it is only necessary to know the rated power of the capacitor and the system reactance to know the voltage rise caused by the capacitors, which is in the range of 4%–5%. Although capacitors raise voltage levels, it is unfeasible to install them in industrial plants only for this purpose. The voltage increase should be considered an additional benefit of the capacitors.

Reduction of losses: RI^2 losses (Joule effect) vary from 2.5% to 7.5% of the kWh of the load, depending on the working hours at full load, conductor's size and length of feeders and distribution circuits. The losses are proportional to the square of the current and, as the current is reduced in the direct ratio of the power factor increase, the losses are inversely proportional to the square of the power factor.

Then, correcting the power factor presents the following advantages:

- Significant reduction in the electricity bill.
- Increasing the company's energy efficiency.
- Increase the useful life of installations and equipment.
- Joule effect reduction.
- Reduction of the reactive current in the electrical grid.

In addition to these advantages for the industry, there are the following benefits for the electric company: reactive power stops flowing in the transmission and distribution system, Joule loss reduces, transmission and distribution system capability increases and there are lower generation costs.

3.10 Methods to Correct the Power Factor

The power factor correction in an installation should be analysed with special care, avoiding immediate solutions that may lead to unsatisfactory technical and/or economic results. Criteria and experience are needed to make an appropriate compensation, analysing individual cases, as there is no standardised solution. In principle, the power factor correction can be achieved:

- By increasing the active energy consumption.
- Using synchronous motors.
- Installing capacitors.

Regardless of the method to be used, the ideal power factor, both for the installation (i.e., for the consumer) and for the electric company, would be the unitary, which means that there is no reactive power in the installation. However, this condition is generally unfeasible from the economic point of view and the value (0.95) is generally considered satisfactory.

The increase in the active energy consumption can be achieved by adding new loads with a high power factor or by increasing the load period whose power factors is high.

In addition to meeting the production needs of the industry, the load that will increase the consumption of real energy must be carefully chosen not to exceed the maximum contracted demand, which would lead to an increase in the electricity bill.

On the other hand, it must not be forgotten that this solution should not opposed with the need to save electricity. Thus, for example, it is not convenient to replace an oil furnace with an electric furnace (whose cosφ is practically equal to 1) only to improve the power factor. However, if the industry has two ovens, one electric and one oil, alternating, extending the periods of use of the electric furnace may be a good option to correct the industry power factor.

Synchronous motors can also work as "generators" of reactive power, either by moving mechanical loads or by running in a no-load condition (in this case, called "synchronous compensators"). This property is controlled by excitement of the magnetic field. When under-excited, they do not generate sufficient reactive power to meet their own needs and, therefore, must receive additional reactive power from the system. When overexcited (normal operation), they supply their own needs for reactive power and also provide reactive power to the system.

It is necessary to consider that correction with this method is not always economically viable, and is only recommended when large mechanical loads are applied with a power exceeding 50 HP (for example, large compressors) and run for long periods (more than eight hours a day). In such cases, the synchronous motor has the dual function of moving the load and correcting the installation power factor.

The use of capacitors is the most common method in industrial plants for the correction of power factor, being, in general, the most economical, allowing greater flexibility. The capacitors used are characterized by their rated power, their manufacture in single- and three-phase units, for high and low voltage with standard values of power, voltage and frequency, for being connected internally in delta and with powers of up to 50 kvar.

3.11 Capacitors Parameters

The capacitor is an electrical device used to apply capacitance into a circuit. It consists of a system of conductors and dielectrics that have a storing electrical energy property when subjected to an electric field. The main parameters of a capacitor are:

Capacitance: Capacitance value assigned by the manufacturer, typically in microFarads.

Voltage: The effective value of the sine-wave voltage between its terminals, for which a capacitor is designed in volts.

Current: The current value corresponding to its rated power, when the nominal voltage is applied to the capacitor at rated frequency.

Rated power: Reactive power under voltage and frequency at which the capacitor is designed in kvar.

The capacitors have a discharge device that is electrically (basically a resistor) connected between or embedded in the capacitor terminals, or connected between the power terminals. It reduces the voltage between their terminals when the capacitor is disconnected from the power supply. Figure 3.5 shows a capacitor used for power factor correction. A capacitor bank is a set of power capacitors with a given capacitance and reactive power connected to supporting structures, with the necessary control, maneuvering and protection devices, assembled to constitute a complete piece of equipment, as shown in Figure 3.6.

Capacitors are usually specified by their rated reactive power given in var (reactive volt-amp).

FIGURE 3.5
Capacitor used for power factor correction. (Courtesy of Weg.)

FIGURE 3.6
Capacitor bank used for power factor correction.

The capacitor rated power is considered when it is submitted to rated voltage and to frequencies at ambient temperature equal to 20°C. Thus, with the following equation, it is possible to determine the capacitance:

$$C = \frac{1000 \cdot P_c}{2 \cdot \pi \cdot f \cdot V_n^2} \tag{3.17}$$

where:

P_c is the rated reactive power in kvar
f is the rated frequency in Hz
V_n is the rated voltage in kV
C is the capacitance in μF

Table 3.1 shows the basic electrical characteristics of typical capacitors at 60 Hz.

TABLE 3.1

Typical Electrical Characteristics of Capacitors at 60 Hz

Voltage (V)	Reactive Power (kvar)	Capacitance (μF) (connection Δ)	Rated Current (A)
220V	1	18.3 × 3	2.62
	2	36.6 × 3	5.25
	3	54.8 × 3	7.87
380V	1	6.1 × 3	1.52
	2	12.3 × 3	3.03
	3	18.4 × 3	4.56
440V	1	4.6 × 3	1.31
	2	9.1 × 3	2.62
	3	13.7 × 3	3.94
480V	1	3.8 × 3	1.20
	2	7.7 × 3	2.41
	3	11.5 × 3	3.61

3.12 Power Factor Correction Using Capacitors

As previously mentioned, the use of capacitors has the function of power factor correction, since all installations that have motors, transformers and so on, have a reactive inductive power that does not perform work. The purpose of the capacitor is to reduce the amount of reactive power that the generators must supply.

Thus, in an installation where it is desired to correct the power factor from $\cos\varphi_1$ (installation power factor) to $\cos\varphi_2$ (desired power factor), i.e., reduce the amount of inductive reactive power from $kvar_1$ to $kvar_2$, we have:

$$Q_1 = P \cdot \tan\varphi_1 \quad (3.18)$$

$$Q_2 = P \cdot \tan\varphi_2 \quad (3.19)$$

The difference between the reactive powers (Q_1 and Q_2) will be given by:

$$Q_1 - Q_2 = P \cdot (\tan\varphi_1 - \tan\varphi_2) \quad (3.20)$$

This results in the following power factor correction are shown in diagram in Figure 3.7.

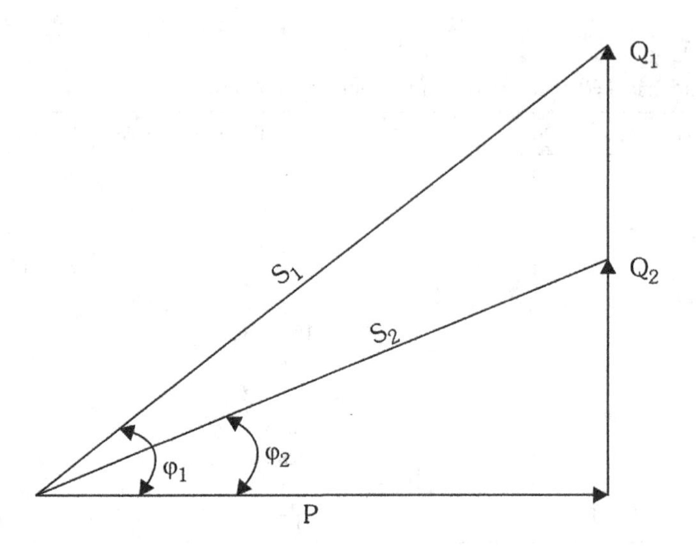

FIGURE 3.7
Diagram of the power factor correction.

In this way, the amount of reactive power (Ckvar) to correct the installation power factor is obtained by:

$$\text{Ckvar} = P.(\tan\varphi_1 - \tan\varphi_2) \tag{3.21}$$

3.12.1 Examples of Power Factor Correction

1. It is desired to correct the power factor of an electrical installation with the real power of 150 kW. It has a power factor of 0.7 and must be corrected to 0.92.

   Solution:

   For the power factors presented, we will have the following angles:

   Installation power factor:

   $$\cos\varphi_1 = 0.7 \text{ and } \varphi_1 = 45.57^0 \text{ and } \tan 45.57^0 = 1.02$$

   Desired power factor:

   $$\cos\varphi_2 = 0.92 \text{ and } \varphi_2 = 23.07 \text{ and } \tan 23.07^0 = 0.42$$

Thus, we find the reactive power required for the correction of the power factor with the following equation:

$$\text{Ckvar} = P \cdot (\tan\varphi_1 - \tan\varphi_2) = 150.(1.02 - 0.42) = 90\,\text{kvar}$$

2. Given the following electrical installation, find the installation power factor (PF) and correct it to 0.92.
   a. 2 motors of 10 HP with power factor = 0.85
   b. 1 motor of 15 HP with power factor = 0.85
   c. 2 motors of 30 HP with power factor = 0.87

Solution:
To solve this problem, you first have to calculate the real and apparent powers of the motor group. The sum of the real powers (PT) will be given by:

$$P_1 = 2.10.746 = 14920\,\text{W}$$

$$P_2 = 1.15.746 = 11190\,\text{W}$$

$$P_3 = 2.30.746 = 44760\,\text{W}$$

$$\text{PT} = 14920 + 11190 + 44760 = 70870\,\text{W}$$

Considering $S = P/\text{PF}$, the sum of apparent power (ST) will be given by:

$$S_1 = \frac{P_1}{\text{PF}_1} = \frac{14920}{0.85} = 17552.94\,\text{VA}$$

$$S_2 = \frac{P_2}{\text{PF}_2} = \frac{11190}{0.85} = 13164.70\,\text{VA}$$

$$S_3 = \frac{P_3}{\text{PF}_3} = \frac{44760}{0.87} = 51448.27\,\text{VA}$$

$$\text{ST} = 17552.94 + 13164.70 + 51448.27 = 82165.91\,\text{VA}$$

In this way, the current power factor of the installation (PF_A) will be obtained by the relation:

$$PF_A = \frac{PT}{ST} = \frac{70870}{82165.91} = 0.86$$

Now, we will obtain the following angles:
For the actual installation power factor:

$$\cos\varphi_1 = 0.86 \text{ and } \varphi_1 = 30.68^0 \text{ and } \tan 30.68^0 = 0.59$$

For the desired power factor:

$$\cos\varphi_2 = 0.92 \text{ and } \varphi_2 = 23.07^0 \text{ and } \tan 23.07^0 = 0.42$$

Thus, we calculate the reactive power that is necessary for the correction of the installation power factor:

$$\text{Ckvar} = P.(\tan\varphi_1 - \tan\varphi_2) = 70870.(0.59 - 0.42) = 12047.9\,\text{kvar}$$

3.13 Location of Capacitors

For power factor correction, capacitors can be located in four positions shown below:

- Correction in the input of the high voltage distribution system.
- Correction the input of the low voltage distribution system.
- Localized correction.
- Mixed correction.

Figure 3.8 shows the location of the capacitors,

- **C1**: capacitor installed directly in the load (individual power factor correction).
- **C2**: capacitor installed in the main low voltage panel.

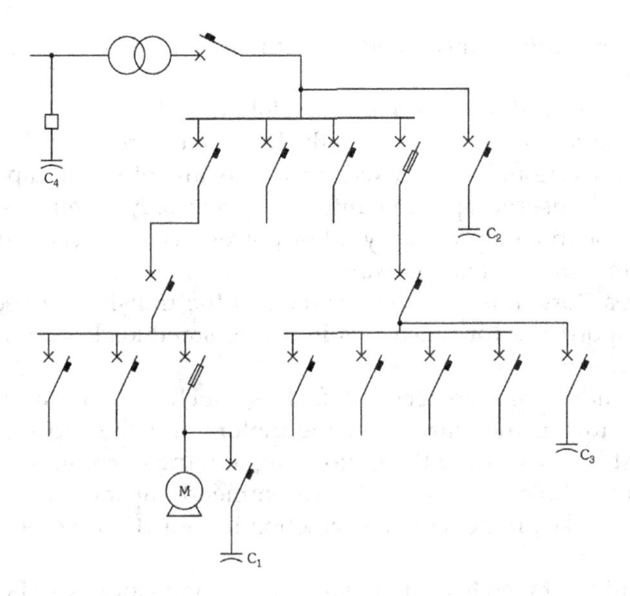

FIGURE 3.8
Location of capacitors to correct power factor.

- **C3**: capacitor installed in a low voltage secondary panel.
- **C4**: capacitor installed in the high voltage power supply.

The most effective method of power factor correction is the installation of capacitors in each motor to make a localized correction, because, with this type of correction, we have:

- Reduction of Joule losses in the whole installation.
- Single drive for motor and capacitor.
- Capacitive reactive power generated only when necessary.

However, due to the high cost, this method becomes economically unfeasible.

The power factor correction at the high voltage power supply only corrects the power factor seen by the energy company, and all the drawbacks mentioned above remain in the installation. In addition, at the moment of maneuvering, a series of problems arise with sparking and great probability of overvoltage, since it is impracticable to install variable capacitors. Therefore, the installation of a high voltage capacitors bank is only recommended in large consumer units, since in addition to all the problems mentioned, the cost is high.

The best technical and economical solution is the correction at the low voltage power supply, together with the localized correction, characterizing the mixed correction. In the mixed correction, an automatic capacitor bank is used, which inserts capacitors into the system only when it is necessary, because, as mentioned previously, a low power factor can cause overvoltage and fines from the electric company.

Fixed capacitors must also be installed in the transformer secondary to correct the power factor when it is in no- or small-load condition, as presented in Table 3.2.

In this individual correction, factors, such as the occurrence of harmonics during the starting of the motors and the capacitor current, which must be lower than the motor magnetizing current, must be taken into account. Table 3.3 shows the recommend capacitors that must be installed to make an individual correction in induction motors in a 60 Hz rated frequency.

Care should be taken to avoid the use of capacitors for power factor correction in motors in the following cases:

- Motors rotating in a reversed operation.
- Subject to excessive starts.
- Activation of elevators or cranes.
- Running at more than one speed.

TABLE 3.2

Recommended Capacitor to Correct Power Factor in Transformers at No-Load Condition

Transformer (kVA)	Capacitor (kvar)
15	0.81
30	1.37
45	1.84
75	2.57
112.5	3.42
150	4.28
225	5.77
300	7.13
750	9.98
1000	12.35
1500	14.25

TABLE 3.3

Typical Capacitor to Make Individual Power Factor Correction in Motors

	Synchronous Speed (rpm)	
	3.600	1.800
Motor (cv)	kvar	kvar
5	2	2
7.5	2.5	2.5
10	3	3
15	4	4
20	5	5
25	6	6
30	7	7
40	9	9
50	12	11
60	14	14
75	17	16
100	22	21
125	27	26
150	32.5	30
200	40	37.5
250	50	45
300	57.5	52.5
400	70	65
500	77.5	72.5

3.14 Real Power in Three-Phase Motors

The real power consumed by the three-phase motor is represented by the following equation, independent of the motor connection:

$$P_n = I \cdot V_L \cdot \cos\varphi \cdot \eta \cdot \sqrt{3} \tag{3.22}$$

where:

P_n is the real power in watts
V_L is the voltage between two phases in volts
I is the current in phase of motor in amperes
$\cos\varphi$ is the power factor
η is the efficiency

3.15 Power Factor in Induction Motors

The same way of any load connected to the power supply, three-phase induction motors have a rated power factor. The power factor tends to be higher as the motor rated power increases. Figure 3.9 shows this relation between cosφ and rated real power (PR).

As load is applied to the motor, the armature current increases and the phase shift between the voltage applied at the terminals and the current circulating in the armature decreases. Thus, as the motor receives load (P = actual real power), its power factor increases, as indicated in Figure 3.10.

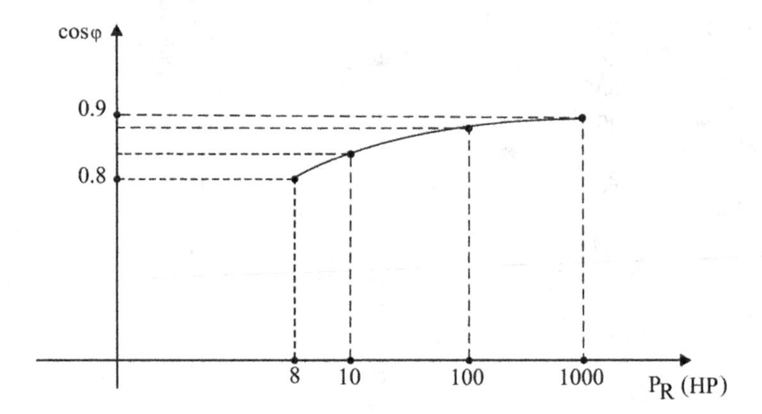

FIGURE 3.9
Power factor by rated real power curve.

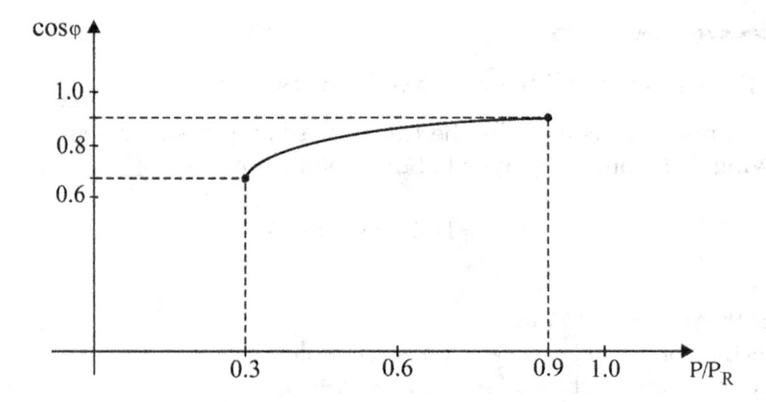

FIGURE 3.10
Power factor curve by load on the motor shaft.

The power factor trend of increasing with the load addition on the motor shaft is observed in practically all three-phase motors. It is more visible on very small power motors and less visible on larger power motors.

3.16 Power Characteristics of a Three-Phase Motor

To understand the power characteristics of a three-phase motor, we will consider the following example.

- Determine the real power and current consumed in a motor of 20 HP, 220 V, with an efficiency of 89.3% and a power factor of 0.79, with a load of 75%.

Solution:
First, it is necessary to convert the real power ($P(w)$) in HP to watts:

$$P(w) = 20.746 = 14920\,\text{W}$$

This is the real power available on the motor shaft.

Taking into account the motor efficiency, the real power consumed from the network is:

$$\eta = \frac{P(w)}{P_{EL}} \tag{3.23}$$

where:
P_{EL} is the real power
η is the motor efficiency

So:

$$P_{EL} = \frac{14920}{0.893} = 16707.72\,\text{W}$$

The apparent power (S) at the motor terminals shall be given by:

$$S = \frac{P_{EL}}{\text{PF}} = \frac{16707.72}{0.79} = 21149.02\,\text{VA}$$

The rated motor current will be given by Equation (3.6):

$$S = \sqrt{3} \cdot V \cdot I$$

So we have:

$$I = \frac{21149.02}{\sqrt{3}.220} = 55.50\,\text{A}$$

Exercises

1. What is the definition of real, reactive and apparent power?
2. Represent the electric powers in the power triangle and show the mathematical relationship between them.
3. Conceive power factor.
4. An equipment has a real power of 500 kW with power factor of 0.5. Find their apparent and reactive powers.
5. What are the causes of a low power factor?
6. List the advantages of power factor correction.
7. Find the installation power factor and correct to 0.96 in the following industry:
   a. 30 motors of 10 HP with power factor = 0.85.
   b. 100 motors of 50 HP with power factor = 0.87.
   c. 3 motors of 150 HP with power factor = 0.88.
8. Describe the methods for correcting the power factor in an installation.
9. Name and describe possible capacitor location points for power factor correction.
10. How does the power factor of a motor change according to its rated power and load in the shaft?

4

Motor Starter Components

In order to make possible the design and implementation of motor starter circuits, it is essential to know the devices that receive electrical circuit signals and drive the electric motors properly. The main functions that must be considered in a motor start circuit are: switching, protection against short circuit and overload, as we can see in Figure 4.1.

In this chapter, we will consider the two main organizations that determine the rules used in electrical motors: the National Electrical Manufacturers Association (NEMA), used preferably in the USA, and the International Electrotechnical Commission (IEC), widely used in Europe, Asia and other parts of the world. NEMA-rated products are generally heavy duty and are able to be employed in a broad range of applications. Typically, the devices are more expensive than IEC-rated ones. On the other hand, IEC-rated devices are often less expensive, smaller in size and are designed for specific motor performance requirements.

In the following are presented the main devices that are employed to start motors.

4.1 Push Buttons

Push buttons are manually operated and have the purpose of momentarily interrupting or establishing, by pulse, a control circuit to start, interrupt or command an automation process. Figure 4.2 shows commonly used push-buttons employed in a start induction motor circuit.

Push-buttons only remain activated by applying external force. When the force stops, the device returns to the starting position.

There are two types of contact: normally open and normally closed.

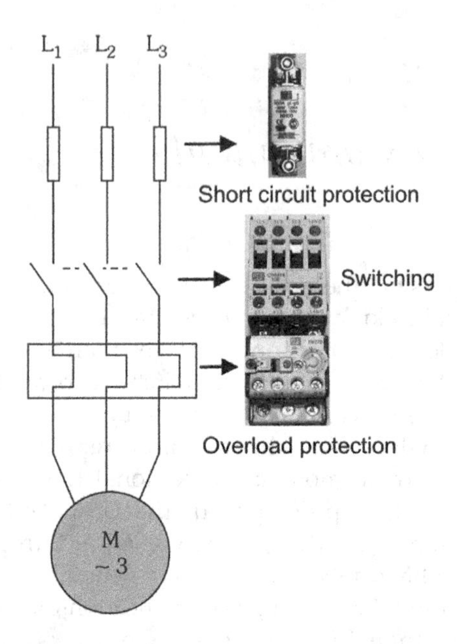

FIGURE 4.1
Functions assigned to starting circuits.

FIGURE 4.2
Typical push-button used in a start induction motor circuit.

Normally open contact (NO): Its original position is open (i.e., it remains open until an external force is applied).

There is a digit that represents the function code for the auxiliary contacts that are normally open, which are 3 and 4. Figure 4.3 shows the graphical representation of a NO contact for IEC and NEMA standards.

Normally closed contact (NC): Its original position is closed (i.e., it remains closed until an external force is applied). There is a digit that represents the function code for the auxiliary contacts for normally closed, which are 1 and 2. Figure 4.4 shows the graphical representation of an NC contact for IEC and NEMA standards.

Figure 4.5 shows in detail the operation of typical contacts that are normally open (a) and closed (b).

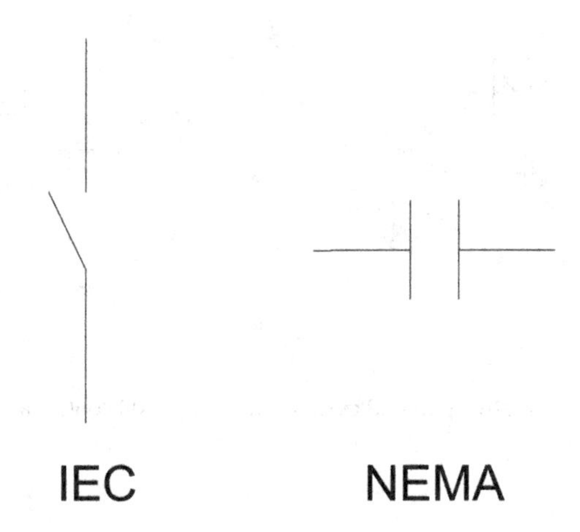

FIGURE 4.3
Graphical representation of NO contact by IEC and NEMA standards.

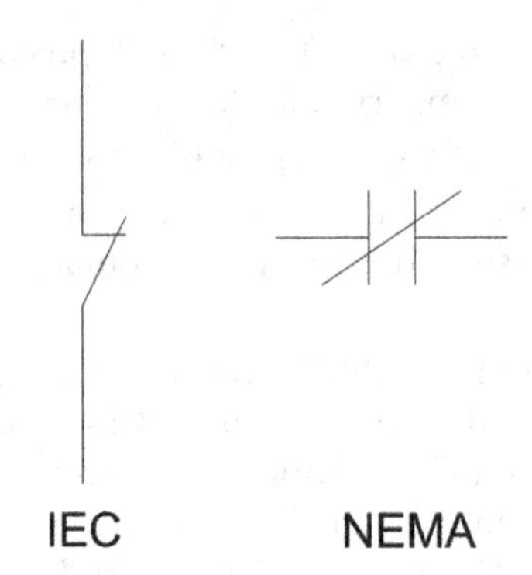

FIGURE 4.4
Graphical representation of NC contact by IEC and NEMA standards.

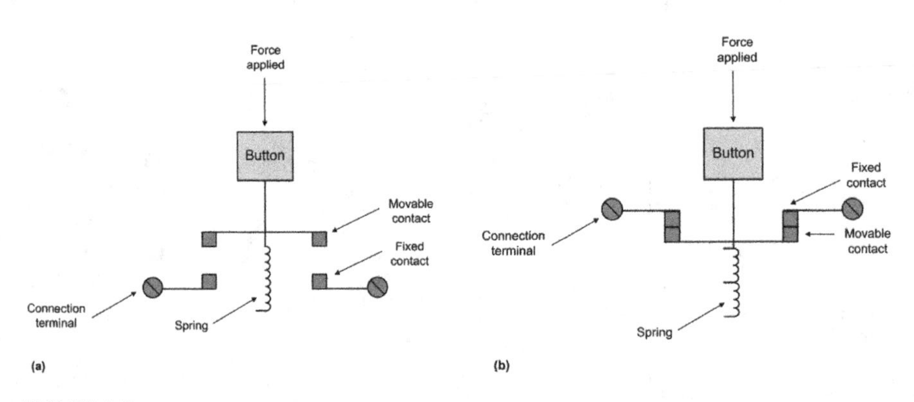

FIGURE 4.5
Detailed operation of normally open and closed contacts. (a) NO contact and (b) NC contact.

4.2 Selector Switch

This switch has two or more positions and allows selection of one of several positions in a given process with a common point of contact (C). This type of switch typically presents an NC and an NO contact, as shown in Figure 4.6.

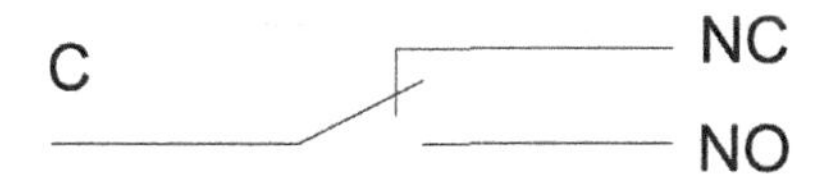

FIGURE 4.6
Representation of a typical selector switch.

4.3 Limit Switches

These switches are considered auxiliary control and actuating devices that operate in a circuit function, such as:

- Contactor control.
- Signalling circuit command to indicate the position of a particular element in movement.

This switch basically consists of a lever or rod, with or without pulleys at the end, which transmits the movement to the contacts that open or close according to their function, which could be:

Control: Signals the start or stop points of a given process.

Safety: Turns off equipment when there is a door opening or equipment alarm.

Figure 4.7 shows a typical limit switch with its main parts.

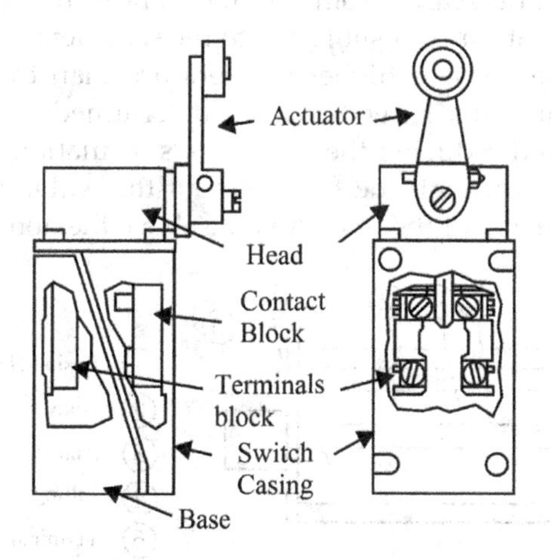

FIGURE 4.7
Limit switch description parts.

4.4 Fuses

This device is responsible for protecting the motor against short circuit. It also protects feeder conductors and control devices in the event of an internal short circuit. Its operation is based on a properly designed fuse element that opens the circuit, interrupting when a fault occurs.

The fuse element is basically a wire or a blade, usually of copper, silver, tin, lead or alloy, allotted within the fuse body, typically of porcelain, and hermetically sealed. Most fuses have an indication to check the fuse integrity. This indicator is made of a wire, usually steel, connected in parallel with the fuse element that releases a spring after its operation, releasing an external indication attached to the fuse body.

Inside the fuses, there is a granular material to help extinguish the electric arc; quartz sand of convenient granulometry is generally used. In Figure 4.8, a typical fuse representation is presented.

The fuse element may assume different forms according to its rated current and may be composed of one or more parallel brands with reduced cross sections. In the fuse, there is a soldering point where the melting temperature is less than that of the fuse element.

When the fuse is operating in a steady state condition, the conductor and fuse element are submitted through the same current that produces heating. The conductor temperature reaches a value of T_1. Because it has a high electrical resistance, the fuse has a higher heating (T_2), resulting in a high temperature at the midpoint of the fuse element, as shown in Figure 4.9.

The temperature decreases from the midpoint to the fuse element ends. The connection points are not subjected to the same temperature of the midpoint, however they have a higher temperature than the conductors. The temperature T_3 must not exceed a value determined by the standards to avoid damaging and reducing the conductor's insulation lifespan. The current that circulates through the fuse without this value being exceeded is called the rated current of the fuse. A value above the nominal causes a fuse

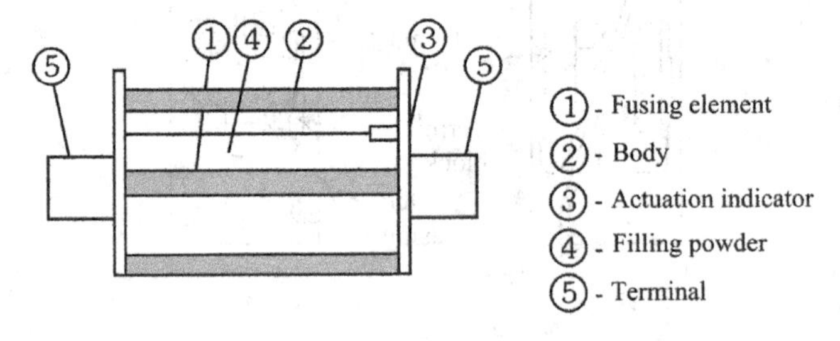

FIGURE 4.8
Constructive elements of fuse.

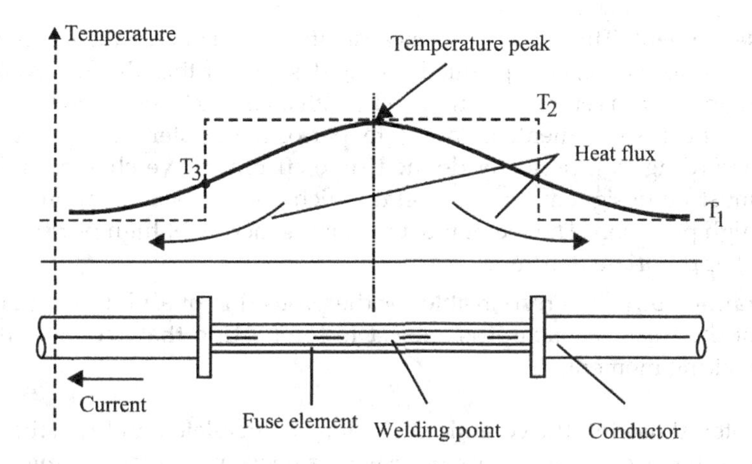

FIGURE 4.9
Internal temperature behaviour in a fuse.

element rupture according to its time-current characteristics curve, causing the circuit to interrupt.

When the current flowing through the fuse is much higher than rated, such as ten times, the reduced cross section of the fuse element melts before the soldering point due to the high current flowing through it.

When the melt occurs, the fuse element is mechanically interrupted, but the current is not interrupted fully and is maintained by an electric arc. The melt and the electric arc cause the fuse element metal to evaporate. The arc is surrounded by the extinguishing element, vaporizes and the metal vapour is pushed against the sand, where much of the arc is extinguished. The sand penetrates and withdraws the heat energy from the arc, extinguishing it.

Thus, it is possible to summarize the mains fuse parts, as described below:

Base: The current enters in the base and circulates to the external contact of the fuse through a contact surface between the base metals and the external contact. Special care must be taken with these surfaces in regard to oxidation; if so, the current flowing through them leads to a temperature rise that masks the time-current curve, which necessarily characterizes a fuse.

It is essential to choose a metal or metal alloy used in the construction of the respective contacts, which do not oxidize, or which oxidize very slowly. One of the solutions is the silver plating of the contact pieces, because silver is a good electric conductor and its oxidation is slow.

Fuse element: This element is encapsulated to guarantee the safety of its performance, as is planned during design. For this, the fuse is surrounded by a ceramic external body, with metallic closure at both ends.

The fuse element, in order to perform its interrupting action according to a perfectly defined time-current curve characteristic, must be designed using a metal which allows its calibration with high precision. The metal must be homogeneous, of high purity and of appropriate hardness.

Ceramic body: It is responsible for the protection of all internal parts of the fuse. As such, it is subject to the heating that occurs at the melting moment.

The material used in the ceramic body must be insulated and remain isolative after the fuse element melts; otherwise, a new current-conducting circuit may form after the fuse element melts. The material must also withstand high temperatures, without changing its insulating properties, since some materials, when subjected to high temperatures, lose their insulating characteristics. It should also withstand the pressures from inside to outside, which appears in the act of melting the fuse element from the thermal expansion of the extinguishing medium and internal gases; porcelain or soapstone type of insulating ceramics are recommended.

4.4.1 Short Circuit Definitions

To better understand the concepts related to the fuse, consider Figure 4.10, which shows what happens in an alternating current system where a short circuit happens.

According to the figure, the main aspects in a short circuit occurrence are present:

Joule integral (I^2t): Is the integral squared current in a specific time interval, such as:

$$I^2t = \int_{t_1}^{t_1} i^2 \mathrm{d}t \, [\mathrm{A}^2\mathrm{s}] \tag{4.1}$$

It expresses a measure of the thermal energy associated with current flow.

Breaking capacity: Also called interrupting rating or short circuit rating, is the maximum current that a fuse can interrupt safely.

Melting time: Amount of time to melt the fuse after a short circuit occurs.

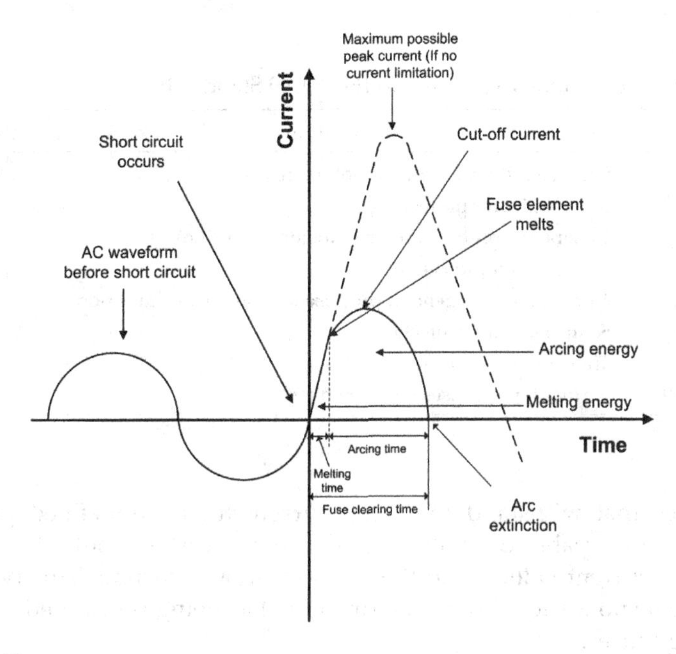

FIGURE 4.10
Alternating current (AC) waveform when a short circuit occurs.

Arcing time: The amount of time after the fuse has melted until the short current is interrupted (cleared).

Clearing time: The time between the short circuit occurrence and the current interrupting. It is the sum of melting and arcing time.

Arcing energy: Is the energy (I^2t) in the arcing time.

Melting energy: Is the minimum energy (I^2t) required to melt the fuse.

Clearing energy: Is the sum of arcing and melting energy representing the total I^2t in a short circuit occurrence.

4.4.2 Operation Classes

In addition to the rated current, the fuses are classified by the IEC 60269 standard, as the operation class is represented by the lowercase letters g and a (the type of equipment are represented by uppercase letters). The operation classes are described below:

g: Fuses that withstand the rated current indefinitely and are able to operate from the lowest overcurrent value up to the rated shutdown current. They are considered full-range fuses.

TABLE 4.1

Typical Fuse Applications according to IEC 60269 Standard

Type	Application	Protection
aM	Short circuit protection of motor circuits	Partial range
aR	Semiconductor protection	
gG	General purpose (mainly conductor protection)	Full range
gM	Motor circuit protection	
gN	North American general purpose for conductor protection	
gR, gS	Semiconductor protection	
gTr	Transformer protection	
gL, gF, gI, gII	Former type of fuses replaced by gG	

a: Fuses that withstand the rated current for an undefined period and are capable of switching off from a certain multiple of the rated current value up to the rated shutdown current. This type of fuse responds to a high overcurrent value, being considered partial range fuses.

The classes are used together to represent fuse applications as presented in Table 4.1.

4.4.3 Types of Fuses (European)

In the following are presented the main fuse systems employed in Europe and other countries under the influence of the IEC standard.

4.4.3.1 D Type

Also called Diazed type, which was derived from "diametral abgestruft," that is "diametral steps." The term "bottle" type is also used, but the official name is "D" type which is derived from Diazed. It is suitable for nominal currents of 2–63 A, breaking capacity of 50 kA and maximum voltage of 500 V. Figure 4.11 shows the D fuse representation (a) and a commercial D Fuse (b).

4.4.3.1.1 D-Fuse Components

Fuse Element: Is the element that melts when driven by a current greater than a reference value for a specified time. The fuse also has metal ends in one of which is located the indicator wire that shows when the melting occurred.

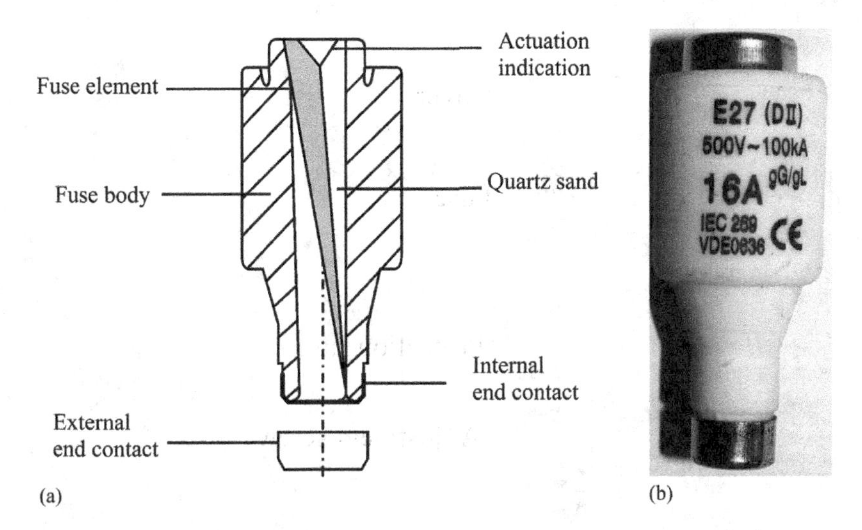

FIGURE 4.11
Representation of a "D" fuse (a) and a commercial D fuse (b).

Base: A fixed part of the porcelain device in which the energy input/output is connected via contacts and terminals and accommodates all fuse safety components.

Screw cap: Porcelain with a threaded metal body. Its function is to fix the fuse to the base.

Protective ring: The porcelain element in a ring shape designed to prevent accidental contact of the fuse.

Indicator wire: As the D-fuse is encapsulated, there is an apparent difficulty in checking whether it is perfect or damaged due to the enclosure. This difficulty is overcome by checking the positioning of the indicator wire. When the fuse indicator is retracted in its mounting position, the fuse is perfect, and when it is ejected (in case of type D), the fuse is "blown" and needs to be replaced. Figure 4.12 shows the D-fuse components.

4.4.3.1.2 D-Fuse Time-Current Characteristic Curve

The time by current characteristic curve shows a graphical representation of the average melting time of the fuse elements by considering the ambient temperature, relative to the rms current, as shown in Figure 4.13.

NOTE: The fuses are typically designed to operate in altitude ranges under 2000 meters and a temperature range up to 25°C ± 5°C. If the installation does not fall within these limits, the derating factor presented by the manufacturers must be considered.

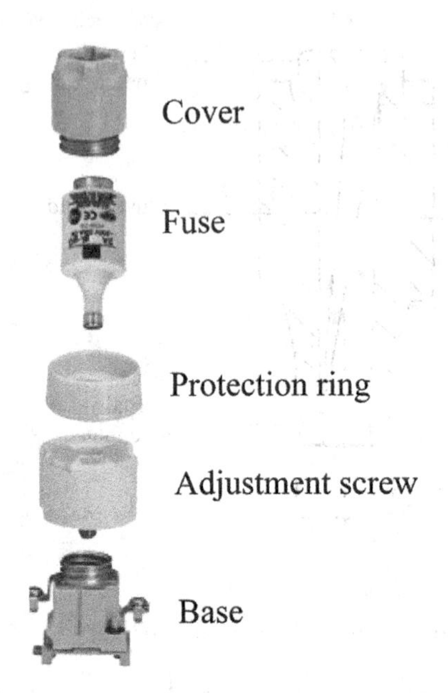

FIGURE 4.12
Components of "D" fuse.

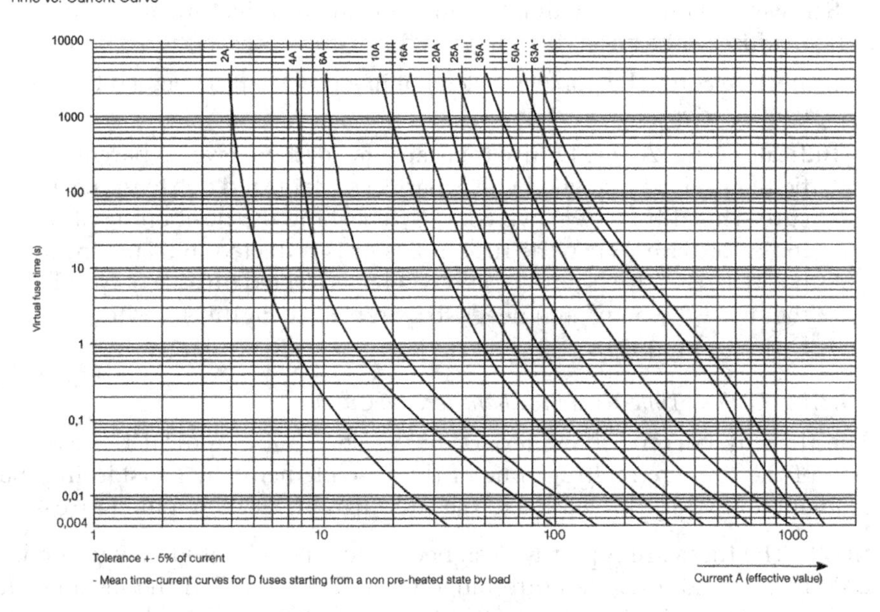

FIGURE 4.13
Typical D fuse time by current characteristic.

4.4.3.2 Type NH Fuse

It is a high interrupting capacity fuse for industrial use. NH is an abbreviation of Niederspannungs Hochleitungs, which is German for low-voltage high-breaking-capacity. It is designed for standard current from 4–630 A, a breaking capacity of 120 kA and maximum voltage of 500 V. Figure 4.14 shows a commercial NH Fuse.

NH fuses are suitable for protecting circuits that, in service, are submitted to short-term overloads, such as during start-up for three-phase motors with squirrel cage rotors, and for maintaining their characteristics according to their time by current curves when submitted to small, long-lasting overloads.

Another important feature is the limitation of the intensity of the short-circuit currents due to its short melting time (<4 ms). Table 4.2 shows NH fuses, their respective codes and typical interrupting rating currents.

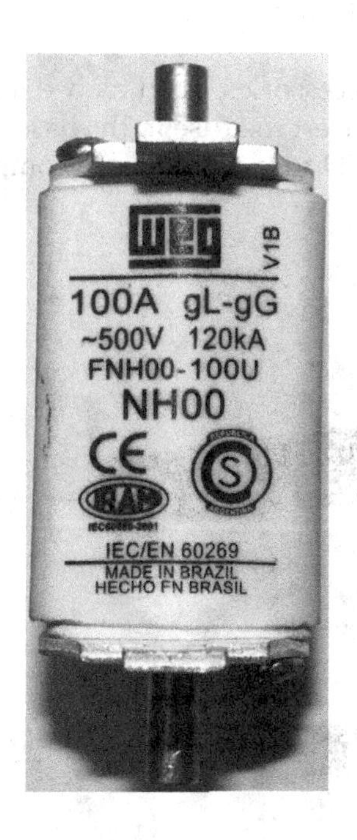

FIGURE 4.14
NF fuse.

TABLE 4.2

NH Code Fuses and Interrupting Rating Currents

Code	Interrupting Rating (A)
NH 00	4–160 A
NH 1	50–250 A
NH 2	125–400 A
NH 3	315–630 A

4.4.3.2.1 NH Fuse Components

The main NH fuse components are presented in the following:

Base: Soapstone-based building material. It has contacts in the form of silver claws pressed by springs.

Fuse: Rectangular porcelain body with knife-shaped metal ends. Inside the porcelain body are the fuse element and the indicator of melting, immersed in special sand of suitable granulometry whose function is the extinction of the arc.

NH fuses also feature accessories, such as a detachable handle, which is made of plastic, used to insert the fuse in the base.

Figure 4.15 shows the NH fuse with it main components.

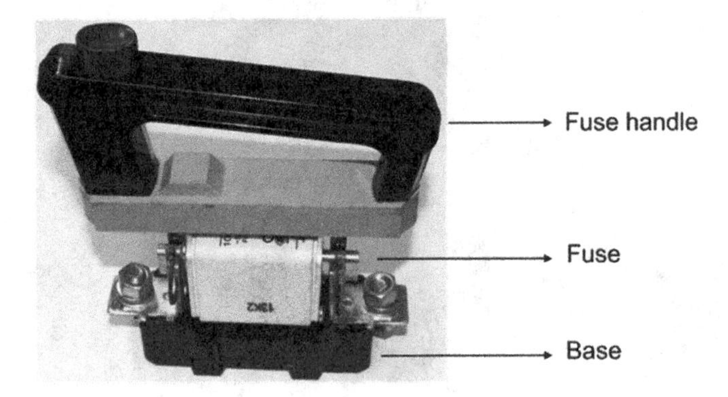

FIGURE 4.15
NH fuse main components.

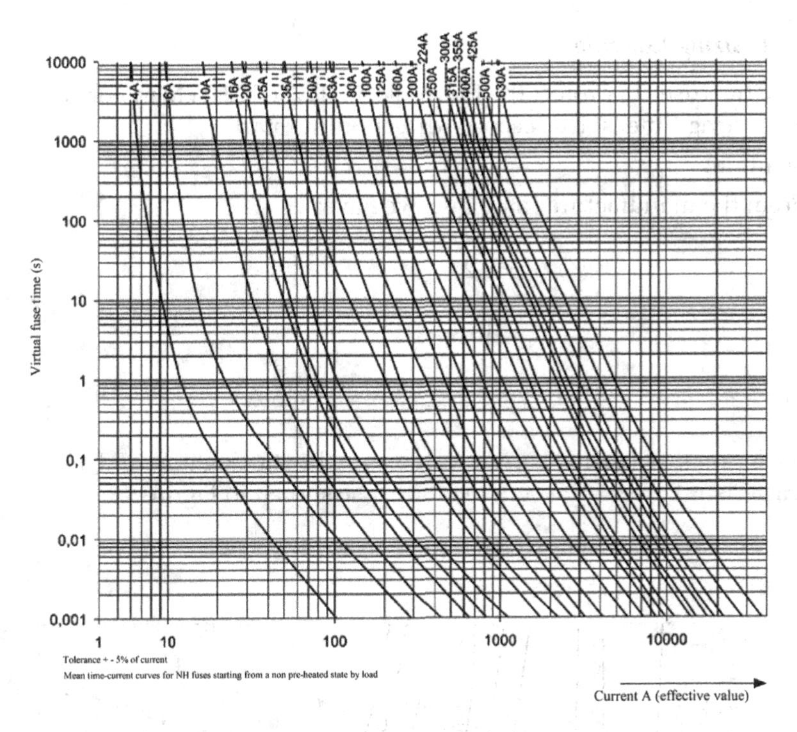

FIGURE 4.16
Typical NH time by current characteristic curve.

4.4.3.2.2 NH Fuse Time-Current Characteristic Curve

Figure 4.16 presents a typical NH Fuse Time-current characteristic curve.

4.4.4 Sizing of Fuses

In the fuse sizing for the European standard, the following aspects must be taken into account:

1. Virtual melting time (starting time and current): The fuses must withstand, without melting, the starting current peak (*Is*) during the motor start time (*Ts*). With the values of *Is* and *Ts*, enter in the time by current curve to size the fuse.
2. $I_{Fuse} \geq 1.2 \cdot FLC$ must be sized to a current at least 20% higher than the rated current of the motor (FLC) that it protects, avoiding premature fatigue, increasing its useful life.
3. The fuses should also protect the contactors and overload relays. This check is done in overload relay and with contactor tables provided by manufactures.

4.4.4.1 Sizing Example

To size the fuses to protect the four-pole 5 HP, 220 V/60 Hz motor, assuming your starting time is five seconds (direct-online start):

Solution:

From the manufacture catalogue, we have:

$$I_{\text{start}}/I_r = 8.2$$

$$I_r = 13.8\text{ A}$$

$$I_{\text{start}} = 8.2 \cdot 13.8\text{ A} = 113.16\text{ A}$$

Considering this starting current value with the starting time and analyse the characteristic current-time curve of the fuse shown in Figure 4.17.

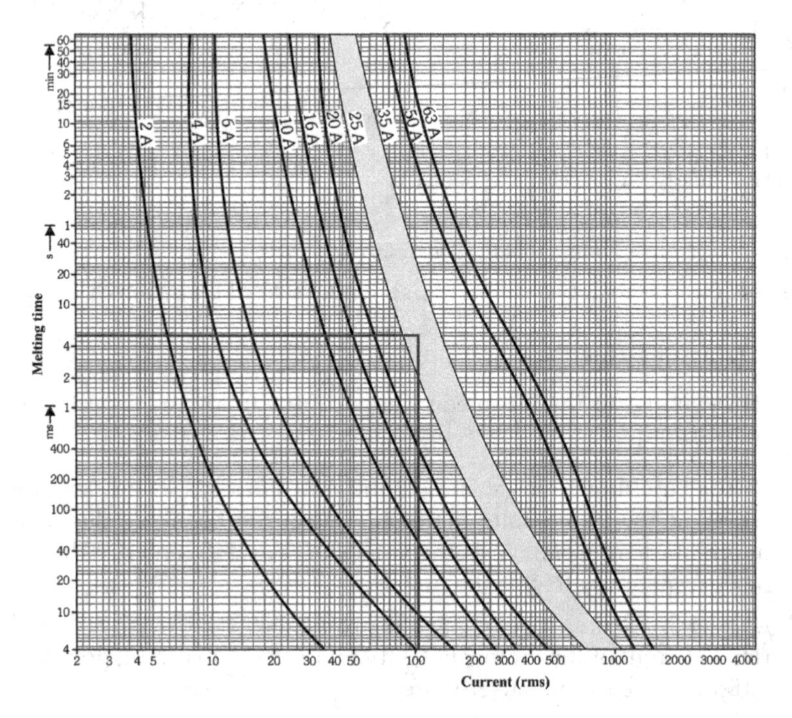

FIGURE 4.17
Characteristic current time used to size the fuse.

With a value of 113.16 A and a starting time of 5 seconds, the fuse chosen is 35 A.

Consider the second criterion:

$$I_{\text{fuse}} \geq 1.2 \cdot \text{FLC}, I_{\text{fuse}}\ (35\text{A}) \geq 16.56\ \text{A} \qquad \textbf{Criterion ok}$$

Thus, the 35 A is the correct fuse correct for this motor.

4.4.5 Types of Fuses (US)

In the in United States, the most common fuse types are presented in the following:

Class H: Type of cartridge fuse, also called an national electrical code (NEC) dimensional fuse, is a general purpose branch circuit, lighting circuit, for the protection of non-inductive equipment, such as electric ovens and resistance heaters. Class H fuses are available in renewable and non-renewable models where renewable type allow the user to replace the internal fusible link when a fault occurs.

Class G: Employed in lighting and appliance panel boards with a special fusible-switch unit. This type is a non-renewable cartridge for use in AC circuits (interrupting ratings to 100 kA rms); it also is available with direct current (DC) ratings.

Class K: Non-renewable fuses available in 250VAC and 600VAC ratings, with current ratings from 0 to 600 A. Class K fuses are also available with DC ratings.

Class R: As with class K fuses are non-renewable fuses made in 250VAC and 600VAC ratings, with current ratings from 0 to 600 A, and are available with DC ratings. They are typically labelled with current-limiting characteristics and could have time delay functions.

Class J: Non-renewable fuses are considered current-limiting, with current ratings from 0A to 600A; a voltage rating of 600VAC is also available with DC ratings. The typical interrupting rating is 200kA rms symmetrical and presents a time-delay characteristic.

Class L: Non-renewable fuses designed for the protection of feeders and service entrance equipment with typical ranges of 601–6000 A.

Class T: Non-renewable fuses are designed for protection of feeders and branch circuits. It has a typical current range from 0 to 1200 A and is also available with DC ratings.

Class CC: Non-renewable fuses are designed for the protection of components sensitive to short-time overloads, non-inductive loads, and short circuit protection of motor circuits. Presents a typical current range from 0 to 30 A.

There is a very large demand in the United States for electric motors installations, with specific rules being applied that dictate the rules for the use of fuses, emphasizing NEC 430.52 (Table 4.3).

The sizing should be done considering the full load amp values shown in Table 4.4 (NEC430.250), not considering motor nameplate values.

NOTE: The dual-element fuse is built by using two fusible elements connected in series and contained in one tube, with the thermal cut-out element and the fuse link element surrounded with arc-extinguishing filler. This type of fuse provides both short circuit protection and overload protection in circuits subject to temporary overloads and surge currents. The fuse has a distinct and separate overload element and short circuit element.

Example: What time-delay fuse is recommended for a 15 HP, 230 Volts, 3 phase motor?

Answer:

Full load Current (15 HP/3 phase motor) = 42 A, considering a 175% (time-delay fuse) = 73.5 A.

TABLE 4.3

(NEC 430.52) Lists the Maximum Ratings for Non-Time-Delay Fuses, Dual Element (Time-Delay) Fuses, Instantaneous Trip Circuit Breakers, and Inverse Time Circuit Breakers

	Percentage of Full-Load Current			
Type of Motor	**Non-Time Delay Fuse**	**Dual Element (Time-Delay Fuse)**	**Instantaneous Trip Breaker**	**Inverse Time Breaker**
Single-phase motors	300	175	800	250
AC poly-phase motors other than wound rotor				
Squirrel cage other than Design B energy-efficient	300	175	800	250
Design B energy-efficient	300	175	1100	250
Synchronous	300	175	800	250
Wound rotor	150	150	800	150
Direct current (constant voltage)	150	150	250	150

TABLE 4.4

Full Load Current, Three-Phase Alternating Current Motors (NEC430.250)

	Squirrel Cage and Induction Motor Current (Amperes)							Synchronous Type Unity Power Factor Current[a] (Amperes)			
HP	115V	200V	208V	230V	460V	575V	2300V	230V	460V	575V	2300V
1/2	4.4	2.5	2.4	2.2	1.1	0.9	_	_	_	_	_
3/4	6.4	3.7	3.5	3.2	1.6	1.3	_	_	_	_	_
1	8.4	4.8	4.6	4.2	2.1	1.7	_	_	_	_	_
2	13.6	7.8	7.5	6.8	3.4	2.7	_	_	_	_	_
3	_	11	10.6	9.6	4.8	3.9	_	_	_	_	_
5	_	17.5	16.7	15.2	7.6	6.1	_	_	_	_	_
10	_	32.2	30.8	28	14	11	_	_	_	_	_
15	_	48.3	46.2	42	21	17	_	_	_	_	_
20	_	62.1	59.4	54	27	22	_	_	_	_	_
25	_	78.2	74.8	68	34	27	_	53	26	21	_
30	_	92	88	80	40	32	_	63	32	26	_
60	_	177	169	154	77	62	16	123	61	49	12

[a] For 90% and 80% power factor, the figures shall be multiplied by 1.1 and 1.25, respectively.

TABLE 4.5

Standard Fuse Sizes according to NEC 240.6

Standard Fuse Sizes								
15	20	25	35	40	45	50	60	70
80	90	100	110	125	150	175	200	225
250	300	350	400	450	500	600	700	800
1000	1200	1600	2000	2500	3000	4000	5000	6000

Note: Additional standard fuse sizes are 1, 3, 6, 10 and 601 amps.

Standard sizes for fuses and fixed trip circuit breakers, according NEC 240.6, are presented in Table 4.5.

So the fuse to this motor must be *80 A*.

NOTE 1: NEC 430.52 allows the user to increase the size of the overcurrent device if the motor is not able to start. All Class CC fuses can be increased to 400%, along with non-time-delay fuses not exceeding 600 amps. Time-delay (dual-element) fuses can be increased to 225%. All Class L fuses can be increased to 300%. Inverse time (thermal-magnetic) circuit breakers can be increased to 400% (100 A and less) or 300% (larger than 100 A). Instant trip circuit breakers may be adjusted to 1300% for other than Design B motors and 1700% for energy-efficient Design B motors.

NOTE 2: It is also important to check the type of protection coordination defined in IEC and NEMA standards:

Type 1: Allows extensive damage to the contactor and overload relays and it may require comprehensive repair or complete replacement.

Type 2: Assures no damage to the contactor or overload relays, allowing easily separable light tack welding of contacts with minimum contact burning is allowed.

4.4.6 High Speed Fuses

Recently, variable frequency drives, soft starters, and other power electronic devices are frequently present in motor circuits. These devices are much more sensitive to the damaging effects of short circuit currents and require a level of protection that is not offered by conventional protection devices (circuit breakers and fuses).

The high speed fuses are indicated for protection of diodes and thyristors, and are, in practice, recommended for rectifiers, frequency converters and soft starters.

Typically, they can be found in two types of construction: blade contact connection type and thread or bar connection type (flush-end), as shown in Figure 4.18.

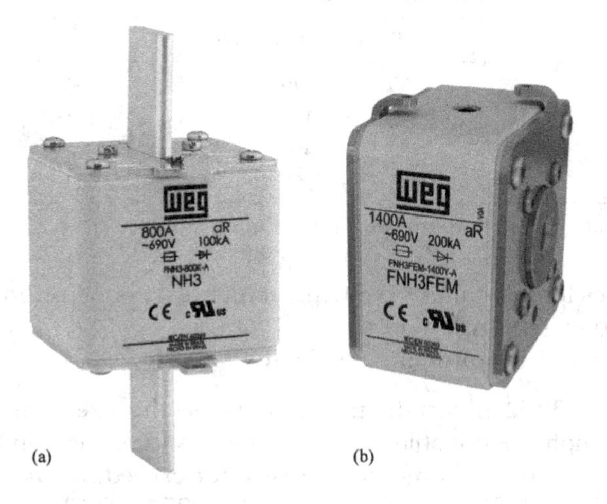

FIGURE 4.18
High speed types of construction fuses. (a) blade contact connection type, (b) thread or bar connection type. (Courtesy of Weg.)

The contact connection type blade type connections have typical 20–1000 A current ratings and are mounted on bases, as shown in Figure 4.18. The bar contact connection type can be directly mounted on the bus bar supporting large nominal current ranges from 450 to 2000 A.

The high speed fuses acting can be due to three factors:

- **Internal short-circuit**: A faulty component produces a short circuit in the power electronic converter.
- **External short-circuit**: A fault in the load can short circuit the power electronic driver.
- **Fault during operation (regenerative braking)**: In the case that a variable frequency drive (VFD) control system faults acts as inverter (causing switching faults).

The fuses must be installed between the power supply and the devices to be protected.

Figure 4.19 shows fuse current (I_f) by average melting time curve for a high speed fuse.

4.4.7 Final Considerations about Fuses

Thus, is possible to summarize the main characteristics of fuses:

- Simple operation.
- Generally low cost.
- Do not have the ability to perform maneuvers, and are associated with appropriated switches if maneuvering is necessary.
- Unipolar devices susceptible to cause damage to motors with the possibility of unbalanced operation.
- They have a non-adjustable time-current characteristic; it is possible to change it only by replacing the fuse with a different nominal current or type of fuse.
- Non-repetitive operation and must be changed after melting.
- Only provides protection against short circuit, being faster than circuit breakers for high overcurrents and slow for small overcurrent.
- They do not have a well-defined time by current curve, but have a likely actuation range.

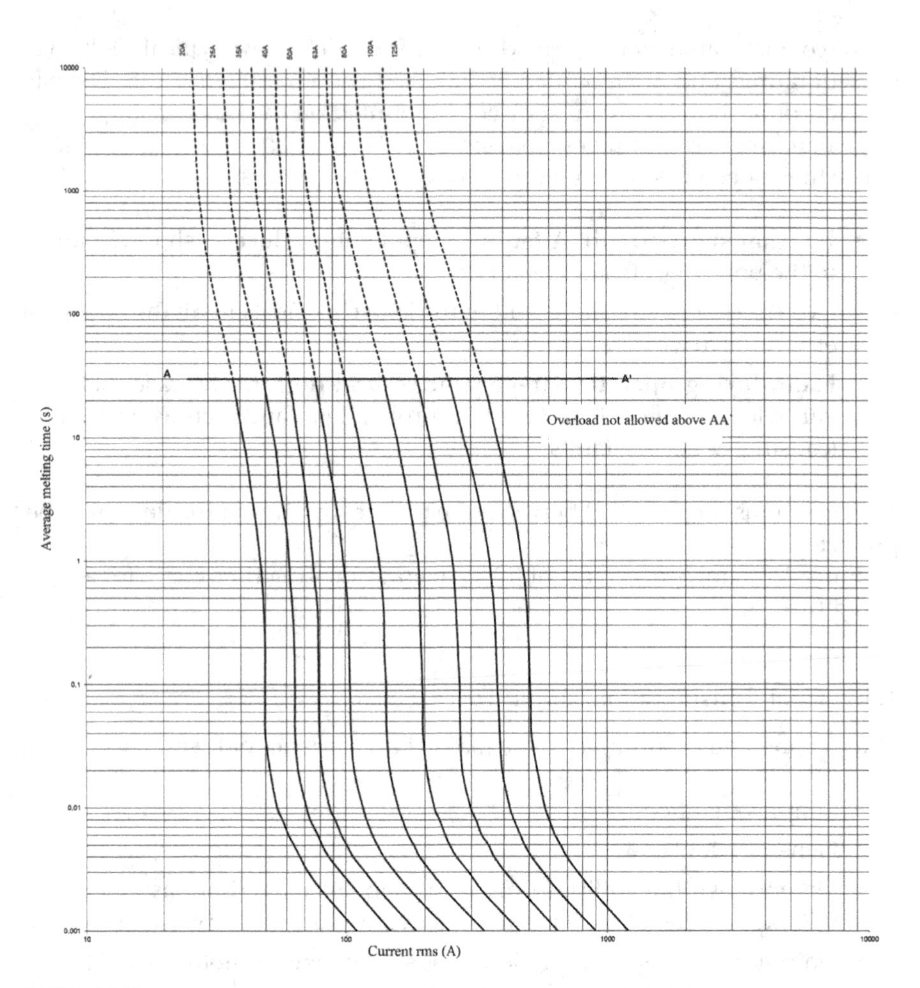

FIGURE 4.19
Fuse current (If) by average melting time curve for a high speed fuse.

4.5 Overload Relays

Overload is the most frequent damage that occurs in electrical machines. Defined as a situation that leads to overheating, the materials used only withstand up to a certain heat value and for a limited time. The determination of both amounts (heat value and time) is done via appropriated technical standards.

This occurs due to an increase in the current consumed by the motor and by thermal effects; whenever the operating temperature is exceeded, the motor expectation life is reduced by premature aging of the insulation.

Therefore, an overload that causes a higher than normal heating will not have immediate effects if it is limited in time and frequency. To do so, it is necessary to stop the motor and restore normal operating conditions.

When temperature values are exceeded, the insulation starts to deteriorate, which means that it loses its initial characteristics, its dielectric strength, which defines the insulation capability. It is the function of the overload relay to operate before these deterioration limits are reached, ensuring an appropriate life for the circuit components. Figure 4.20 shows a commercial overload relay.

The overload relay can be defined as a protection device whose operation is based on an indirect method of overload detection in motors, in which a thermal model of the motor to be protected by a thermal element is created. A three-pole overload relay has three bimetallic strips, each consisting of two metals bonded by rolling with different coefficients of expansion and a heating winding around each bimetallic. Each of the heating windings is connected in series with one of the motor phases. The heating windings cause a deformation in the bimetallic presented in Figure 4.21.

In addition to the power contacts, the relays have auxiliary contacts for current interruption in the control circuit. The overload relay operation occurs when the auxiliary contacts opens when an overcurrent in a motor happens, as presented in Figure 4.22.

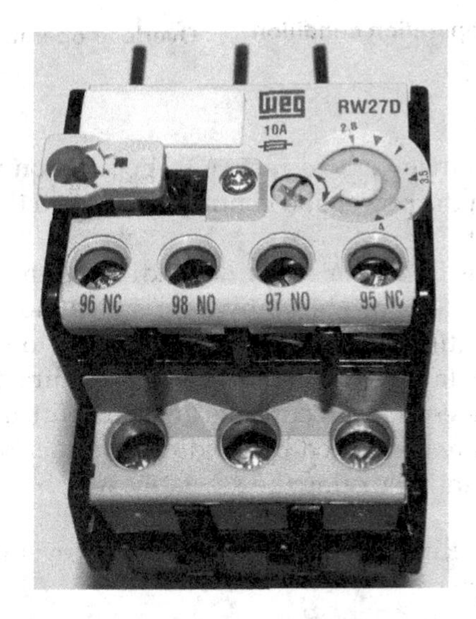

FIGURE 4.20
Overload relay and its symbol.

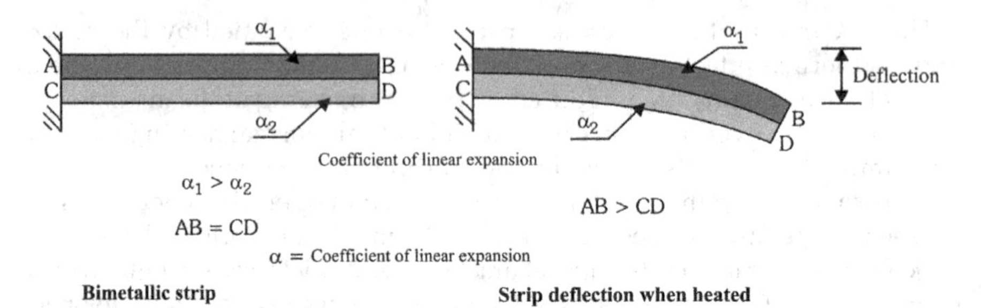

FIGURE 4.21
Principle of operation of overload relay.

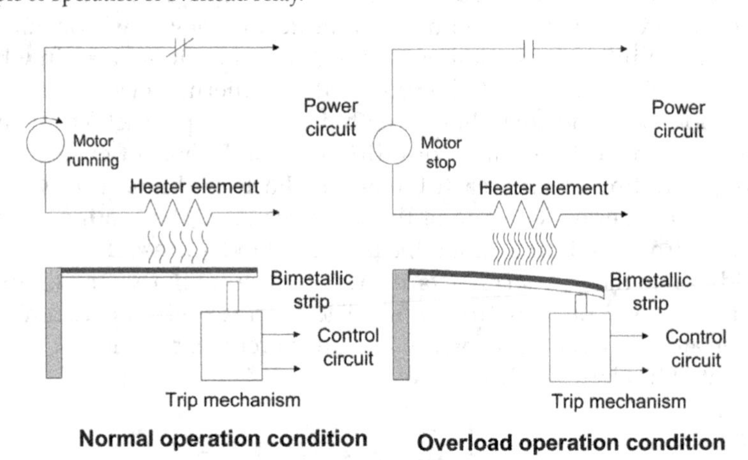

FIGURE 4.22
Overload relay operation.

Figure 4.23 shows the overload relay (OL) connection with the main and auxiliary contacts in the power and control circuits of a three-phase motor in a direct-on-line start.

The deformation is higher or lower depending on the current value. The actuation of the device is produced by the relative movement of mechanical elements with different coefficients of expansion (thermocouples—expansion of the metals according to the temperature change), under the action of certain values of input currents (overcurrent), in which two blades of different metals are connected by welding or under pressure and, when heated, dilate differently, curving the assembly, which causes two effects:

Releasing the locking device: This causes the overload relay power contacts to open.

Opening an NC contact: This causes the opening of the control circuit of a motor.

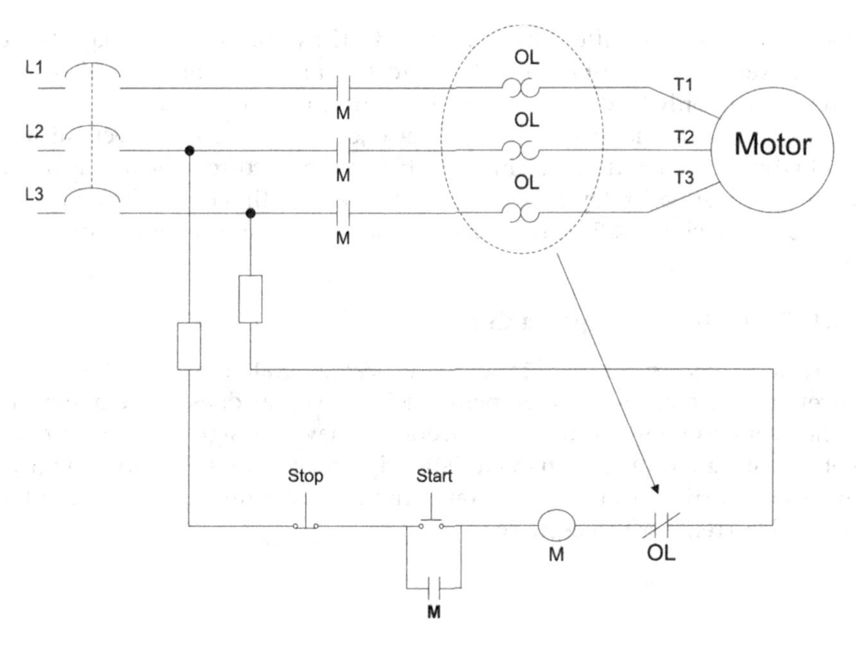

FIGURE 4.23
Overload relay connection in a direct online starter.

They are used to protect motors from possible overheating caused by:

- Mechanical overload.
- Too high starting time.
- Locked rotor.
- Phase loss.
- High frequency of switching.
- Voltage and frequency deviation.

To supervise the motor, a thermal element is placed in each phase conductor, so the thermal model of three-phase motors consists of three thermal elements, one for each phase.

The thermal relay does not protect the motor in the event of a short circuit and must be associated with fuses to provide complete motor start protection.

A relay tripped once typically does not automatically return to its resting position and must be reset manually. This is critical to avoid an unexpected motor power supply after being switched off by the thermal relay. There are also devices that allow remote shutdown and reset of thermal relays. Resetting can only be done when bimetals are sufficiently cold.

Contactor manufacturers already offer overload relays that mechanically fit into the contactors that they manufacture, with the three line inputs of the respective overload relay phases being connected directly to the contactor load contacts. Figure 4.24 shows an overload relay connected with contactor.

4.5.1 Characteristic Tripping Curve

Overload relays are designed to sense the heat generated in the motor. As the current motor increases, its temperature increases, as does the temperature of the overload relay thermal. The overload relay is design to reproduce the motor heating curve. The characteristic tripping curves states how the tripping time, starting from the cold state, changes according to multiples of the full-load current for three-phase motors.

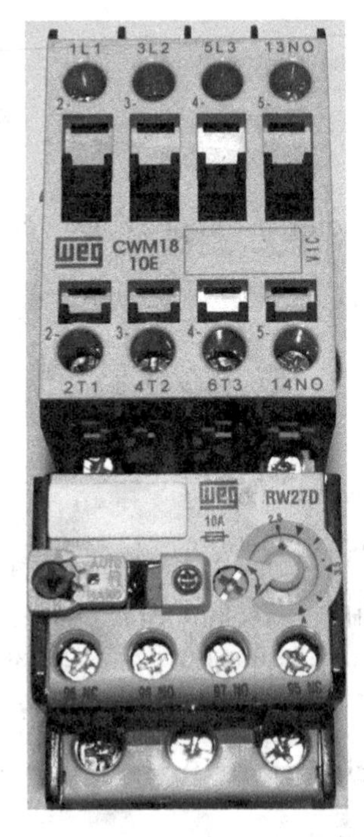

FIGURE 4.24
Contactor and overload relay connection.

4.5.1.1 Trip Class

As previously presented, the overload relays is designed to protect the motors against overload, but at start-up, the motor presents a current peak and the relays must not trip with this temporary overcurrent and must trip only if that peak, i.e., the starting time, is extended. Depending on the applications, the normal starting time of the motors can change from a few seconds (no load start condition, low resistant torque, among others) to a few tens of seconds (motors with a great inertia). Therefore, overload relays must be adjusted to the starting time according to the following tripping classes:

- **Trip class 10**: Applications with a starting time of less than 10 seconds.
- **Trip class 20**: Applications with starting time of up to 20 seconds.
- **Trip class 30**: Applications with starting time of up to 30 seconds.

Figure 4.25 shows the overload relays tripping class.

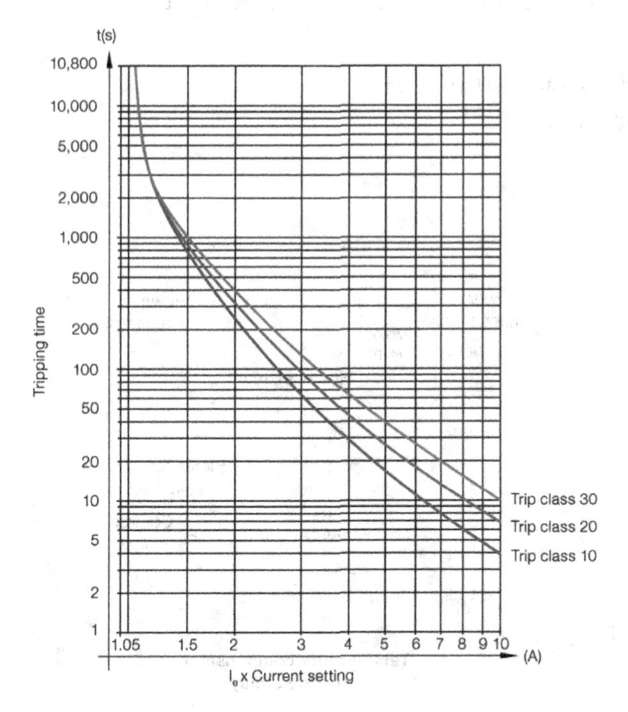

FIGURE 4.25
Trip classes of overload relays.

4.5.2 Ambient Temperature Compensation

Bimetallic strip deformation results from the heating caused by the current circulating in the phases and changes in the ambient temperature. To minimise this effect, there is a compensating bimetallic, influenced only by variations in ambient air temperature. Only the deformation originated by the current can modify the position of the bimetals and cause the relay to trip. In general, a compensated thermal relay is insensitive to ambient temperature variations between −40°C and +60°C. Figure 4.26 shows the ambient temperature compensation in an overload relay.

4.5.3 Forms of Operation

The overload relay has the following parts, as presented in Figure 4.27

1. Reset button.
2. Auxiliary contacts.
3. Test button.
4. Bimetallic strip for ambient temperature compensation.
5. Slide bar.
6. Main bimetallic strip.
7. Current adjust.

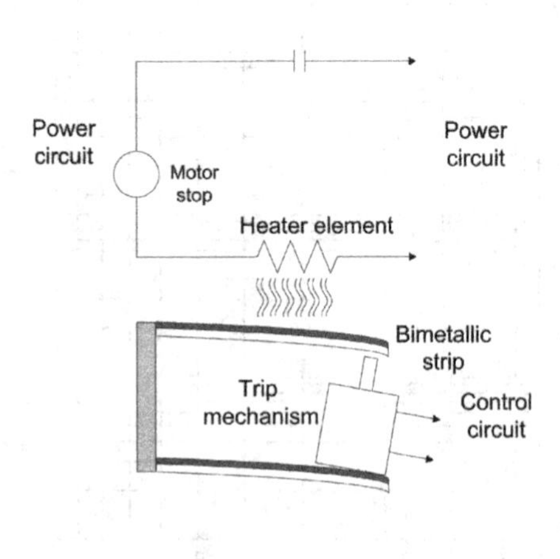

FIGURE 4.26
Ambient temperature compensation in an overload relay.

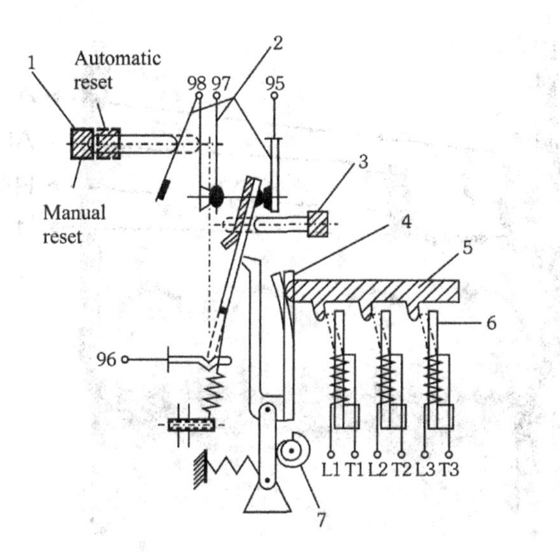

FIGURE 4.27
Overload relay elements.

By means of a button located on the front of the relay, it is possible to parameterize its operation according to the following functions, as shown in Figure 4.28.

The thermal overload relay has a multifunction RESET/TEST button that can be set in four different positions (Table 4.6).

In HAND and AUTO positions, when the RESET button is pressed, both NO (97–98) and NC (95–96) contacts change states. Operation of this button occurs as follows:

In H (manual RESET only) or A (automatic RESET only) position, the test function is blocked. However, in the position HAND (manual RESET/TEST) or AUTO (automatic RESET/TEST) it is possible to simulate the test and the trip functions by pressing the RESET button.

When set in the H or HAND position, the RESET button must be pressed manually to reset the overload relay after a tripping event. On the other hand, when set in A or AUTO position, the overload relay will reset automatically after a tripping event.

Table 4.7 summarizes the operation of this button.

NOTE: Overload relays have thermal memory. After tripping due to an overload, the relay requires a certain period of time for the bimetal strips to cool down. This period of time is a so-called recovery time. The relay can only be reset once it has cooled down. The recovery time depends on the characteristic tripping curves and the level of the tripping current. After tripping due to overload, the recovery time allows the load to cool down.

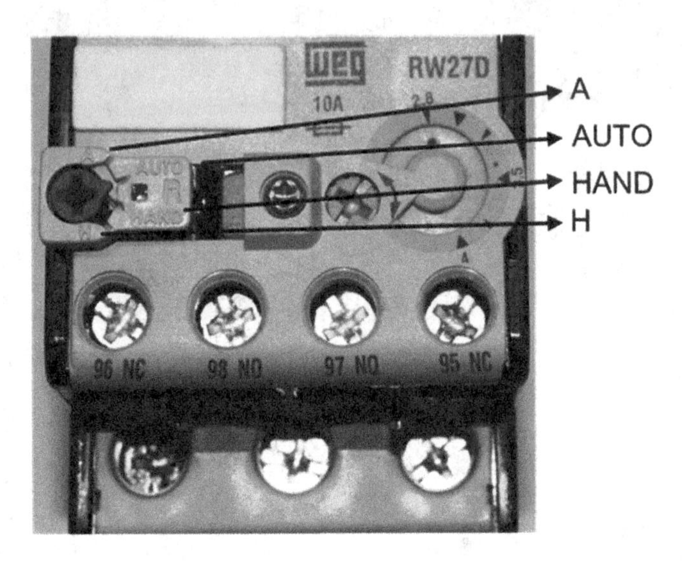

FIGURE 4.28
Types of operation in an overload relay.

TABLE 4.6
Overload Relay Types of Operation

Representation	Function
A	Automatic Reset/Test
AUTO	Automatic Reset/Test
HAND	Manual Reset/Test
H	Manual Reset only

TABLE 4.7
Overload Relay Forms of Operation

Operation	H	HAND	AUTO	A
Overload relay reset	Manual	Manual	Automatic	Automatic
Auxiliary contact trip test 95–96 (NC) and 97–98 (NO)	Disabled	Allowed	Allowed	Disabled

4.5.4 Sizing

The overload relays must be sized considering that they contain in their adjustment range the rated motor current (I_r) that circulates through the section where they are connected, and the adjustment being made by a knob that rotates acting on the elongation or on the curvature of the bimetallic strips. Each relay covers only a certain current range; thus, each manufacturer provides a wide variety of protection relays.

The relay must not be sized with the rated current of the circuit at the upper end of its adjustment range, because, if the motor needs to be used with a service factor above 1, the relay will not allow such a current even if the motor can endure this situation. The current setting in the relays should be done as follows:

$$I_{\text{relay}} = 1.15 \text{ to } 1.25.I_r \tag{4.2}$$

where:

I_r is the rated motor current
I_{relay} is the overload relay current adjust

In the case of motors with a service factor equal to or greater than 115%, or motors with a permissible temperature rise of 40°C, the setting may be up to 125% of the rated current ($1.25.I_r$). In other cases, the thermal relays must be set to 115% of the rated current ($1.15.I_r$).

This analysis should also be made regarding the beginning of the actuation range where the relay would have difficulty detecting loss of phase (Figure 4.29) when the motor was operating below 60% of its rated current.

4.5.5 Solid-State Overload Relays

In solid-state overload relays, the motor current is obtained using the current transformers and converted into an electronic signal for the instrument.

This principle of operation is different from the overload relay where a significant amount of energy is wasted in the bimetal strips. The solid-state overload relay has low heat losses because it is an electronic circuit, reducing the power consumption and cooling problem of cabinets. Typically, there is a reduction in consumption compared to overload relays of up to 87%. It is worth mentioning that the solid-state relay is self-powered, that is, no additional external power is required for operation; thus, it can be directly applied to the contactor. Figure 4.30 shows a commercial solid-state relay.

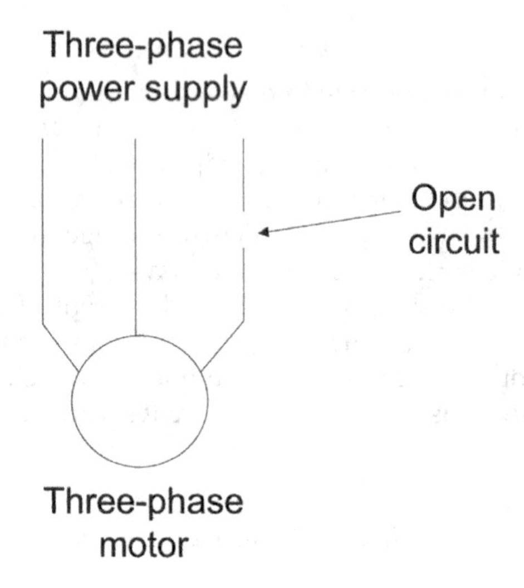

FIGURE 4.29
Phase loss in a three-phase motor.

FIGURE 4.30
Solid-state overload relay.

This device features built-in auxiliary contacts that, when properly wired in series with the coil of the contactor, assures the motor is switched off when a failure occurs and can also be used for monitoring purposes.

It also has a RESET push-button and a TEST switch. Both functions allow the checking of proper wiring and the status of the auxiliary contacts. The status window (TRIP) that displays the current operation status is also located on the front side, where there is a dip switch to choose a trip class.

In applications where frequent motor start-ups (intermittent duty) take place, the increase of heating behaves slightly different in the bimetal strips than in the motor windings, and undesired early trippings are common. In such situations, the thermal capacity of the motor is not properly utilised, and thermal overload relays are not the most suitable solution. In these cases, the use of solid-state relays is recommended.

Another important issue that must be taken into account is the large current range of a solid-state relay that has a ratio of 5:1, between the maximum and minimum values of adjustable current, while, for overload relays, this ratio is approximately 1.5:1. It allows a drastically reduced amount of equipment in stock with the use of a solid-state relay.

There are also overload relay options that offer a greater number of functions with communication through industrial networks (DeviceNet, Profibus DP, etc). In these devices, in addition to overload protection, there are the following functions of monitoring:

- Phase imbalance
- Lockout
- Underload
- Current imbalance
- Earth leakage current
- Thermistor input (PTC)
- Individual phase currents
- Average current value
- Percentage of thermal capacity used
- Percentage of current imbalance

These parameters can be obtained online by connection with industrial networks, as shown in Figure 4.31.

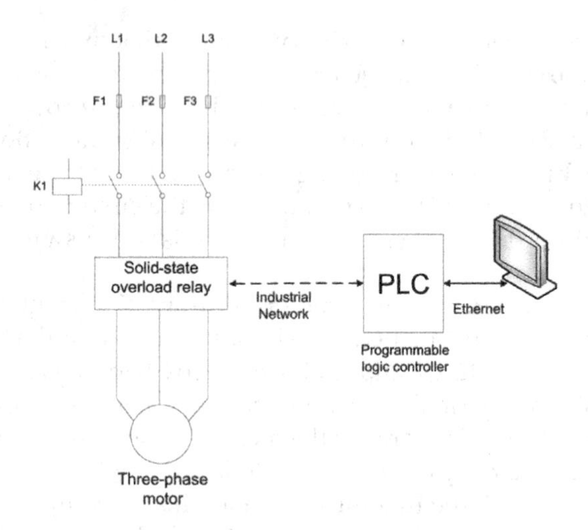

FIGURE 4.31
Solid-state overload relay connected with industrial networks.

4.6 Motor Protective Circuit Breakers

The motor protective circuit breaker is built in a moulded-case circuit breaker. This device simultaneously offers overload and short circuit protection and also allows switching (approximately 15 operations/hour) according to IEC 60947 and Underwriters Laboratories (UL) 508. Figure 4.32 shows the functions that a circuit breaker can perform when connected to a direct online starter on a three-phase induction motor.

Figure 4.33 shows a commercial motor protective circuit breaker. At the front of the circuit breaker there is a current setting dial to adjust the overload protection. It is also possible to lock the current setting so as to ensure the reliability of the current setting on the circuit breakers installed in the field on electrical panels and machines. The circuit breaker also has an optional accessory to signal the trip (overload or short circuit) occurrence by means of auxiliary contacts or mechanical indicators. The circuit breakers can also be locked by means of padlocks installed on the handle, ensuring safety in stoppages for maintenance.

4.6.1 Advantages of Using Motor Protective Circuit Breakers

- Unlike the fuses, promotes multipolar operation, avoiding unbalanced operation in three-phase equipment, as in the case of the fuse, if a single element is melted.

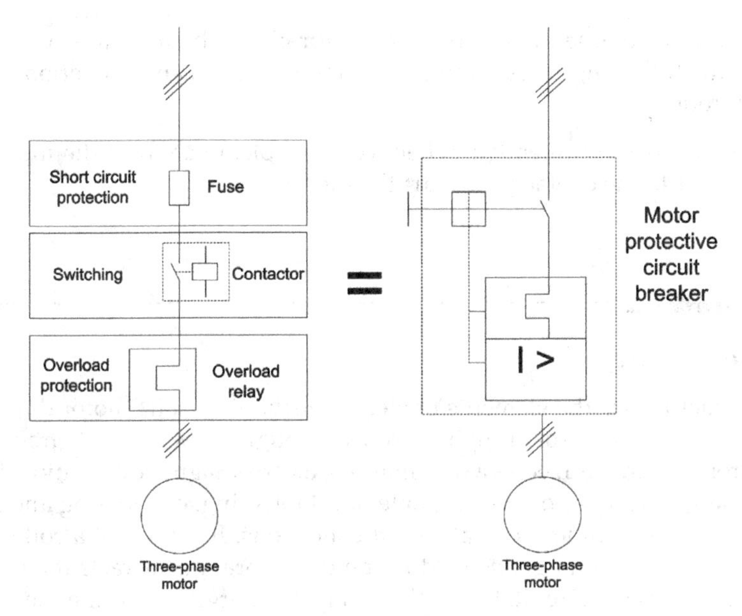

FIGURE 4.32
Motor protective circuit breakers with their functions.

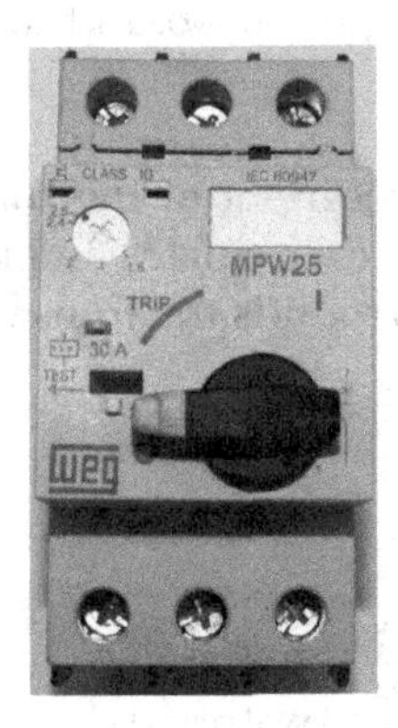

FIGURE 4.33
Commercial motor protective circuit breakers with their functions.

- Offers a wide range of rated currents, simplifying coordination with other protection devices.
- Repetitive operation, that is, can be restarted after tripped, with no need for replacement.
- Its characteristic time by current is adjustable.
- In some cases, it allows remote control.

NOTE: It is recommended to use the motor circuit breaker to start motors only in the following cases: Command should be local and operation should be low frequency.

Therefore, it is recommended, whenever possible, to connect the motor circuit breaker to the contactor to start the motor.

4.7 Contactors

The contactors are the elements of electromechanical start motor diagrams, which allow the control of high currents through a low current circuit. The contactor is characterized as a non-manual, electromagnetic operation device with a single resting position, capable of establishing, conducting and interrupting currents under normal circuit conditions. It is built of a coil which, when fed, creates a magnetic field in the fixed core that attracts the moving core that closes the circuit. When the coil is de-energised, the magnetic field is interrupted, causing the return of the core to a rest position by springs actuation, as presented in Figure 4.34a; a commercial contactor with its auxiliary contacts is shown in Figure 4.34b. In the following are presented the four main contactor parts.

Coil: Represents the control input of the contactor which, when connected to a voltage source, circulates an electric current that creates a magnetic field that surrounds the iron core.

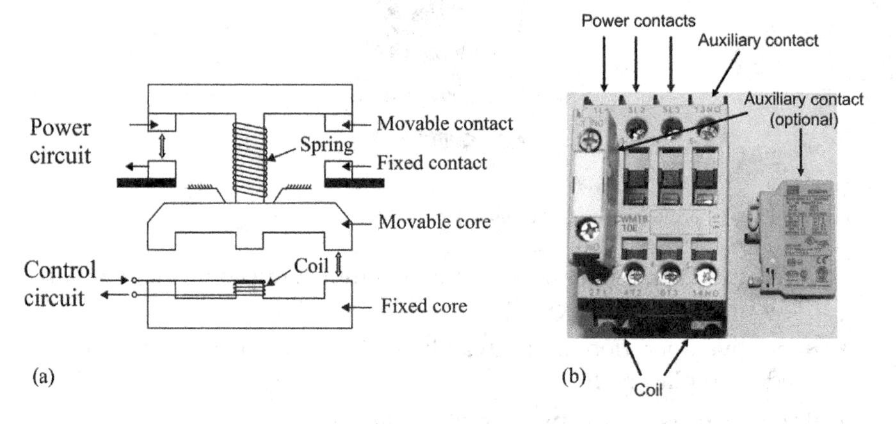

FIGURE 4.34
(a) Contactor parts and commercial contactor. (b) Commercial contactor.

The coils must be chosen according to the voltage (24–660 V) and power supply type (DC or AC) of the control circuits.

The energy consumption of the coils is relatively low. A current peak occurs at the energisation time (approximately ten times the retention current), with the coil consumption estimated at 6.5 and 25 VA, depending on the type of contactor. The coil has a power factor of approximately 0.3.

Iron core: Attracted to the coil by the magnetic field, it is coupled to the contact and, consequently, the movement of the core close the movable contact.

Movable contact: It is driven by the iron core and is coupled to a spring that tends to take it to the rest position, but when the coil is energised, the magnetic field strength is greater than that of the spring, causing the fixed core to attract the movable core.

Spring: Element responsible for bringing the contact back to the rest position when the coil is disconnected from the source when the magnetic field ceases and the spring force becomes stronger than the core.

Figure 4.35 shows the symbols of a contactor used in multi-wire diagrams according to NEMA and IEC standards.

4.7.1 Contactor Ratings

As mentioned previously, there are two main standards that rule the specification and application of electrical equipment according to NEMA and IEC, which have significant differences that will be presented below.

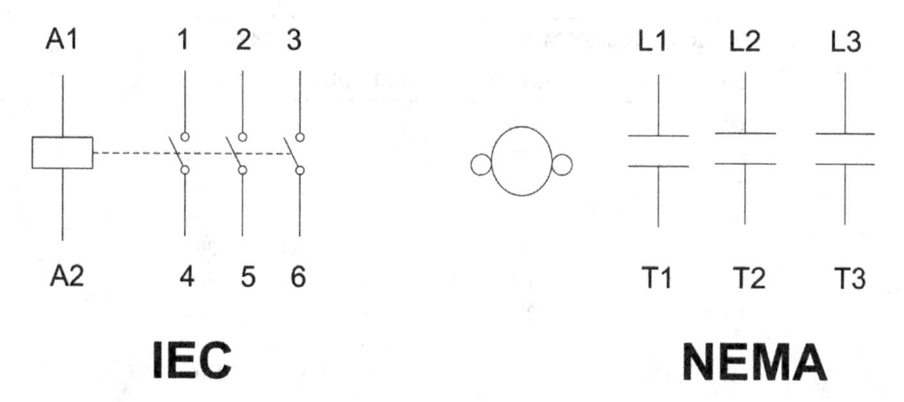

FIGURE 4.35
US (NEMA) and European (IEC) contactors.

4.7.1.1 NEMA

The NEMA standard is fundamental for the interchange between manufacturers of a determined specification. This standard assumes that the end user does not know more specific details of the load such as duty cycles, based on their specification by current, voltage and motor power. The NEMA contactor has a reserve to meet these variations while maintaining its performance over a wide range of applications.

Rated current and motor voltage are taken into account in the NEMA size ratings specification, which are shown in Table 4.8 for AC and DC.

4.7.1.2 IEC

When comparing NEMA and IEC contactors, it is noted that the IEC contactor typically has sizes smaller than 30%–60% relative to the NEMA sizes. IEC contactors are not defined by standard sizes, as in NEMA, but by the application according to utilization categories for AC and DC, which are shown below.

4.7.1.2.1 Utilization Categories

Contactors must be compatible with load power, because the ability to conduct and switch currents is their main function. Therefore, the contactors must be chosen according to the type of load they are going to drive; otherwise, their lifespan will be reduced. The classification of the contactors is made according to utilization categories, which depend on the following factors:

TABLE 4.8

NEMA AC/DC Contactor Rating

AC 60 Hz 600V Max		DC 600V Max	
NEMA Size	**Current**	**NEMA Size**	**Current**
00	9	1	25
0	18	2	50
1	27	3	100
2	45	4	150
3	90	5	300
4	135	6	600
5	270	7	900
6	540	8	1350
7	810	9	2500
8	1215		
9	2250		

- Type of load to be driven: squirrel cage motor or wound rotor, resistances, etc.
- Operational conditions: motor in steady state or with locked rotor, reversal of direction of rotation.

Electric or electromechanical loads, connected to an electric circuit, have different electrical characteristics.

The loads are classified into three large groups of which one always predominates in each component/equipment:

Inductive loads: Such as electric motors. Although there is the presence of a certain resistive portion, manifested by the existence of the Joule losses, exist the predominance of the inductive effect.

This type of load has a characteristic current that is greater than nominal. The current reduces its intensity as time passes; in the case of the motor, when it begins to acquire speed.

In Figure 4.36, in the time axis, the unit of measure is the second, and the axis of the currents is the multiple of the rated current ($x.I_r$). This greater current is a consequence of the need for a higher power at the beginning of the motor operation to overcome the mechanical inertias connected to its axis, which are presented by the mechanical machine that the motor must move. As soon as the inertia is overcome, the motor reduces the current and reaches its rated value (I_r).

Due to the fact that the starting current is greater than nominal, there are electrical losses and fluctuations in the grid, which need to be controlled.

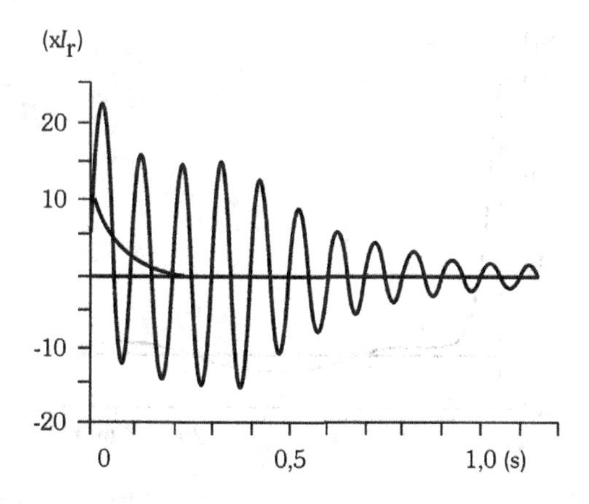

FIGURE 4.36
Inductive load characteristic.

Resistive loads: Examples can be found in electric furnaces and heating systems.

The evolution of the relation time by current of this load type is in a totally different way than that occurs in inductive loads.

According to Figure 4.37, in the time axis the scale is milliseconds, which shows that the duration of an initial peak of current is much lower, so the resulting effects are smaller, as is the case of heating, on the axis of the current, which, again, is the multiple of the rated current ($x.I_r$). However, the peak current is much higher, reaching values in the order of 20 times the rated value. Despite this, the current by time product is much less critical than in the case of inductive loads.

Capacitive loads: In this load type, the time and current references are similar to the previous case; some are larger, but with short duration, overcurrent peaks. Thus, the heating and dynamic effects on the components of the device are relevant, with a peak of $60.I_r$ (Figure 4.38), which may compromise a switching in this load type. For this reason, the capacitor switching devices must be of special type, or the user must consult the manufacturer about the appropriate switching device.

The different utilisation categories, according to IEC 947 standard, are referred to as AC in alternating current and DC for direct current and are following presented.

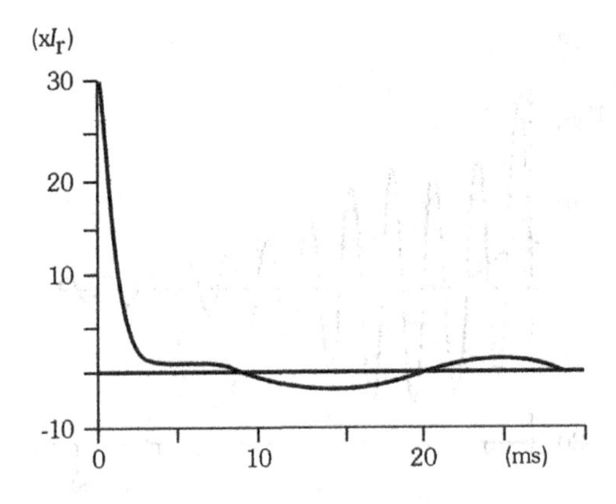

FIGURE 4.37
Resistive load characteristic.

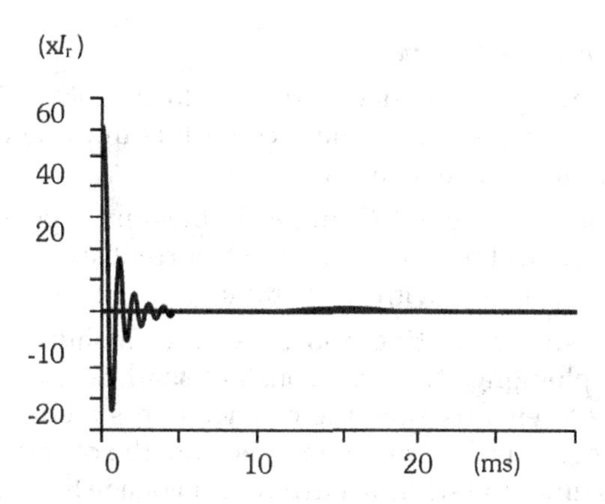

FIGURE 4.38
Capacitive load characteristic.

4.7.1.2.2 Alternating Current Categories

AC-1: Applies to AC devices with power factor ≥0.95. It is used for slow switch frequency with resistive or low inductive loads. Examples of applications: heaters, incandescent lamps, and fluorescent lamps with corrected power factor.

AC-2: This category should be used for motors with low switch frequency and can be used to control starting and stopping in steady state running conditions and counter current braking, as well as in starting for wound rotor motors.

At the closing, the contactor can set the starting current close to two and a half times the rated current of the motor, which in the opening is able to interrupt the starting current with a voltage close to the grid. Application examples: winches, pumps, and compressors.

AC-3: Applied to squirrel cage induction motors, which stop with the motor running. In closing, the contactor withstands the starting current, which is five to seven times the rated motor current. In the opening, it interrupts the rated current absorbed by the motor when, at this moment, the voltage at the terminals of its poles is approximately 20% of the line voltage. Application examples: pumps and fans.

AC-4: Used for heavy switching, such as starting motors at full load, intermittent control, reversing at full load, and counter-current braking. On energisation, the contactor closes on an inrush current approximately 5–8 times the nominal current. On de-energisation, the contactor breaks the same magnitude of nominal current at a voltage that can be equal to the supply voltage.

4.7.1.2.3 Direct Current Categories

DC-1: This category is intended to operate loads with a time constant (L/R) of 1 ms or less. Thus, these contactors must be used to drive resistive or low inductive loads.

DC-2: Used for the drive of DC motors with shunt type excitation. On closing, the contactor makes the inrush current around 2.5 times the nominal rated current with easy braking.

DC-3: Applies to the starting and braking of a shunt motor during inching or plugging. The time constant shall be less than or equal to 2 msec. On energisation, the contactor sees a current similar to that in Category DC-2. On de-energisation, the contactor will break around 2.5 times the starting current at a voltage that may be higher than the line voltage. This category considers severe braking.

DC-5: Contactors of this category are applied for starting series motors with series excitation during inching or plugging. The time constant is less than or equal to 7.5 msec. On energisation, the contactor receives about 2.5 times the nominal full load current. On de-energisation, the contactor brakes the same amount of current at a voltage which can be equal to the line voltage. In this category, braking is severe.

NOTE 1:

Plugging: Stopping a motor rapidly by reversing the incoming power connections.

Inching: Energising a motor repeatedly for short periods to obtain small incremental movements.

NOTE 2: In this section, we have shown the most common utilisation categories; however, there are others for specific applications that should be consulted in the standard for more details.

4.7.2 Lifespan of Contactor

The lifespan of the contactor is directly related to the electrical life of its contacts, which depends on the current intensity and is determined by its number of operations. According to the number of operations estimated for a certain time range, it is possible to estimate the useful life of the contactor, expressed according to mechanical and electrical aspects.

The mechanical lifespan is a fixed value, defined by the design and wear characteristics of the materials used. Its value varies from 10 to 15 million operations for small contactors and is indicated in the manufacturer's catalogue.

Considering the electrical aspects, the lifespan is a variable value, being a function of the operations frequency of the load to which the contactor is connected, the total number of operations that the contactor can perform, its utilisation category and the effects of electric arc, which depend on the voltage and the electric current.

Typically, considering the rated current shutdown conditions in the AC-3 utilisation category, this value varies from 1 to 1.5 million of operations, as we can see in Figure 4.39, where it is possible to observe the following aspects:

- On the horizontal axis is the braking current, which is not necessarily the rated current, so its value must be determined or measured at each load connected to the contactor.
- On the vertical axis is the total value of the operations for each size of contactor that is able to perform in AC-3, which is most used in industrial installations. In other words, the electrical durability of the contactor.

This information is essential if an industry maintenance plan is to be set up to adequately plan for the purchase of spare parts and the best period for the exchange of some parts.

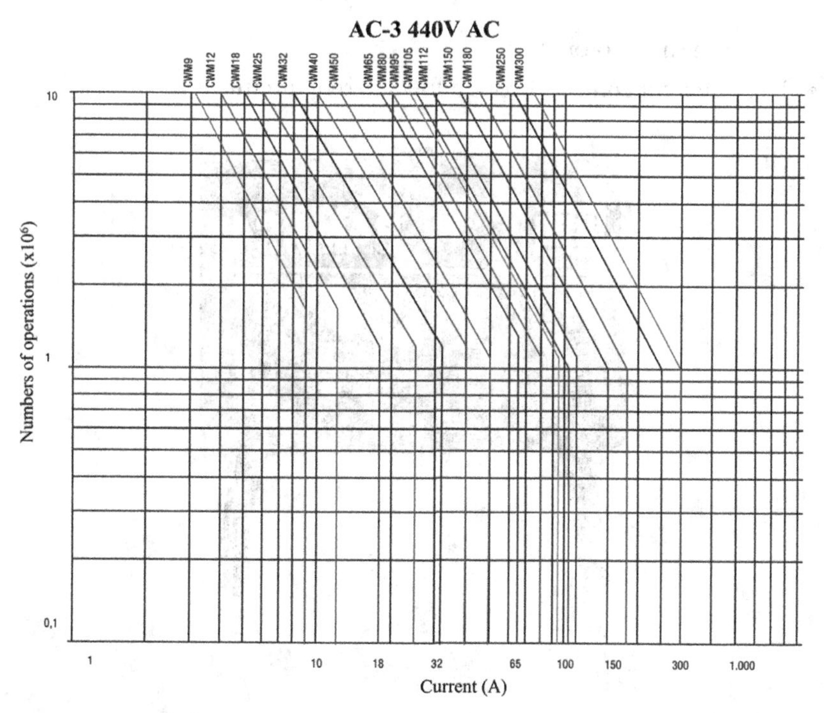

FIGURE 4.39
Lifespan of AC-3 contactor.

4.7.3 Suppression Modules

The contactor coil represents an inductive reactance when carrying current. Switching from switched-on to switched-off mode represents a very abrupt change in current intensity, causing excess voltage.

An electric arc is created at the contact that switches the coil current, damaging it. In addition, undesired electromagnetic signals are emitted that interfere with electronic equipment, as well as programmable logic controllers (PLCs).

To reduce this problem, suppression modules are connected in parallel to the contactor coils. These blocks are an optional part in commercial contactors. The suppression module is nothing more than a resistor from 100–220 Ω connected in series with a capacitor of 0.1–0.22 μF. This association must be placed in parallel with the coil, using as short conductors as possible. Figure 4.40 shows a commercial suppression module and a wiring diagram with its connection in a contactor.

4.7.4 Main Features of Contactors

The main contactor features are summarized below:

- Fast and safe motor start.
- High current control through low current circuit.

FIGURE 4.40
Suppression modules applied in a contactor. (Courtesy of Weg.)

- Local or remote control.
- Possibility to use different logics to start the motor.
- Provides effective operator protection.
- Ensuring motor shutdown in case of overload.
- Possibility of simplification of operation, supervision and installation.

4.8 Auxiliary Relays

Considering the motor drive circuits, it is common to use relays for control, alarm, protection, timing etc. The following are the most common types of relays used in practice.

4.8.1 Time Relays

This type of relay is used in timer functions, the most common types are shown below.

4.8.1.1 ON-Delay

This type of relay switches its contacts after the selected time has elapsed. When a supply voltage is connected on terminals A1–A2 the pre-set time delay begins, and, at the end of the delay time, the output relay is ON and remains energised until the supply voltage is removed. The timing diagram is presented in Figure 4.41.

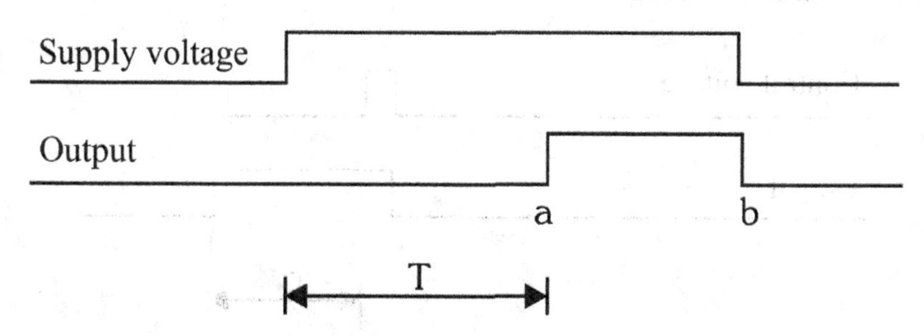

FIGURE 4.41
ON-delay time relay timing diagram.

where:

a: commutation
b: return to initial state
T: pre-set time
Reset: Remove supply voltage resets the time delay and the output

4.8.1.2 OFF-Delay

This relay requires continuous supply voltage timing, which is controlled by a command contact (control voltage). If the command contact is energised, the output relay is ON, and after the command contact is removed the selected time delay begins. When the selected time elapses, the output relay is de-energised. Figure 4.42 shows an OFF-delay relay timing diagram.

4.8.1.3 Star-Triangle

It has been specially developed for use in star-delta starters. It has two separate timing circuits. One is to control the star contactor, and the other, with a fixed time of approximately 100 ms, is to control the contactor that connects the motor windings in a delta (triangle). This relay works as follows: by applying voltage to terminals A1–A2, the Y contact switches. After the time has elapsed and the output contact of the star connection is selected, it returns to its initial state, starting the fixed time count of 100 ms, which, after elapsing, closes the triangle output contact. The schematic of Figure 4.43 shows the time instants of each timing step.

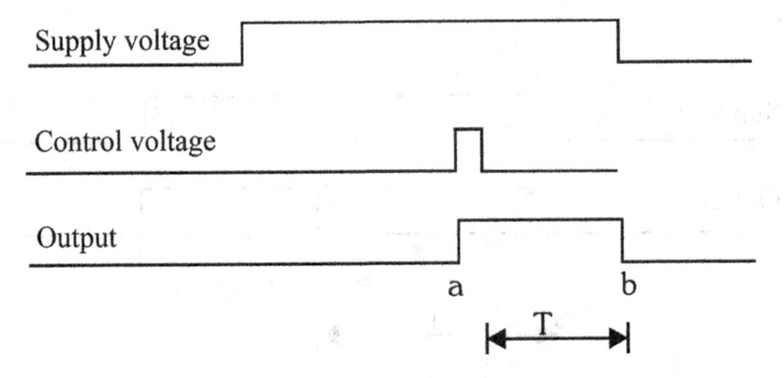

FIGURE 4.42
OFF-delay time relay timing diagram.

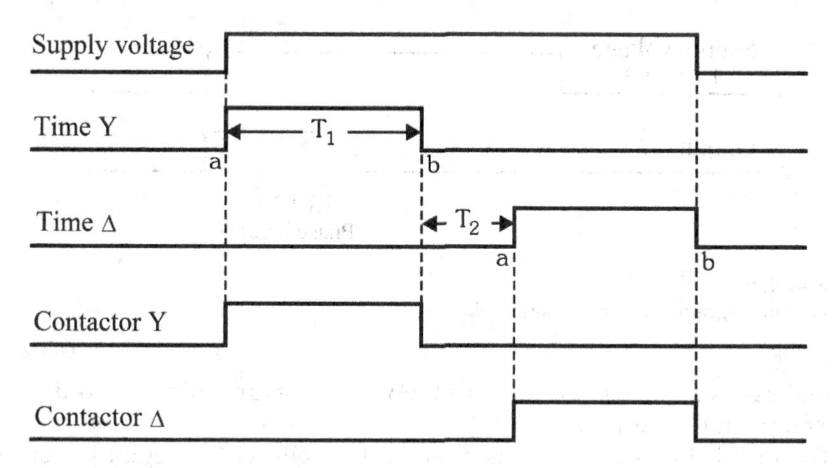

FIGURE 4.43
Star-triangle relay timing diagram.

where:

a: start of switching
b: return to rest
T_1: adjustable time for star connection
T_2: fixed time of 100 ms for delta connection

4.8.2 Monitoring Relays

These relays are intended for the power supply monitoring and motor operating conditions. There are several types of relays intended for monitoring below in which will be presented some of the most used in the industry.

4.8.2.1 Phase Sequence

Used in three-phase systems to detect the inversion in the phase sequence. The relay acts as follows:

When the relay is connected to the supply with the phase sequenced correctly, the output relay is not activated. If a phase sequence inversion occurs, the output contacts will turn ON. Figure 4.44 shows a diagram with this operating principle.

4.8.2.2 PTC Thermistor

Used in motors that use positive temperature coefficient (PTC) probes, a sensor whose resistance rises sharply to a defined temperature value.

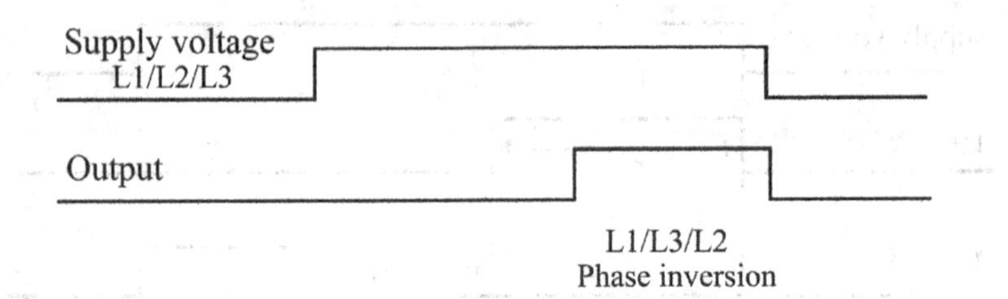

FIGURE 4.44
Operation diagram of phase sequence relay.

It undergoes a sudden variation in the value of its resistance, according to the graph in Figure 4.45.

The installation of the PTC is made in the coils, at the beginning of the coils, always on the opposite side of the fan. Generally, a PTC is installed per phase and is connected in series (Figure 4.46).

The PTC relay must be sized according to the protection temperature. It compares a reference signal with the signal sent by the PTC. Due to heating, the temperature of the probe increases beyond normal and so its resistance also increases. The PTC relay protects the motor against any type of heating in the winding, such as overload, ambient temperature rise, ventilation failure, too high starting frequency, too long starting, etc.

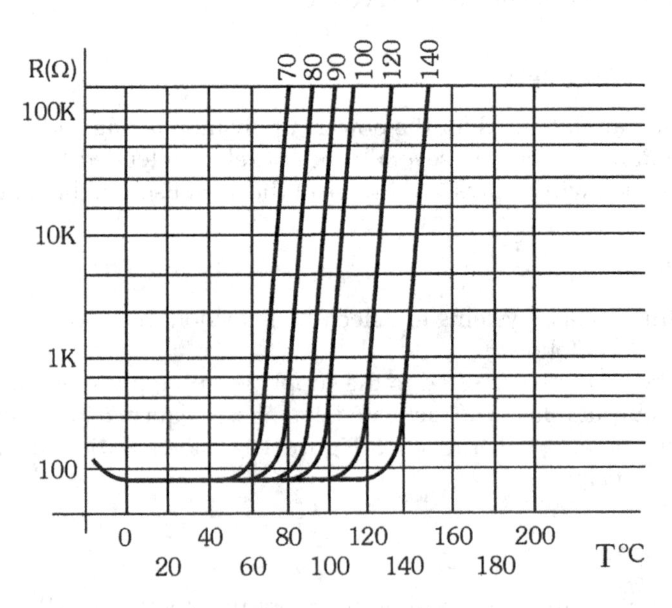

FIGURE 4.45
Resistance by temperature thermistor operation curve.

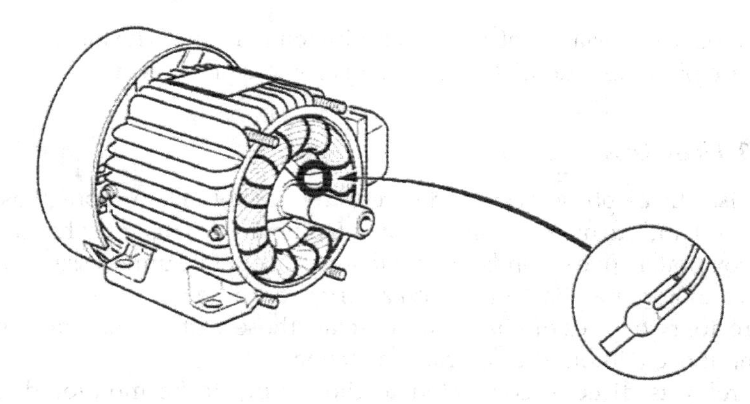

FIGURE 4.46
Location of the PTC in the motor.

In this relay, when the temperature reaches the threshold value, this resistance value rapidly increases and the relay switches its output contacts OFF.

Some relays have testing feature in case of the failure of the PTC detectors, indicating that the PTC detectors are either open or in short circuit condition (20 Ω or less).

The PTC relay acts by causing the output contact to open, and only returns if the temperature returns to normal, decreasing its resistance, as we can see in Figure 4.47.

NOTE: When a temperature control is desired in which it is possible to visualise the temperature range and act before it reaches working limits, a calibrated resistance that varies linearly with temperature, called PT-100, should be used. These thermoresistors have their operation based on the

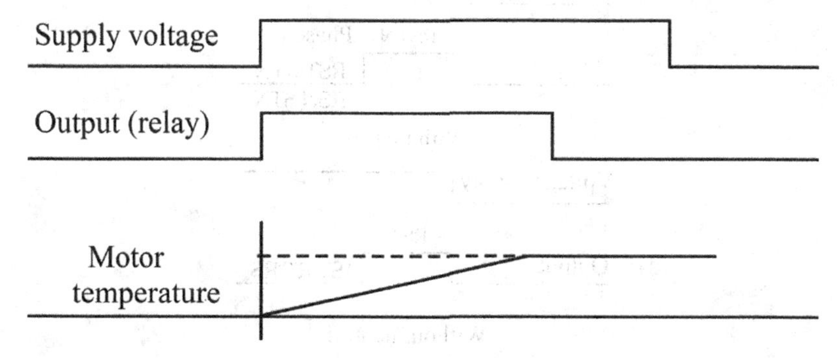

FIGURE 4.47
Operation diagram of the PTC thermistor relay.

characteristic of variation of resistance with temperature. The most used elements are platinum and nickel, having resistances of 100 Ω at 0°C.

4.8.2.3 Phase Loss

Supervises three-phase grids and detects the loss of one or more phases of the neutral and turns off a contact when fault occurs. This relay has a delay of approximately five seconds, so that it does not operate unnecessarily at the start of a motor or a phase fault with a very short time.

There are two types of phase-failure relay: those with the neutral connection and those without the neutral connection.

The relay is directly connected to the supply to be monitored, feeding the three-phases with a phase amplitude within the selected limits. The output relay switches the contacts to operation position. When a loss occurs in one of the phases in relation to the others to a value below the percentage limit selected, the output contact will become de-energised (OFF). The diagram that shows the operation of this relay is presented in Figure 4.48.

NOTE: The user sets a percentage that varies approximately from 70%–90%, which will define the loss percentage of a phase in relation to the others.

4.8.2.4 Undervoltage and Overvoltage

Used as protection in single-phase and three-phase networks. If a voltage change in the grid occurs outside the limits established in two dials with typical variations of 3%–15%, one for maximum voltage and another for

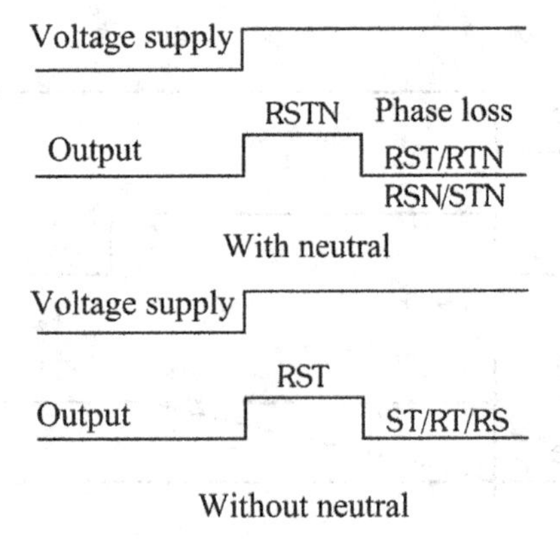

FIGURE 4.48
Phase loss relay operation diagram.

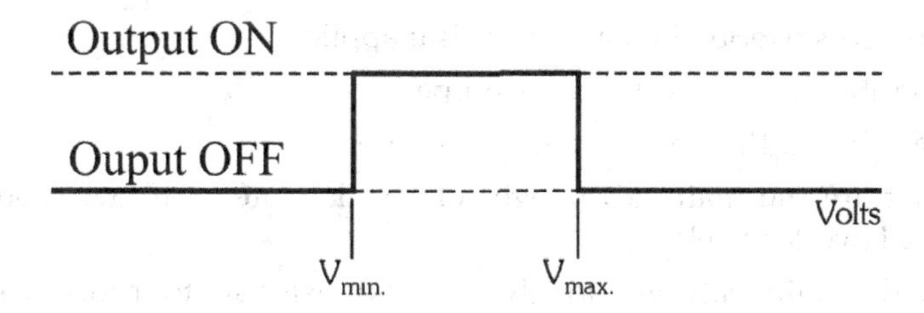

FIGURE 4.49
Undervoltage and overvoltage relay operation diagram.

minimum voltage, they close an output contact for signalling. Thus, the relay is energised within the set range and de-energised below it. They also act for phase failure and may have a delay of approximately five seconds, such as phase loss relays.

This relay is directly connected to the three phases to be monitored (L1, L2 and L3) the relay switches the contacts to the operation position (ON).

When any failure occurs in the system causing under or overvoltage, or even phase loss, relay de-energisation will occur (OFF) protecting the monitored equipment. Figure 4.49 shows a timing diagram of this relay.

Exercises

1. What are normally open (NO) and normally closed (NC) contacts? Present your symbols according to IEC and NEMA standards.
2. What is the main function of the fuses? List the main fuse parts.
3. With the help of a picture, present the following definitions:

   Joule Integral:

   Breaking capacity:

   Melting time:

   Arcing time:

   Clearing time:
4. Define the fuses operation classes.
5. Make a comparison of type D and NH fuses, taking into consideration constructive aspects and operation.
6. What aspects should be taken into account in fuses by IEC and NEMA standards?
7. What is the application of high speed fuses?

8. Conceives overload relay. Where is it applied?
9. Describe the overload relay tripping curves.
10. How should the overcurrent relays be sized?
11. What are the main advantages of a solid-state relay over conventional overload relays?
12. List the main features and advantages of using motor protective circuit breakers.
13. Describe the operation of a contactor and its main parts.
14. Show the differences between IEC and NEMA contactor ratings.
15. What are contactor suppression modules? When should they be used?
16. Describe the operation of the following auxiliary relays:
    a. ON-delay
    b. OFF-delay
    c. Star-triangle
    d. Phase sequence
    e. PTC thermistor
    f. Phase loss
    g. Undervoltage and Overvoltage

5

Starting Methods of Induction Motors

The starting of an electric motor is a critical moment, because the motor requests a much higher current than when in continuous operation, due to the change of the state of inertia of the motor. In this chapter will be presented the circuits used to start the induction motor correctly and safely.

5.1 Starting Current of Induction Motor

At the time of starting an induction motor, the current usually ranges from six to eight times the rated motor current. The amplitude and time of the initial current peak depends on the starting conditions. If it is starting under load, the peak will be higher than the no-load condition, reaching up to 10 times the rated current. This high current can even trip the protection devices of the circuits. In addition, it could overload the power supply in a damaging way. Figure 5.1 shows a curve that relates the starting current (I_{start}) to the motor speed (N).

It may be noted that, at start, when the motor speed is practically zero, there occurs the maximum current, locked rotor current (I_{lr}), which remains at this value up to close to the rated motor speed (N_r) and rated current (I_r) considering a synchronous speed (N_s). With this, it is possible to state that the current in a motor is a function of the voltage applied to it, as Figure 5.2 shows. Thus, the starting method functions as a voltage reduction during the motor start and then the rated voltage applies when the motor is already at the rated speed.

There are different methods of starting induction motors that reduces the start current, which will be presented below. Before starting the study of the starting methods, it is fundamental to know the electrical diagrams used, as well as the symbols employed, considering the NEMA and IEC standards.

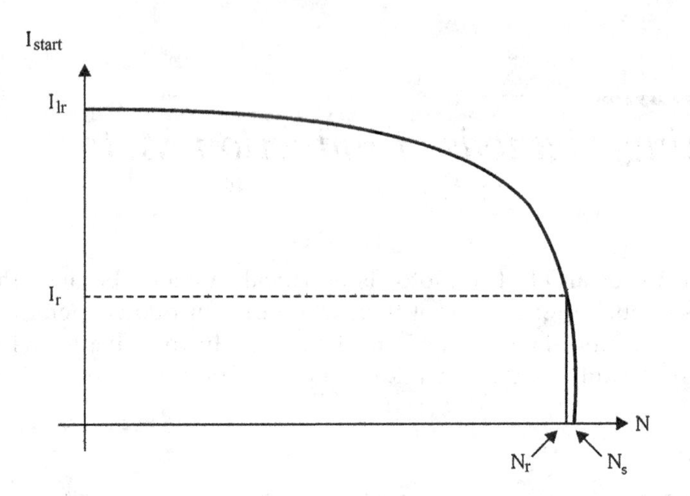

FIGURE 5.1
Relation between starting current and motor speed.

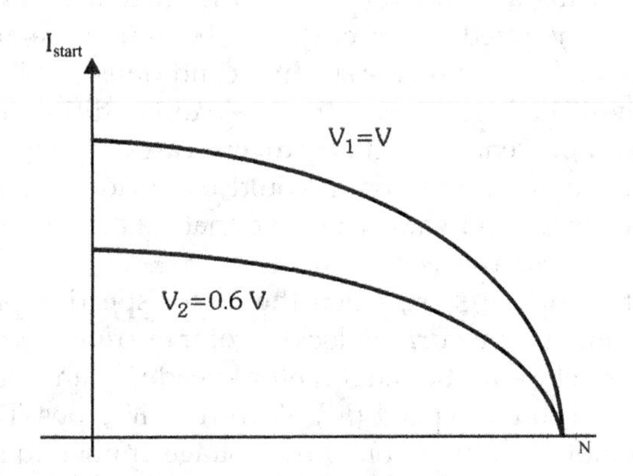

FIGURE 5.2
Motor current change due to voltage reduction.

5.2 Basic Electrical Symbols

In order to make easy the understanding and design of the motor starter electrical diagrams, the symbols that are most frequently used in electrical drawings will be presented, considering IEC and NEMA standards in Table 5.1.

TABLE 5.1

Electrical Symbols Considering NEMA and IEC Standard

Device	NEMA	IEC
Circuit Breaker		
Magnetic only		
Thermal-magnetic		
Thermal overload relay	or	
Fuse		
Coil		

(*Continued*)

TABLE 5.1 (*Continued*)

Electrical Symbols Considering NEMA and IEC Standard

Device	NEMA	IEC
Normally open momentary pushbutton	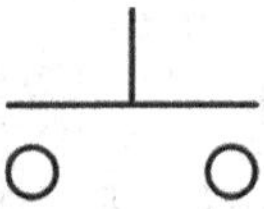	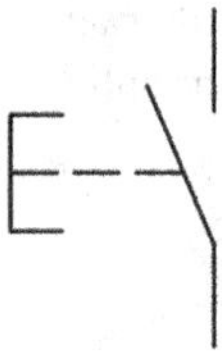
Normally closed momentary pushbutton	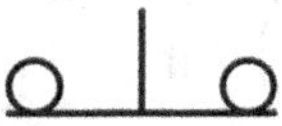	
Pilot indication light	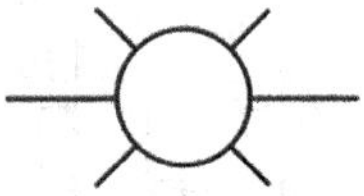	
Magnet coil		
Normally open contact	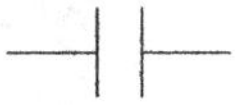 or 	
Normally closed contact	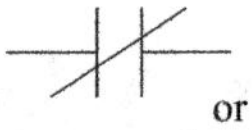 or 	
Single-phase induction motor	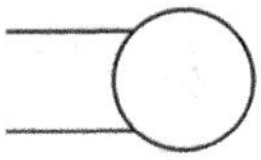	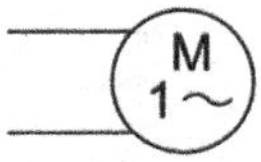 M 1~
Three-phase induction motor	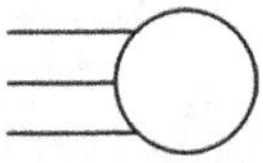	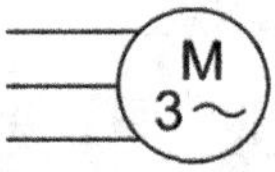 M 3~

5.3 Typical Diagrams for Starting Induction Motors

The following are the circuits most commonly used to start an induction motor: direct on line (DOL), primary resistance, star triangle and autotransformer.

5.3.1 Direct On Line Starter

Direct on line starter is the simplest way to start an electric motor, in which all three phases are connected directly to the motor.

The direct on line starter is recommended in the following cases:

- Low power motor in order to limit the disturbances caused by the current peak in the start.
- The motor does not need a progressive acceleration and is equipped with a mechanical device (gearbox) that prevents a very fast start.
- The starting torque is high.

In direct on line starting, the starting current (I_s) is directly proportional to the supply voltage and decreases as the speed increases, as shown in Figure 5.3.

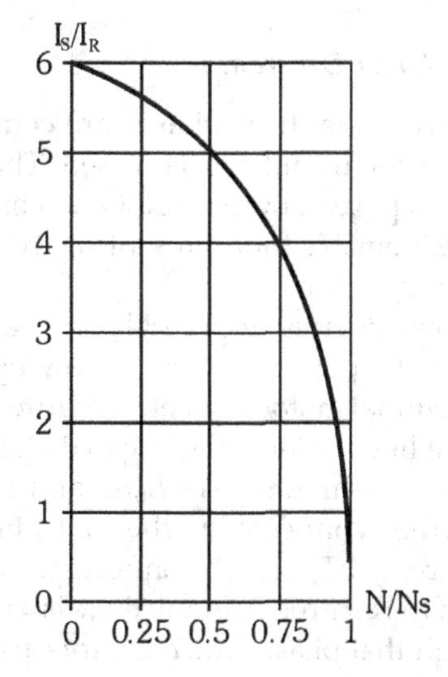

FIGURE 5.3
Relation between start current and speed in DOL starter.

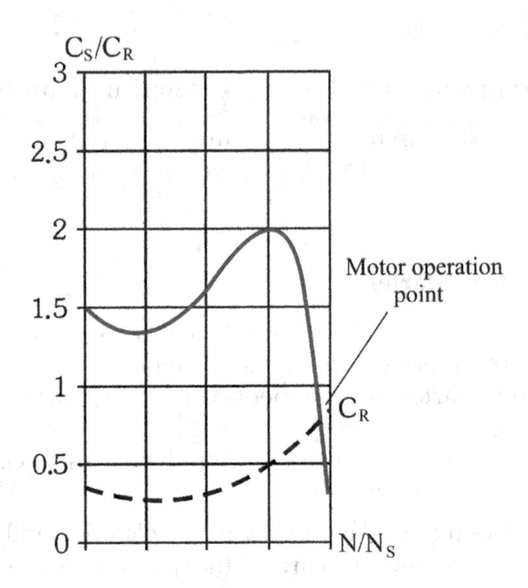

FIGURE 5.4
Torque change as a function of the supply voltage.

Figure 5.4 shows the start torque (C_S) in a DOL starter in the function of motor speed (N). It also shows that the motor torque must be higher than the load resistance torque (C_R) when the torque curve crosses the point of motor operation.

5.3.1.1 Direct On Line Starter Diagram

In the direct on line starter, the three phases are connected directly to the motor at the moment that a start button is pressed. The electrical diagram is typically divided into two parts: power circuit and control circuit.

The power circuit is responsible for representing the components that handle the rated current of the motor.

The control circuit shows devices responsible for the motor drive logic; the same ones being used in the power circuit are now operating with currents much smaller than the nominal motor current. In Figure 5.5, the power and control circuit of the direct on line starter according to IEC standard is represented.

In the power diagram, the three phases L_1, L_2 and L_3 are protected by the fuses (F_1, F_2, F_3) and are then connected to the contactor K_1, which is directly coupled to the overload relay FT_1, which connects the cables to the motor.

In the representation of the control diagram, there is a power supply (LN), and the energy comes through that phase which is protected by a fuse and, below it, an overload relay auxiliary contact (95/96-FT_1) is employed to interrupt the circuit in the case of fault. Thus, the logic part of the circuit works as follows.

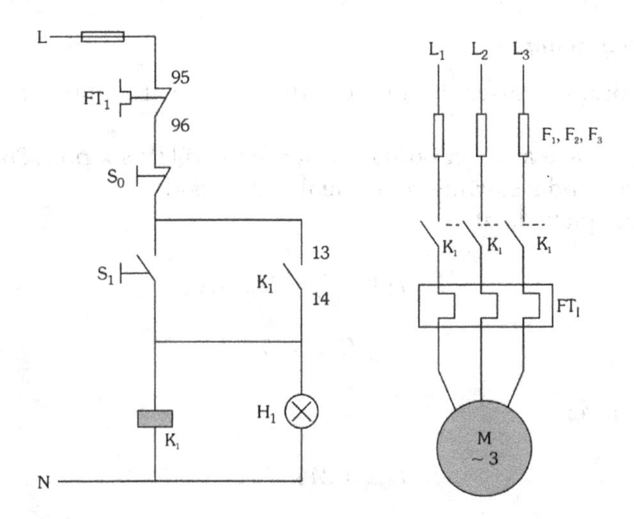

FIGURE 5.5
Control and power circuits of direct on line starter (IEC).

When pressed, the push-button S_1 energizes the contactor coil K_1 which closes the auxiliary contact 13–14 (K_1), performing the contactor seal K_1 keeping the contactor energised. The motor is powered by the three phases and starts to run. In parallel with the contact 13–14 (K_1), there is a signalling lamp (H_1), indicating that the contactor is energised. When the push-button S_0 is pressed, the contactor coil circuit K_1 is turned off and de-energised.

Figure 5.6 shows the control and power circuits of a direct on line starter according to the NEMA standard.

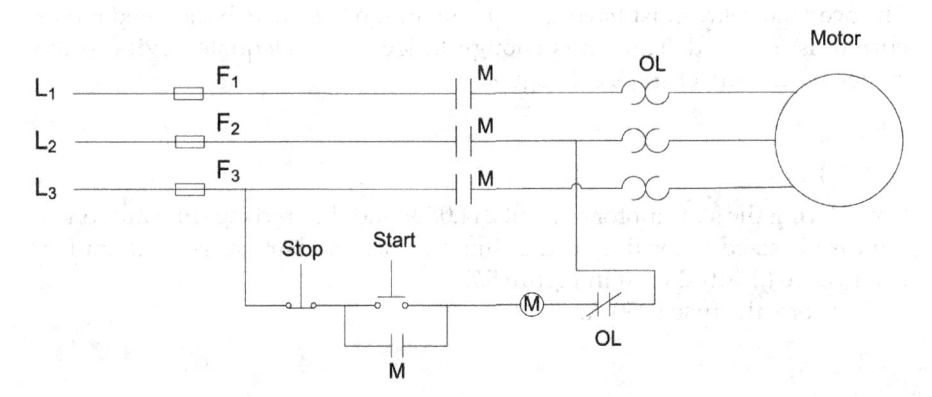

FIGURE 5.6
Control and power circuits of direct on line starter (NEMA).

5.3.1.2 Sizing Example

The direct online starter design according to the IEC standard is shown below.

Example: To size a direct online starter for a 20 HP, six pole, 380 V/60 Hz motor control, and a starting time equal to 2 seconds.

Motor nameplate data:

$$I_{\text{rated}}\ (380\ \text{V}) = 32.35\ \text{A}$$

$$I_{\text{start}}/I_{\text{rated}} = 7.5$$

Thus, we have:

$$I_{\text{start}} = 244.07\ \text{A}$$

5.3.1.2.1 Contactor K_1

To size the contactor K_1, we must take into account the rated current of the circuit (I_r) to make the sizing according to the contactor rated current (I_e). In this way:

$$I_e(K_1) \geq I_r \Rightarrow I_e \geq 32.35\ \text{A}$$

Considering this current value, it is sufficient to locate in the manufacturer's catalogue the contactor that has this current range.

5.3.1.2.2 Overload Relay FT_1

The overload relay must have an adjustment range in which the rated motor current is included. Thus, it is enough to locate an adequate device in the respective manufacturer's catalogue.

5.3.1.2.3 Fuse

Considering the start motor current 244.07 A and the starting time 2 seconds, the fuse is sized according to the time by current characteristics manufacturer curve (draft) shown in Figure 5.7.

Therefore, the fuse is 50 A.

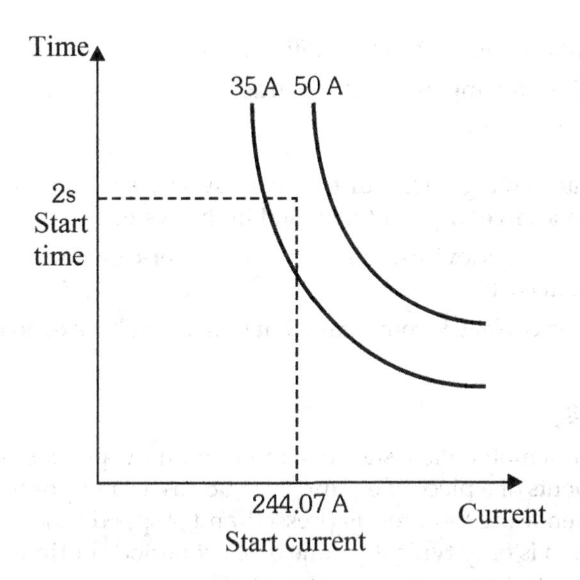

FIGURE 5.7
Sizing fuse according to time by current curve.

The fuse sized (IF) must also meet the following conditions:

IF $\geq$ **1.2** $\cdot I_r = 50 \geq 1.2 \cdot 32.35 = 38.82$ A

IF $\leq$ **Ifmax**K_1: The rated current of the fuse must be less than the maximum fuse current rated for the contactor (checking the manufacturer's catalogue).

If $\leq$ **IfmáxFT**$_1$: The rated current of the fuse must be less than the maximum current of the fuse rated for the overload relay (checking the manufacturer's catalogue).

5.3.1.2.4 Advantages of Direct On Line Starter

Thus, it is easy to list the advantages of a direct on line starter:

- Simple equipment and easy construction and design.
- High starting torque.
- Quick start.
- Low cost.

5.3.1.2.5 Disadvantages of Direct On Line Starter

In direct on line starting, the high starting current of the motor presents the following disadvantages:

- Significant voltage drop in the main supply system, which causes interference in equipment installed in the system.
- Drive systems (devices, cables) must be over-dimensioned, increasing system costs.
- Imposition of electric companies that limit the voltage drop in the grid.

5.3.2 Jogging

It is defined as a momentary start of a motor with the purpose of promoting small movements of a piece of equipment, such as a conveyor belt. The motor is started when a push-button is pressed and stopped when it is released. This application is only typically done in short periods of time. In Figure 5.8, a jogging start electric diagram is presented.

This circuit has a standard start/stop start and also an extra button to promote the jog function. The circuit has two forms of operation:

Start/stop: Pressing the start button will energise the coil M by starting the motor and causing the auxiliary contact M to keep the circuit switched on.

Jog: When the coil M is de-energised and the jog button is pressed, the coil is re-energised, but the circuit is not sealed in due to the action of the closed contact (jog button).

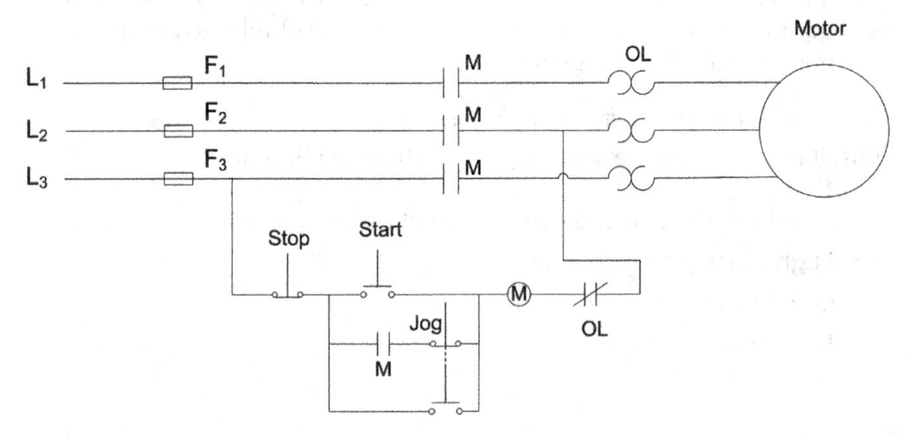

FIGURE 5.8
Jogging starting electric diagram.

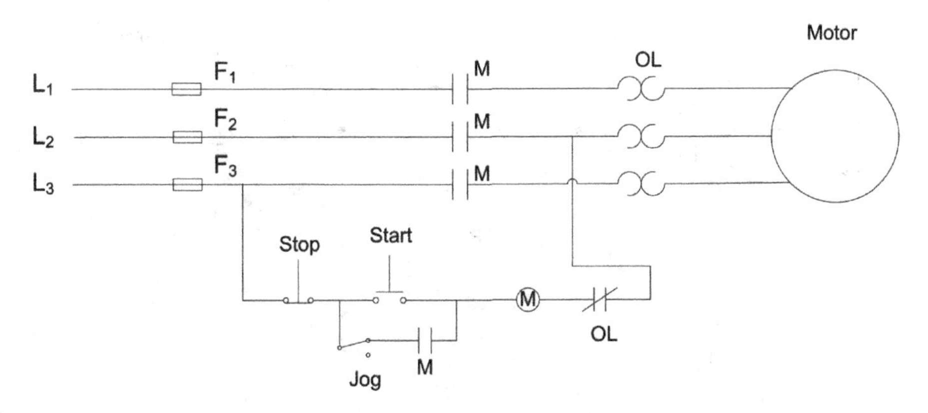

FIGURE 5.9
Jogging starting electric diagram variation.

This circuit may have problems in very fast operations and a diagram variation can be made using a selector circuit with the same button to jog and start/stop functions as shown in Figure 5.9.

This circuit has a selector switch to choose the conventional start/stop start or a start with the jog button.

5.3.3 Forward/Reverse Starter

There are several situations in the industry where reverse rotation is required. As was seen previously, to do this operation, it is sufficient to reverse two phases that feed the motor.

The control and power circuit for this switch are shown in the following Figure 5.10.

The power circuit has two contactors: one responsible for the forward start motor and another for the reverse start coupled in an overload relay for protection of the motor. The connection between the contactors can be made by wires or jumper wires, already commercially available, as shown in the Figure 5.11.

When contactor F is activated, L_1, L_2 and L_3 will be connected directly to the terminals T_1, T_2 and T_3 of the motor. On the other hand, when the contactor R is energised, we have L_3 connected with T_3, L_1 connected T_2 and L_2 to T_1 causing the motor run in the opposite direction. Regardless of the direction of rotation, only one overload relay will be used.

A very important issue is that when the motor is reversed the contactors will never be energised at the same time; if this happens, it will cause a short circuit due to the reversion of the phases in one of the contactors. To avoid this problem, mechanical interlocking is provided and consists of a system of levers that prevents both contactors from being triggered at the same time (this interlocking is shown in the dashed line of Figure 5.11.) So, if the coil of

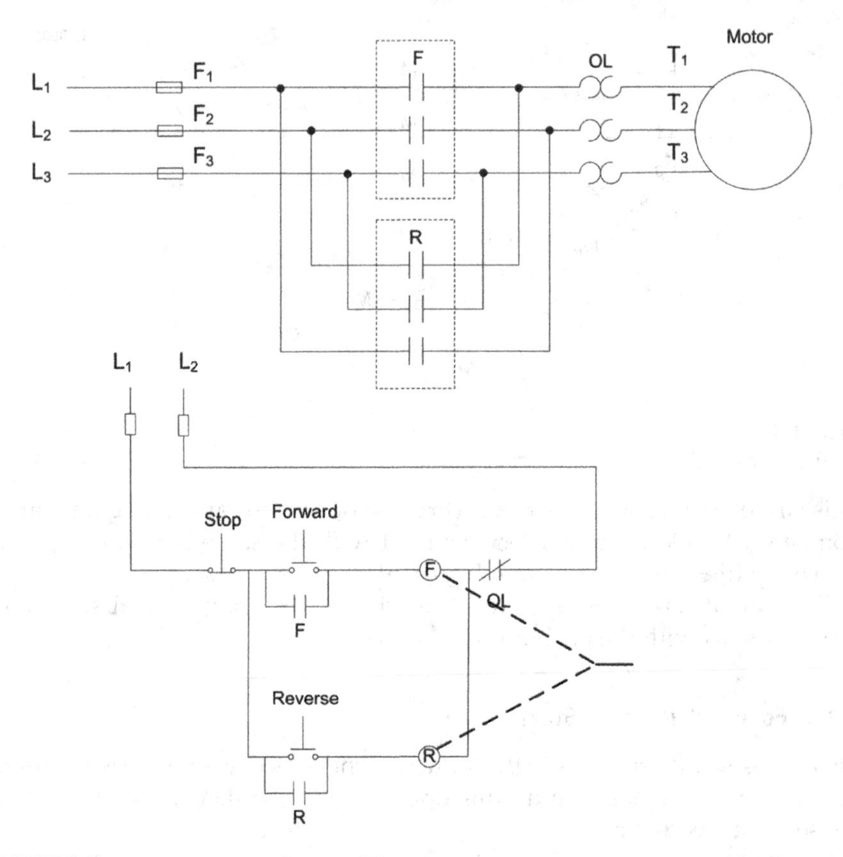

FIGURE 5.10
Control and power circuit to forward/reverse starter.

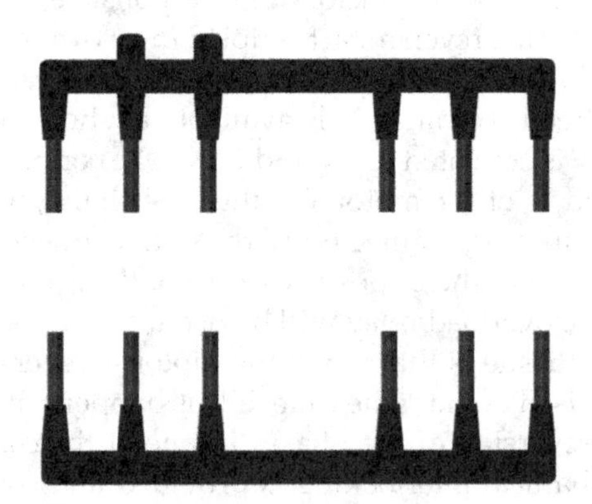

FIGURE 5.11
Jumper wires to reversing motor.

the forward contactor is activated and in sequence, the reverse contactor will never be energised because of this interlock. So to the reverse contactor coil be energised, it is necessary that the forward coil is turned off.

To avoid mechanical interlock failure, additional electrical interlocking is used through NC auxiliary contacts. The forward contactor NC auxiliary contact is placed in series with the reverse coil and the NC auxiliary reverse contact is placed in series with the forward contactor coil. So when the direct contactor is energised, it opens the NC auxiliary contact that is in series with the reverse coil, preventing and energising its coil, and so does the reverse contactor when energised.

In the previously presented circuit to reverse the motor, the stop button must always be pressed first. If you want to do the reversal directly, you must use break-make push-buttons for starting and reversing the motor, as shown in the Figure 5.12.

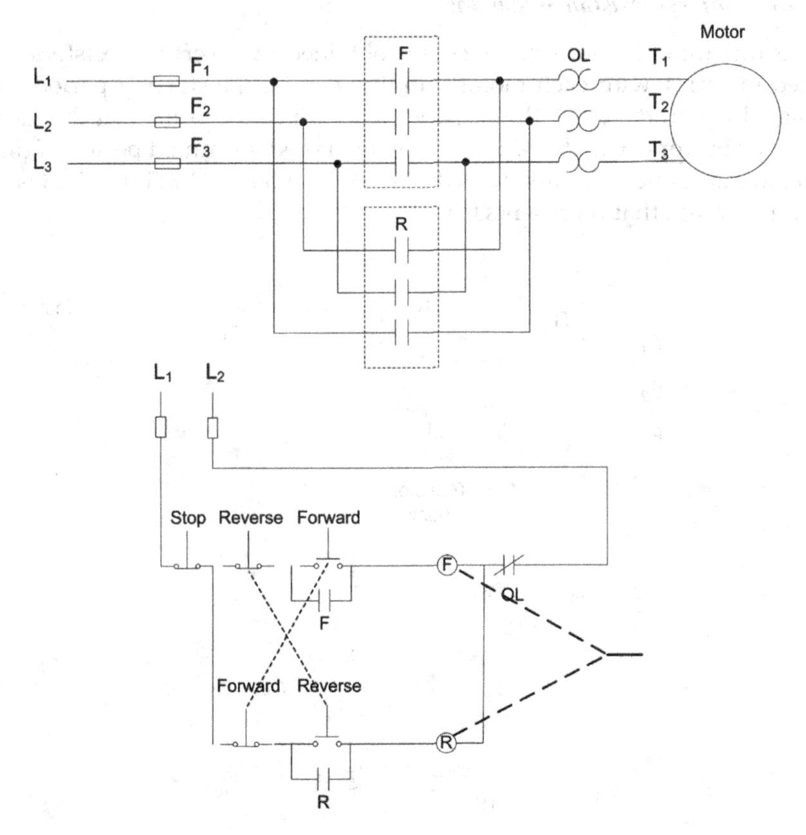

FIGURE 5.12
Forward/reverse start direct electric diagram.

In this circuit, each of the forward/reverse buttons has an NC contact and NO contact, causing it to energise the coil of one contactor and, at the same time, the shutdown of the other contactor, allowing a direct start without the need to press the button to stop.

5.3.4 Reduced Voltage Starters

In applications with high power motors and high inertia loads, it is not possible to start the motor using a direct on line starter due to high starting current. The best way to reduce the starting current in an induction motor is to reducing its voltage at start-up and, under normal operating conditions, apply the rated voltage to the motor. The following are the most-used circuits in the industry to start the motor with reduced voltage.

5.3.4.1 Primary-Resistance Starting

In this diagram, the reduced voltage is obtained by inserting resistors connected in series, with each motor winding during the starting period. The voltage drop produced in the resistors will reduce the voltage at the motor terminals by approximately 75%–80%. After a predetermined period of time, the motor is connected directly to the rated line voltage. In Figure 5.13 is the wiring diagram that represents this starter.

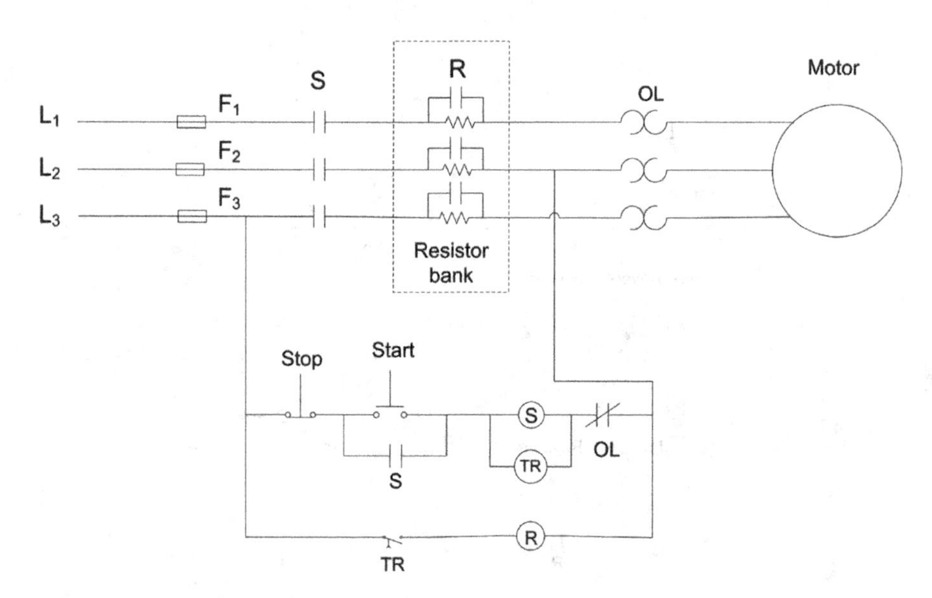

FIGURE 5.13
Primary-resistance starter diagram.

The operation of the circuit will be as follows:

- When the start button is pressed, the coil *S* and the timer TR will be energised, and the contacts will be closed, and the motor will run with low voltage.
- When a predetermined time is reached, the TR contact will close and energise the contactor R, which is in parallel with the resistors, causing the entire current to be diverted from the resistors by applying the nominal voltage to the motor.

NOTE: This type of start will have a loss due to power dissipation in the resistors.

5.3.4.2 Star-Triangle Starter

Also called a Wye-Delta starter, it consists of powering the motor with a voltage reduction in the winding during the start. The motor starts with 58% of the rated voltage, and after a defined time, the connection is converted into a delta, assuming the rated voltage.

This starter provides a reduction in starting current of approximately 33% of its value, as shown in Figure 5.14. It is recommended in applications that have a resistant (load) torque of up to one-third of the starting torque.

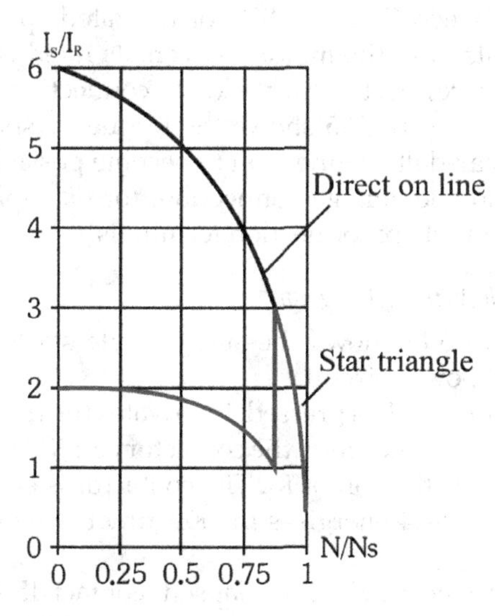

FIGURE 5.14
Starting current in the star-delta starter.

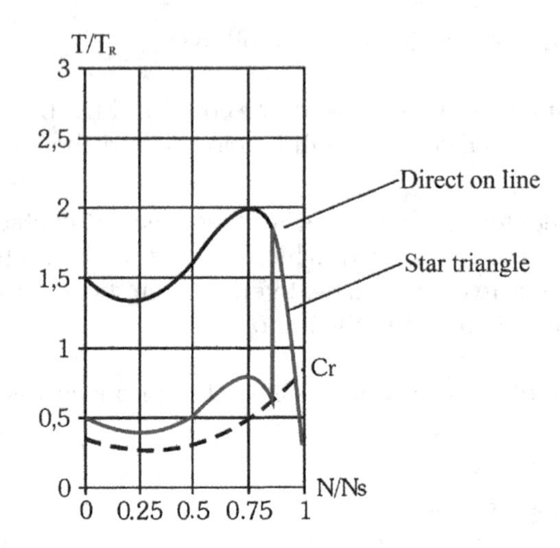

FIGURE 5.15
Start torque in a star-triangle starter.

The star-delta starter, in practice, is used almost exclusively for no-load motor starts. Since the starting torque is proportional to the square of the feed voltage, it will have a torque of more or less 20%–50% of the rated torque. Only after the rated voltage has been reached can the load be applied in the motor.

The motor speed stabilises when the motor and resistant torques are balanced, generally between 75% and 85% of the rated speed. The windings are connected in delta, and the motor recovers its rated characteristics. The change from the star connection to the delta connection is controlled by a timer. The following Figure 5.15 shows the torque by speed curve for this type of start. For a star-delta connection to become possible, the motor must be built-in with a double-voltage connection, for example: 220/380 V, and must have a minimum of six connection terminals.

5.3.4.2.1 Star/Triangle Wiring Diagram

The following Figure 5.16 shows the star-triangle starter wiring diagram with the control and power circuit.

When the push-button SH_1 is pressed, it activates the time relay KT_1, which, through its contact 18–15, energises the contactor coil K_3, which, with its open contact 13–14, energises the coil of K_1. The contactor is energized by contact 13–14. As the contact 13–14 energises the KT_1 timer, the motor starts its star connection.

After the time selected in KT_1 has elapsed, contact 15–18 is opened and, after a fixed time of 100 ms, the time relay contact 25–28 closes, energising the contactor K_2, which opens its contact 21–22, interrupting the power

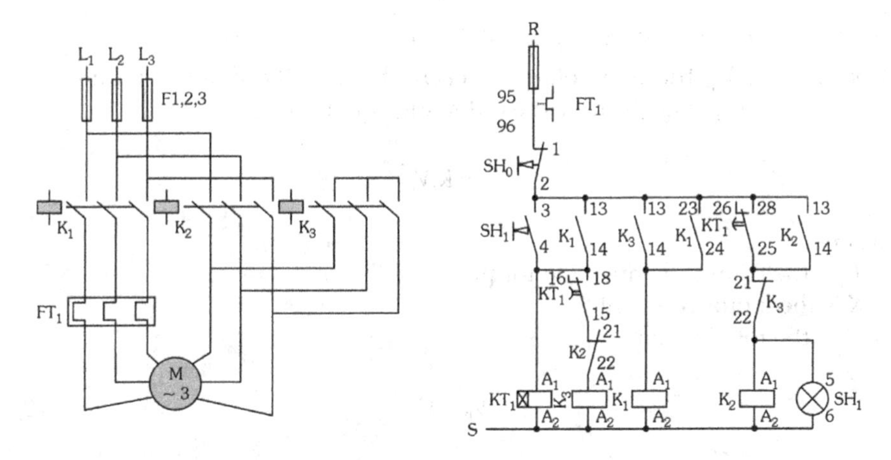

FIGURE 5.16
Star-triangle starter wiring diagram.

supply of the contactor K_3. Through its contact 13–14, the contactor K_2 is kept energized. When the SH_0 pushbutton is pressed, the coil circuits of contactors K_1 and K_2 are switched off and the circuit is de-energised.

Figure 5.17 summarizes the star-delta starter operation.

NOTE: The transition from the star connection to the triangle connection is controlled by the star-delta relay. The closing of the delta contactor occurs with a delay of 30–100 ms (fixed time) to avoid a short circuit between the phases, since the contactors cannot be closed simultaneously.

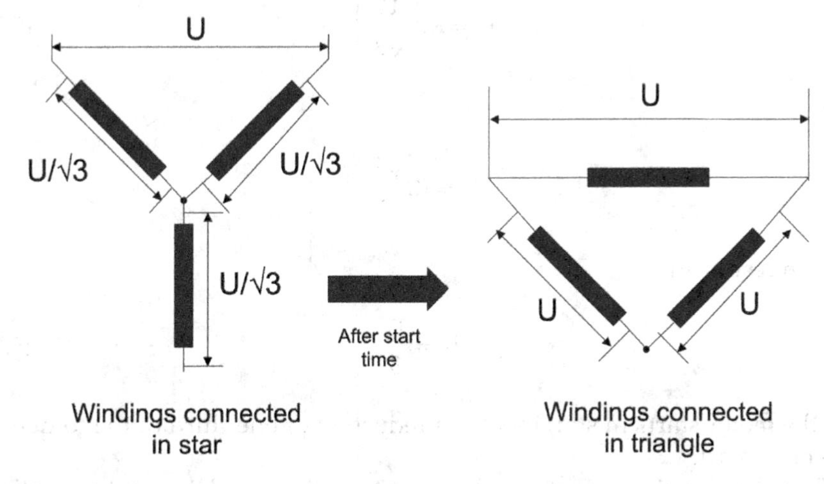

FIGURE 5.17
Star-triangle operation.

5.3.4.2.2 Mathematical Description of the Star-Triangle Starter

Considering V_n, the line voltage of each phase of the motor winding, the developed torque is given by the following equation:

$$T_\Delta = K.V_n^2 \quad (5.1)$$

where:

T_Δ is the triangle connection torque
K is the motor constant
V_r is the rated voltage

$$V_n = V_r \quad (5.2)$$

So:

$$T_\Delta = K.V_r^2 \quad (5.3)$$

When the star connection is made, we have the phase voltage (V_f) applied to the motor and this voltage is given by:

$$V_f = \frac{V_r}{\sqrt{3}} \quad (5.4)$$

So:

$$T_Y = K.\left(\frac{V_r}{\sqrt{3}}\right)^2 \quad (5.5)$$

and

$$T_Y = K.\frac{V_r^2}{3}$$

Which results in:

$$T_Y = \frac{T_\Delta}{3} \quad (5.6)$$

As the motor starts in star, there is a reduction of one-third of the torque at the rated start.

To determine the currents in the circuits of the star-delta starter, we first consider the single-line diagram of the power circuit, as shown in Figure 5.18.

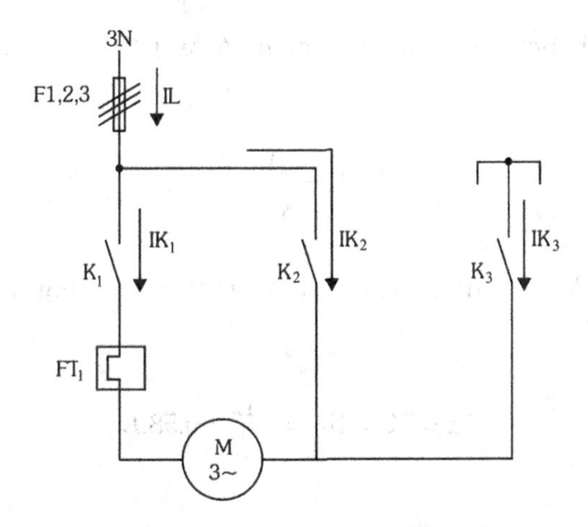

FIGURE 5.18
Single-line diagram star-triangle starter.

Then, consider IK_1, IK_2 and IK_3, the currents carrying in the contactors, K_1, K_2 and K_3, respectively. To analyze the currents circulating the contactors we will consider the motor connection in a triangle to obtain its current values, as shown in Figure 5.19.

The analysis begins, considering:

$$IL = I_r \tag{5.7}$$

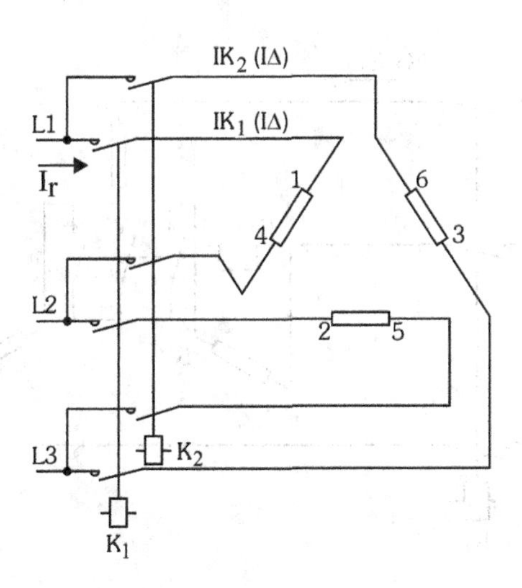

FIGURE 5.19
Motor connection in triangle.

The relationship between the current in Δ and the line current (grid) is given by:

$$I_{\Delta} = \frac{\text{IL}}{\sqrt{3}} \tag{5.8}$$

Since the triangle current (I_{Δ}) is the same as that carrying in the contactors K_1 and K_2, then:

$$I\Delta = \text{IK}_1 = \text{IK}_2 = \frac{\text{IL}}{\sqrt{3}} = 0.58.I_r \tag{5.9}$$

And its impedance is given by:

$$Z = \frac{U_r}{\frac{I_r}{\sqrt{3}}} = \frac{U_r \cdot \sqrt{3}}{I_r} \tag{5.10}$$

In order to determine the current in the contactor K_3 (IK_3), we must consider the star connection, since it only starts in the star connection of the motor. Figure 5.20 shows the power circuit of the star connection with its respective currents.

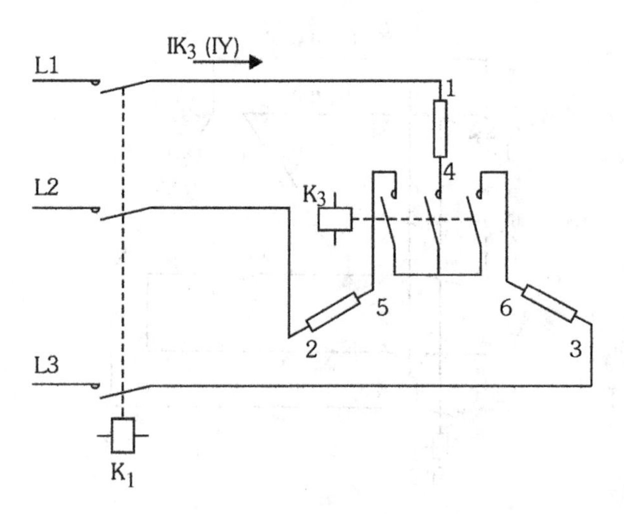

FIGURE 5.20
Motor connection in star.

In this way, the star current is given by the voltage divided by the impedance:

$$I_Y = \frac{\frac{U_r}{\sqrt{3}}}{Z} = \frac{\frac{U_r}{\sqrt{3}}}{\frac{U_r \cdot \sqrt{3}}{I_r}} \tag{5.11}$$

$$I_Y = \frac{I_r}{3} = 0.33 \cdot I_r \tag{5.12}$$

Resulting in:

$$IK_3 = 0.33 \cdot I_r$$

The current in the overload relay FT_1 is the same as that of the contactor K_1, since the relay is connected in series with this contactor and the current flowing in it is the same as that of the contactor K_1. With this, we have all the currents in the circuit:

- $IK_1 = IK_2 = 0.58 \cdot Ir$
- $IK_3 = 0.33 \cdot Ir$
- $IFT = 0.58 \cdot Ir$

The starting current has a reduction of 33% in relation to the direct online start due to the star-delta connection. Thus:

$$I_S = \left(\frac{I_S}{I_r}\right) . I_r \cdot 0.33 \tag{5.13}$$

NOTE: It is important to observe that as the contactor current is reduced, the size of the contactors for this starter will be less than that of the direct on line starter.

5.3.4.2.3 *Example of Star/Triangle Starter Sizing*

In order to apply the mathematical description to size a star/triangle starter, consider the following motor: 100 HP, 380 V/660 V and a start time of 10 seconds.

Considering nameplate data:

$$\text{FLC (380 V)} = 134.44\ \text{A}$$

$$\text{Starting current } (I_s) = 1102.49\ \text{A}$$

5.3.4.2.3.1 Number of Auxiliary Contacts Typically, for a star-delta starter, it is necessary for the contactor K_1 to have two NO auxiliary contacts, and for the contactors K_2 and K_3 to have one NO and one NC contacts.

5.3.4.2.3.2 Contactors K_1 and K_2

$$I_e \geq 0.58 \,.\, \mathrm{Ir} \Rightarrow I_e \geq 78 \text{ A}$$

5.3.4.2.3.3 Contactor K_3

$$I_e \geq 0.33.\mathrm{Ir} \Rightarrow I_e \geq 44.4 \text{ A}$$

5.3.4.2.3.4 Overload Relay The overload relay must have an adjustment range in which the current passing through the contactor K_1:

$$I(\mathrm{K}_1) = 0.58 \,.\, \mathrm{Ir}$$

$$I_e \geq 0.58 \,.\, \mathrm{Ir} \Rightarrow 78 \text{ A}$$

5.3.4.2.3.5 Fuses In the start moment:

$$I_{\mathrm{start}} = 0.33.I_s$$

So:

$$I_{\mathrm{start}} = 363.8 \text{ A}$$

Consider the start time is equal to 10 seconds and considering the time by current fuse curve (draft) presented in Figure 5.21.

According to Figure 5.21, the recommended fuse would be 100 A, however, the following conditions must be met:

- If $\leq \mathrm{If}_{\mathrm{max}}\mathrm{K}_1$ => check manufacturer catalogue
- If $\leq \mathrm{If}_{\mathrm{max}}\mathrm{FT}_1$ => check manufacturer catalogue
- If $\geq 1.2 \cdot \mathrm{Ir}$

Considering the rated current $I_r = 134, 44$ A, one must take into account the criterion (If ≥1.2 Ir), thus, we have If ≥161.32 A. In this way, it is recommended to use the fuse found in the time by current curve, as seen in Figure 5.21, which is greater than 161.32 A, whereby the 200 A fuse is chosen.

5.3.4.2.4 Advantages of the Star-Triangle Starter

- Low cost (in relation to the autotransformer starter).
- Small space occupied by components.
- No start limits.

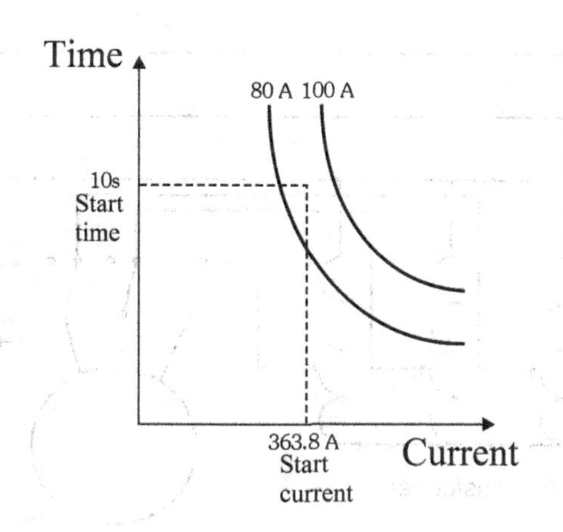

FIGURE 5.21
Time by current NH fuse (draft).

5.3.4.2.5 Disadvantages of the Star-Triangle Starter

- If the motor does not reach at least 90% of its rated speed when switching to the triangle, the peak current is almost the same as that of the direct on line start.
- The motor must have at least six terminals accessible for connection.
- The grid voltage value must match the Delta motor connection voltage.

5.3.4.3 Autotransformer Starter

This starter energises the motor with a reduced voltage at start. The voltage reduction is done by connecting an autotransformer in series with the windings. After starting, the motor is submitted to a rated voltage, as shown in the Figure 5.22.

5.3.4.3.1 Autotransformer

The autotransformer has a flat magnetic core, consisting of three columns of silicon steel plates closed at the top. Three windings are located on the columns. The lower terminals of these windings are connected in star, forming a centre that is suspended. Throughout the autotransformer winding, operational taps with 50%, 65% and 80% of the voltage are applied in the phase, having transformation ratios (k) of 0.5, 0.65 and 0.8, respectively. Sensors are placed (thermal probes) to monitor the autotransformer temperature windings rise and to prevent the start, if its temperature exceeds a certain value.

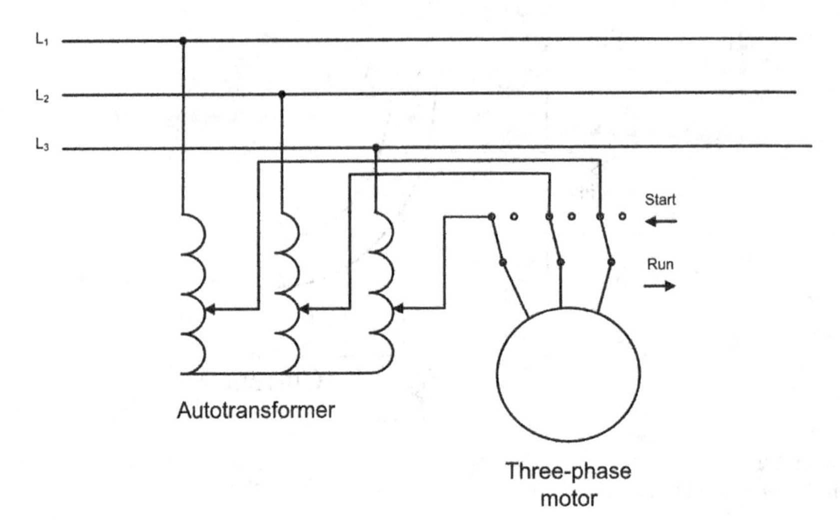

FIGURE 5.22
Autotransformer starter.

In this way, the motor torque and the current carrying in the motor windings are reduced by factors corresponding to the tap chosen for the operation.

The motor current during the starting process is reduced as a result of the application of a voltage less than the rated voltage at the starting motor terminals. As a direct consequence of this fact, the torque is also reduced and the torque versus rotation characteristic curve has a value lower than the full voltage characteristic curve. This new positioning of the characteristic curve depends on the tap chosen in the autotransformer.

To motors that start at no-load or with very small load, the 50% tap (rated voltage) may be adequate. For other situations, the designer must choose a tap that results in higher starting voltage, higher starting current, and, consequently, larger torque throughout the entire starting process, generating more acceleration to the motor in the starting process.

The motor start is made in three stages:

1. First, the autotransformer is connected in star, and then, the motor is connected to the grid through a part of the autotransformer windings. The start is made with a reduced voltage depending on the transformation ratio. The autotransformer has derivations (taps) that make it possible to choose the transformation ratio and the most appropriate reduced voltage. Figure 5.23 shows the autotransformer with its taps.

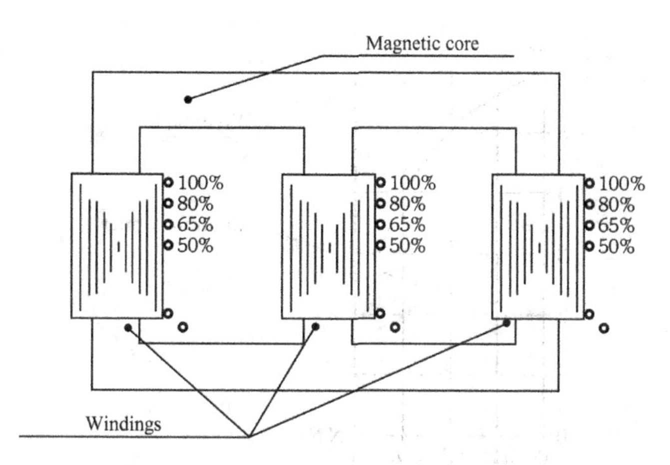

FIGURE 5.23
Autotransformer with its taps.

2. In the second stage, before switching to rated voltage, the star connection is opened. This operation is performed when the equilibrium speed is reached at the end of the first period.
3. The connection to the rated voltage is made after the second stage, in which the autotransformer is disconnected from the circuit. The current and starting torque change in the same proportions, so the following start current (I_s) range of values is obtained:

$$I_s = 1.7 \text{ to } 4.I_r \tag{5.14}$$

In Figure 5.24, the graph illustrates the behaviour of the starting current of the autotransformer start compared to the direct online start.

The autotransformer start has a starting torque from 50% to 85% of the rated torque, depending on the tap to be used, according to Figure 5.25.

Thus, the voltage reduction applied to the motor is according to the autotransformer tap in which it is connected:

- Tap 65% reduces to 42% compared to direct on line starter.
- Tap 80% reduces to 64% compared to direct on line starter.

The autotransformer starter can be used to start motors under load, which can be single-voltage and have only three connection terminals.

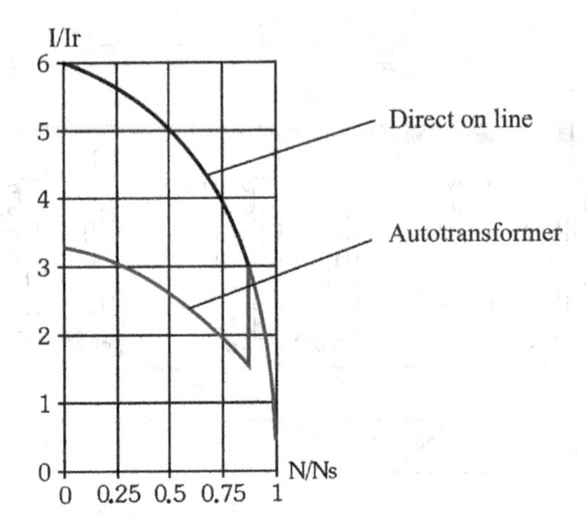

FIGURE 5.24
Starting current in autotransformer starter.

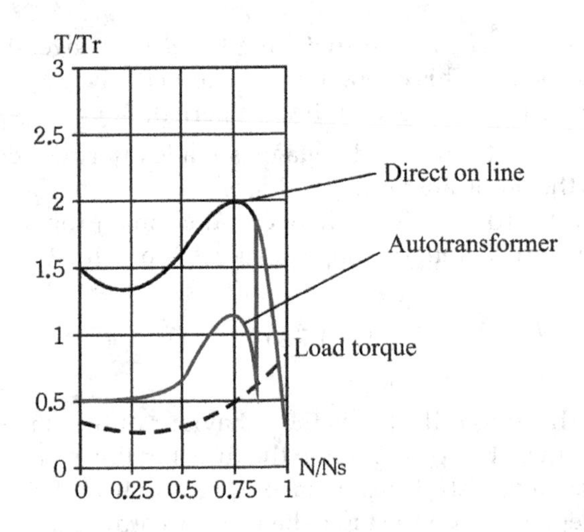

FIGURE 5.25
Torque in an autotransformer starter.

5.3.4.3.2 Electric Diagram of Autotransformer Starter

In Figure 5.26 are presented the control and power circuit of an autotransformer starter.

Pressing the SH_1 button activates the contactor K_3, which closes the secondary winding of the autotransformer, and, through its contact 13–14, energises the coil of the contactor K_2, connecting the autotransformer

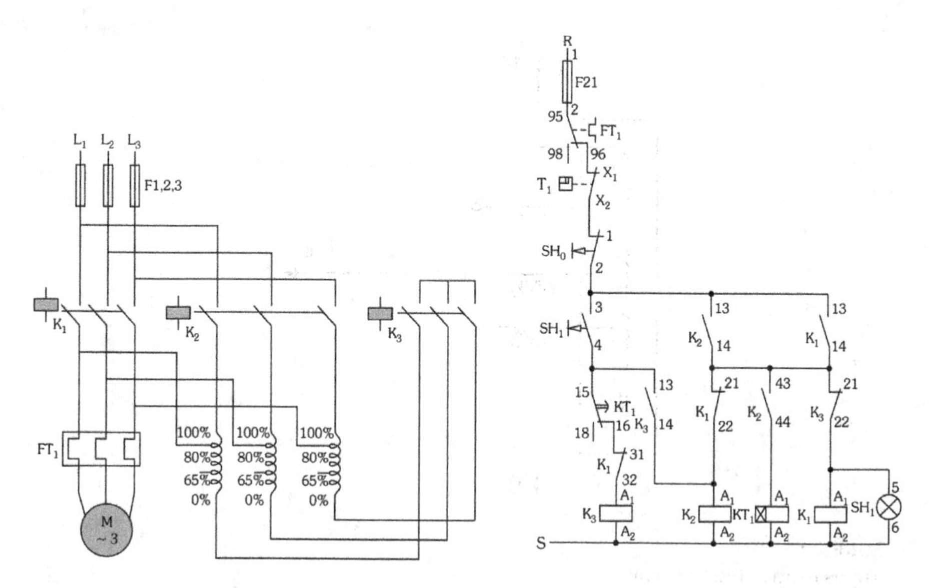

FIGURE 5.26
Control and power circuit of autotransformer starter.

to the grid. This is sealed by its contact 13–14, in the same way that the contactor K_3 is sealed by the contact 13–14 (K_2) and 13–14 (K_3). Thus, the motor starts at reduced voltage.

By the auxiliary contactor 43–44 (K_2), the timer KT_1 is energised. After the preset time programmed in KT_1 has elapsed, the contact 15–16 switches, and the contactor K_3 is de-energised, closing its contact 21–22, and, through the contact 13–14 (K_2), energises the K_1 coil. When energising K_1, its contact 21–22 opens, de-energising K_2, and contactor K_1 remains activated by its contact 13–14. The motor will then receive the rated voltage from grid. When the SH_0 push-button is pressed, the contactor coil K_1 is turned off, and the circuit is de-energised.

5.3.4.3.3 *Mathematical Description of the Autotransformer Starter*

Since an autotransformer is used to start the motor, we should take into account the equations that relate its voltages and currents according to Figure 5.27.

$$\frac{V_{in}}{N_{in}} = \frac{V_{out}}{N_{out}} = k \tag{5.15}$$

$$I_{in}.N_{in} = I_{out}.N_{out} \tag{5.16}$$

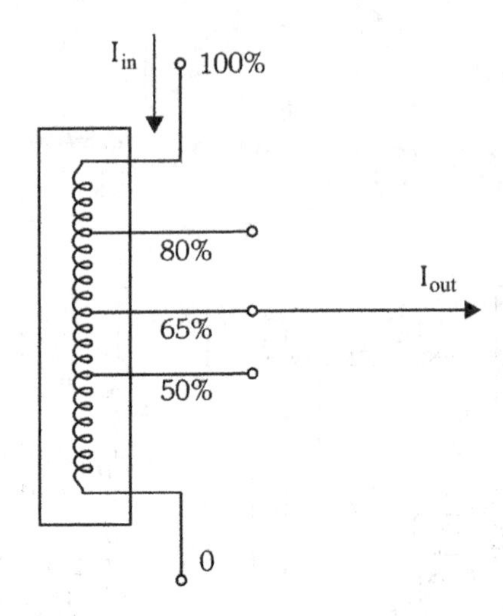

FIGURE 5.27
Currents in an autotransformer.

where:
V_{in} is the autotransformer input voltage
V_{out} is the autotransformer output voltage
I_{int} is the autotransformer input current
I_{out} is the autotransformer output current
N_{in} is the number of turns in autotransformer input
N_{out} is the number of turns in autotransformer output

So we have:

$$I_{in} = I_{out}.k \tag{5.17}$$

The torque is obtained by the equation:

$$T = C.V_r^2 \tag{5.18}$$

where:
T is the motor torque
C is the motor constant
V_r is the motor voltage

So, the rated start torque ($T_{s(r)}$) is:

$$T_{s(r)} = C.V_r^2 \tag{5.19}$$

Considering: $V_{out}/V_{in} = k$

The torque with autotransformer ($T_{s(a)}$) is:

$$T_{s(a)} = C.V_{out}^{2} \quad (5.20)$$

So:

$$T_{s(a)} = C.k^{2}V_{in}^{2} \quad (5.21)$$

The autotransformer input voltage is the same as the grid voltage; we have:

$$T_{s(a)} = C.k^{2}.V_{r}^{2} \quad (5.22)$$

and

$$T_{s(a)} = k^{2}\left(C.V_{r}^{2}\right) \quad (5.23)$$

Associating with the torque equation of the direct on line start, we have:

$$T_{s(a)} = k^{2}\left(T_{s(r)}\right) \quad (5.24)$$

Equation 5.24 shows that the autotransformer torque is the product of the rated torque (obtained by applying the rated motor voltage) by the ratio of the number of turns to the square. Thus, if a transformer ratio of $k = 0.5$ is applied, the torque is reduced to 25% of the rated torque. Table 5.2 shows the relationship between the torque and taps most used in autotransformers.

It is fundamental to know the resistance torque imposed by the load in the starting process to choose the tap that will be used, because if the starting torque reduction is very large, the motor may not have enough torque to overcome the load resistant torque and not start.

To find the currents in the autotransformer start, first consider the single-line diagram of the power circuit, as shown in Figure 5.28.

TABLE 5.2

Relationship between Motor Torque and taps in an Autotransformer

Tap	k^2	Rated Torque (%)
0.5	0.25	25
0.65	0.4225	42.25
0.8	0.64	64

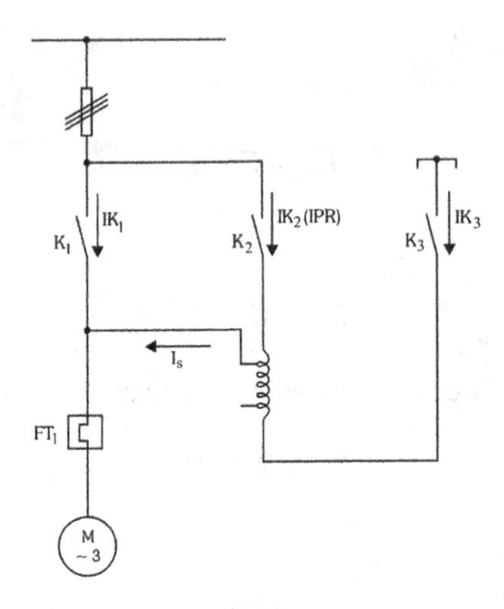

FIGURE 5.28
Single-line diagram of power circuit of autotransformer starter.

In the contactor, the rated current is applied:

$$IK_1 = I_r \tag{5.25}$$

Since the motor impedance is constant, with rated voltage and current are applied:

$$Z = \frac{U_r}{I_r} \tag{5.26}$$

Thus, when applying reduced voltage:

$$Z = \frac{k.U_r}{I_S} \tag{5.27}$$

Since the impedance (Z) is constant:

$$\frac{U_r}{I_r} = \frac{k.U_r}{I_S} \tag{5.28}$$

The primary power (P_1) of the autotransformer is the same as that of the secondary (P_2):

Thus, the following equations are present in the primary:

$$U_1 = U_r \tag{5.29}$$

$$P_1 = U_r.\text{IPR} \tag{5.30}$$

where U_1 is the primary voltage.

The equations of the secondary are as follows:

$$P_S = U_S.I_S \tag{5.31}$$

$$U_2 = k.U_r \tag{5.32}$$

$$I_S = k.I_r \tag{5.33}$$

where U_2 is the secondary voltage.

By making the active power equal in the primary and secondary and considering the current of the primary (I1) the current carrying in the contactor K_2 (IK_2), we have:

$$P_1 = P_2 \tag{5.34}$$

$$(k.U_r).(k.I_r) = U_r.\text{IK}_2 \tag{5.35}$$

So:

$$\text{IK}_2 = k^2\,\text{Ir} \tag{5.36}$$

To obtain the current in contactor K_3:

$$I_S = \text{IK}_2 + \text{IK}_3 \tag{5.37}$$

$$\text{IK}_3 = I_S - \text{IPR} \tag{5.38}$$

$$\text{IPR} = \text{IK}_2 = k^2.\text{Ir} \tag{5.39}$$

We also have:

$$I_S = k.\text{Ir} \tag{5.40}$$

$$\text{IK}_3 = k.Ir - k^2.Ir \tag{5.41}$$

Giving the following current value to the contactor K_3:

$$IK_3 = Ir.(k - k^2) \tag{5.42}$$

The current in the overload relay FT1 is the same as that of the contactor K_1, since the relay is connected in series with this contactor and the current carrying in it is the same as that of the contactor K_1. With this, we have all the currents in the circuit:

$$IK_1 = Ir$$

$$IK_2 = k^2.Ir$$

$$IK_3 = (k - k^2).Ir$$

$$FT1 = Ir$$

The following Table 5.3 shows the current values in the contactors K_2 and K_3 for the respective taps of the autotransformer.

The reduction of the starting current is proportional to the square of the reduction factor k. Thus, the equation below gives the starting current (I_s) as a function of the reduction factor k:

$$I_s = \left(\frac{I_s}{I_r}.I_r\right).k^2 \tag{5.43}$$

5.3.4.3.4 *Autotransformer Starter Sizing Example*

In order to size an autotransformer starter, consider the following example: 30 HP motor, tap of 80% and starting time of 15 seconds. The nameplate FLC is 77.1 A and the starting current is 617 A.

TABLE 5.3

Currents in K_2 and K_3 with Different Autotransformers Taps

Autotransformer Tap (% V_n)	k	IK2	IK3
85	0.85	$0.72.I_r$	$0.13.I_r$
80	0.80	$0.64.I_r$	$0.16.I_r$
65	0.65	$0.42.I_r$	$0.23.I_r$
50	0.50	$0.25.I_r$	$0.25.I_r$

- **Contactor K_1**

  The contactor K_1 must be sized considering the motor FLC. Then:

$$I_e \geq I_r$$

$$I_e \geq 77.1 \text{ A}$$

- **Contactor K_2**

  The current in K_2 depends on the tap in which it is connected; thus, the current in K_2 is given by:

$$I_e \geq k^2.\text{Ir}$$

$$I_e \geq 49.3 \text{ A}$$

- **Contactor K_3**

$$I_e \geq (k - k^2).I_r$$

$$I_e \geq 0.16.I_r$$

5.3.4.3.4.1 Overload Relay Is chosen considering FLC:

$$I_e \geq I_r$$

$$I_e \geq 77.1 \text{ A}$$

Thus, a relay that has its adjustment range within the found current must be chosen.

5.3.4.3.4.2 Fuses The starting current is reduced by factor k^2, and, considering $k = 0.8$, we have $k^2 = 0.64$.

So: $I_s(a) = k^2.I_s = 394.9$ A

Considering starting time of 15 s in the fuse characteristic curve (see Figure 5.29), we have:

Thus, the 125 A fuse is chosen.

Then, it is necessary to check the conditions:

- If $\geq 1.2\ I_r = 125 \text{ A} \geq 1.2 \cdot 71 \text{ A} = 85.2 \text{ A}$
- If $\leq$ IfmaxK_1 => Check contactor manufacturer catalogue
- If $\leq$ IfmaxFT1 => Check overload relay manufacturer catalogue

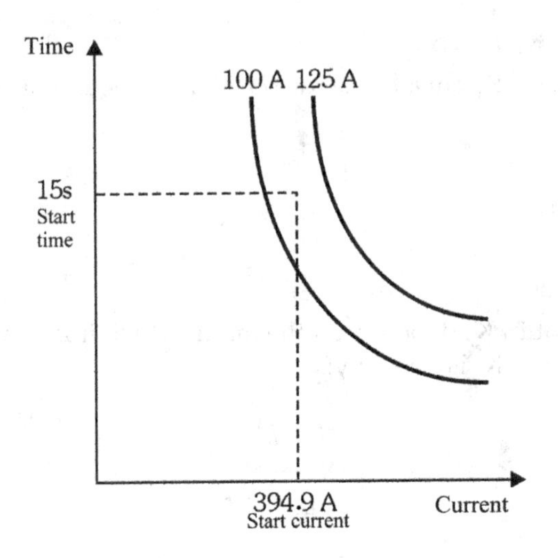

FIGURE 5.29
Characteristic fuse curve to size autotransformer starter fuse (draft).

5.3.4.3.5 *Advantages of Autotransformer Starter*

In the following are listed the main advantages of using an autotransformer starter:

- When switching from tap to grid voltage, the motor is not switched off and the second peak is reduced.
- In order to start the motor satisfactorily, it is possible to change the tap of 65%, 80% or even 90% of the rated voltage.
- The grid voltage can be equal to the voltage value of the motor star or triangle connection.
- The motor only requires three external terminals.

5.3.4.3.6 *Disadvantages of Autotransformer Starter*

In the following are listed the main disadvantages of using an autotransformer starter:

- Limitation of number of starts.
- Higher cost due to autotransformer.
- Larger space occupied in the panel due to the size of the autotransformer.

Exercises

1. Why is not possible to start a motor by applying directly the grid voltage?
2. What is the purpose of a jogging motor starter? Describe the control circuit operation.
3. How is it possible to reverse the rotation of a three-phase induction motor? Describe the control and power circuits to perform it.
4. Compare start currents and torque on the direct on line, star-triangle and autotransformer starters.
5. With the help of the control and power diagrams, explain the operation of the following starters:
   a. Direct on line
   b. Primary-Resistance
   c. Star-triangle
   d. Autotransformer
6. What should be taken into account to determine the tap to be used in the autotransformer starter?
7. What are the advantages and drawbacks of the following starters:
   a. Direct on line
   b. Star-triangle
   c. Autotransformer
8. You were hired to design circuits to start a 50-HP induction motor with a starting time of 10 seconds. Find the contactors, overload relays and fuses for the following starters:
   a. Direct on line
   b. Star-triangle
   c. Autotransformer

6

Solid-State Starters: Soft Starter

The solid-state starter employs silicon-controlled rectifiers (SCRs) to apply reduced voltage in the starting period. Solid-state starters, often referred to as soft starters, are made so that the current flowing into the motor at start is reduced thanks to a reduction in the output starter voltage that occurs electronically. In addition, a soft starter could provide advanced motor control, adjustable reduced voltage, diagnostics and protective features.

With the advent of power electronics, the use of solid-state starters becomes increasingly economically feasible and practical. The following is a description of one of the most commonly used electronic starters: soft starters.

6.1 Soft Starters

These devices are intended for motor control, ensuring progressive acceleration and deceleration and allowing for speed adaptation to operating conditions. The main difference between conventional reduced voltage starters (primary resistor, star-triangle and autotransformer) and the soft starter is the smooth ramp for starting the motor, bringing, as a main benefit, the reduction of the peak current at the motor start. In Figure 6.1, the voltages applied in direct on line starters, star-triangle and soft starters are presented.

These voltage changes will provide the following current and torque curves presented in Figure 6.2.

The progressive voltage ramp can be controlled by the acceleration ramp or dependent on the value of the limiting current, or connected to these two parameters. Thus, the soft starter ensures:

- Control of operating characteristics, especially during start and stop periods.
- Thermal protection of the motor.
- The mechanical protection of the machine moved by suppression of the mechanical blows and reduction of the starting current.

Figure 6.3 shows a commercial soft starter used in industry.

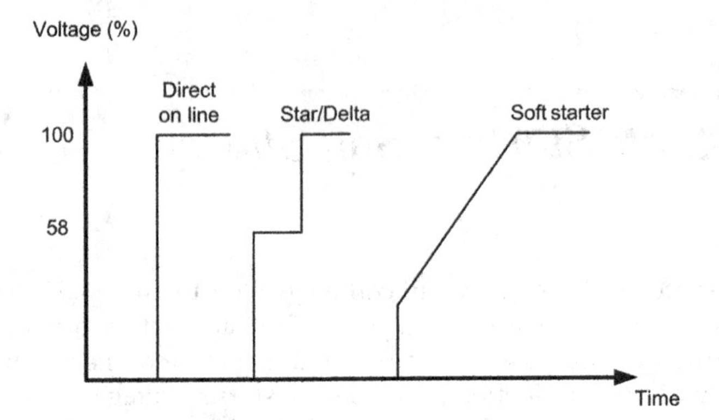

FIGURE 6.1
Voltage change comparison between conventional reduced voltage starters and soft starter.

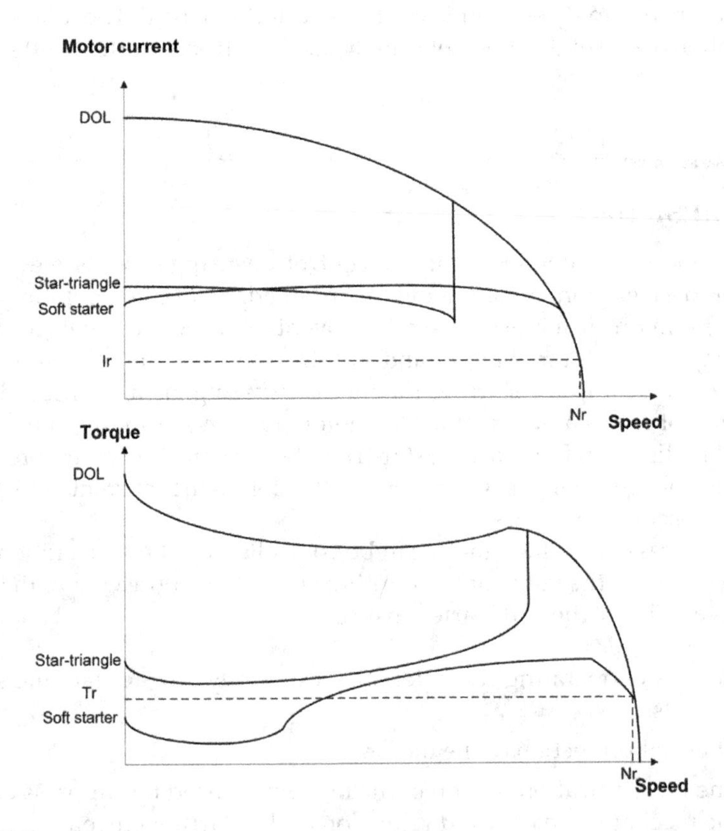

FIGURE 6.2
Current and torque curves comparison between conventional reduced voltage starters and soft starter.

FIGURE 6.3
Commercial soft starter. (Courtesy of Weg.)

6.1.1 Operation Principle

The soft-starter's operation is based on the use of silicon controlled rectifiers (SCR), or rather, a thyristor bridge in the anti-parallel configuration mounted two-by-two in each phase, which is controlled by an electronic control board to adjust the output voltage, according to the programming done by the user, as shown in the diagram of Figure 6.4.

As the diagram shows, the soft-starter controls the grid voltage through the power circuit consisting of six SCRs. By varying its firing angle, the effective voltage value applied to the motor will change. Each part of the structure will be analysed separately, dividing it into two parts: power circuit and control circuit.

6.1.1.1 Power Circuit

This is the circuit where the rated motor current is applied. It consists basically of SCRs and their protections and by current transformers (CT). The SCRs (thyristors) are fundamental components of the soft starter, and the control of its trigger angle to the voltage applied to the load can be controlled, thus controlling its current and power. As we are dealing with alternating current, the voltage is applied in both current directions. In this way, an anti-parallel configuration of the SCRs in each phase (Figure 6.5) is implemented so that each half-cycle has voltage control.

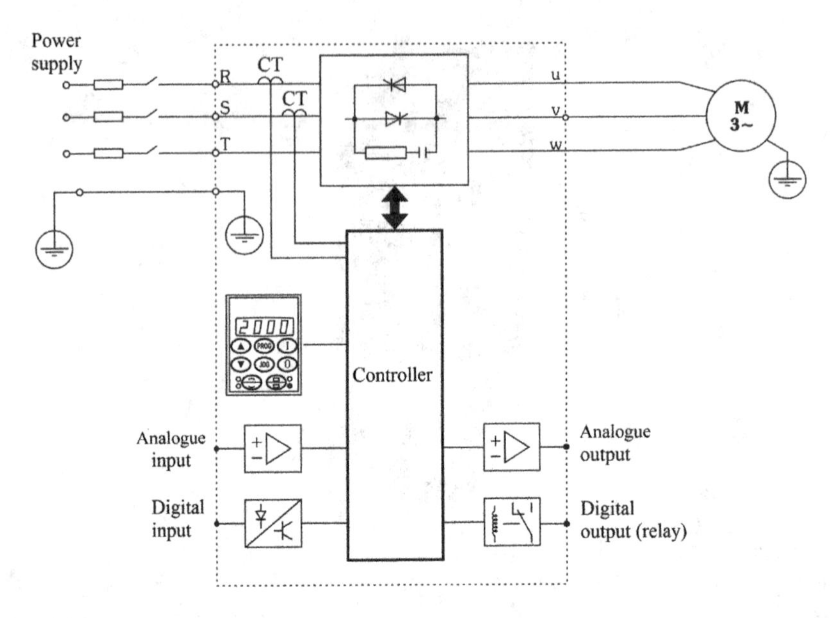

FIGURE 6.4
Simplified block diagram of soft starter.

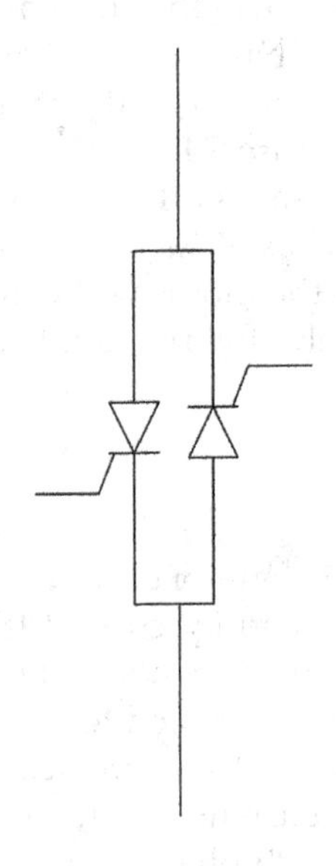

FIGURE 6.5
Antiparallel configuration of the SCRs.

Figure 6.6 shows that the higher the SCR trigger angle, the lower the voltage applied to the motor.

To control the motor voltage, a thyristor trigger circuit is implemented, so the voltage applied to the motor will increase linearly, thereby controlling the motor starting current and applying the grid voltage at the end of start. Antiparallel thyristor pairs are used in each phase of the motor, as shown in the Figure 6.7.

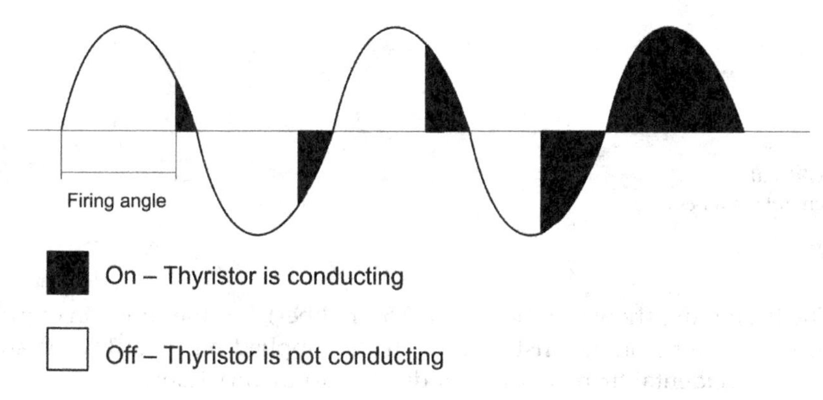

FIGURE 6.6
Relation between trigger angle and voltage applied to motor.

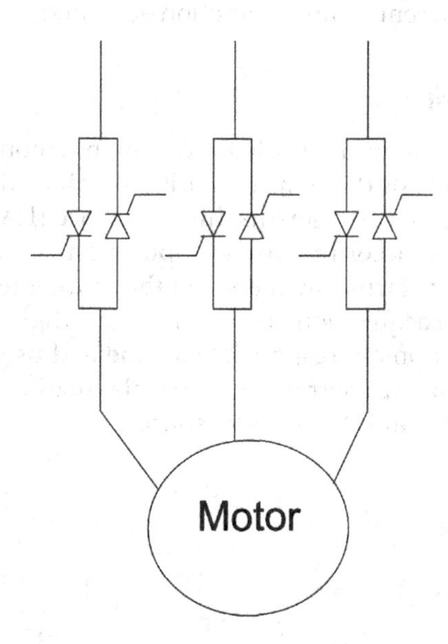

FIGURE 6.7
Antiparallel configuration of the SCRs in each phase of motor.

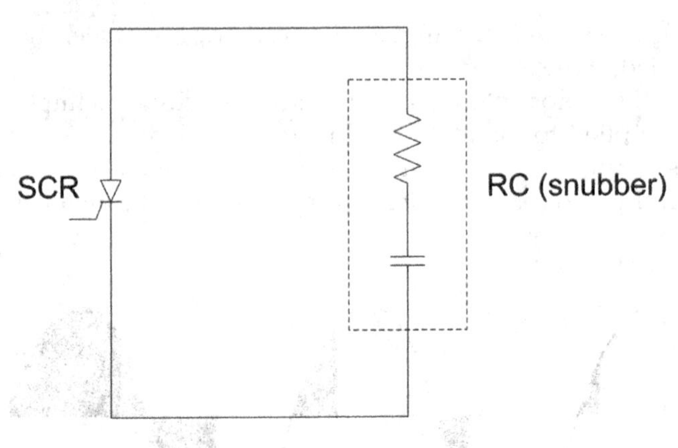

FIGURE 6.8
RC circuit (snubber).

The RC circuit, shown in the Figure 6.8 (snubber), has the function of protecting the thyristors against voltage change applied on them (dv/dt) and avoiding accidental firing of the SCR due to abrupt variations.

In addition, the current transformers monitor the output current, allowing the electronic control to protect and maintain the current value at predefined levels (current limiting function activated).

6.1.1.2 Control Circuit

This is where the circuits responsible for command, monitoring and protection of the components of the power circuit, as well as the circuits used for command, signalling and human/machine interface (HMI), are located and configured by the user, according to the application.

Currently, most soft starters available on the market use microprocessors to perform these data acquisition and control tasks digitally.

The HMI is the equipment responsible for the end user interface where it is possible to perform parametrisation, variable monitoring, indication, etc. Figure 6.9 shows a typical HMI of a soft starter.

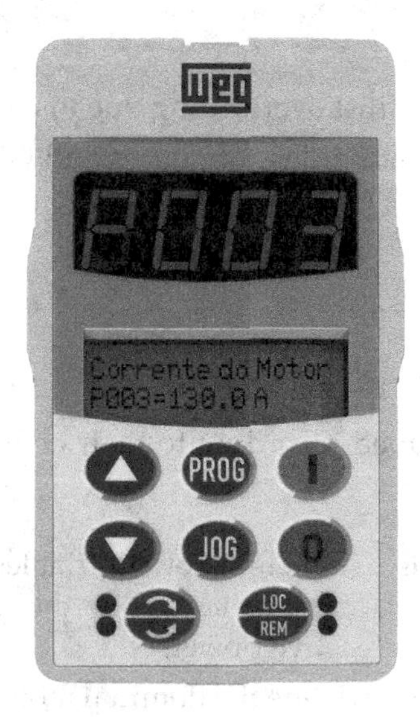

FIGURE 6.9
Typical industrial HMI used in soft starter.

6.2 Main Functions of Soft Starter

In addition to the features previously mentioned, the soft starter also has programmable functions that allow the configuration of the drive system according to the user needs. The thyristors are controlled by a microprocessor that provides the following functions:

- Control of acceleration and deceleration ramps.
- Adjustable current limitation.
- Start torque.
- Direct current injection braking.

- Overload protection.
- Motor protection against overheating due to over-frequent starting.
- Detection of imbalance or loss of phases and failure in thyristors.

6.3 Parameters Description

The parameters are grouped according to their characteristics and particularities, as shown below:

Read parameters: This parameter shows variables that can be viewed on the display, but cannot be changed by the user, such as current, torque, power factor, reactive power, etc.

Motor parameters: These define the nominal motor characteristics, e.g., motor current adjustment, service factor.

Protection parameters: It includes the parameters related to the levels and time of operation of the motor protections.

Regulation parameters: These are adjustable values to be used by the soft starter functions, for example, initial voltage, acceleration ramp time, deceleration ramp time, etc.

Configuration parameters: These define the characteristics of the soft starter and the functions to be performed by the inputs and outputs.

NOTE: There is a soft starter parameter that loads the factory default settings. The parameters are chosen to suit the largest number of applications, minimising the need for reprogramming during commissioning.

The following is a description of some of the typical parameters available in soft starters.

6.3.1 Voltage Ramp in Acceleration

The soft starters can be adjusted in the voltage module so as to have a suitable initial starting voltage, responsible for the initial torque that drives the load. By adjusting the starting voltage by a value, Vs, and a starting time, Ts, the voltage increases from the value of Vs until it reaches the line voltage of the system in a time interval, Ts, also subject to parameterisation, as shown in Figure 6.10. The voltage waveform applied to the motor is presented in Figure 6.11.

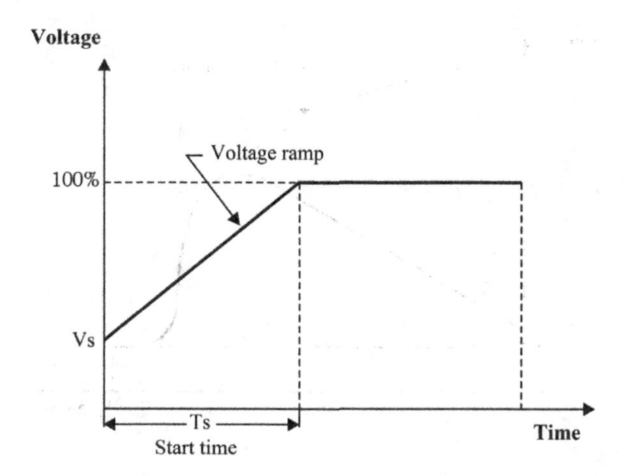

FIGURE 6.10
Acceleration ramp time.

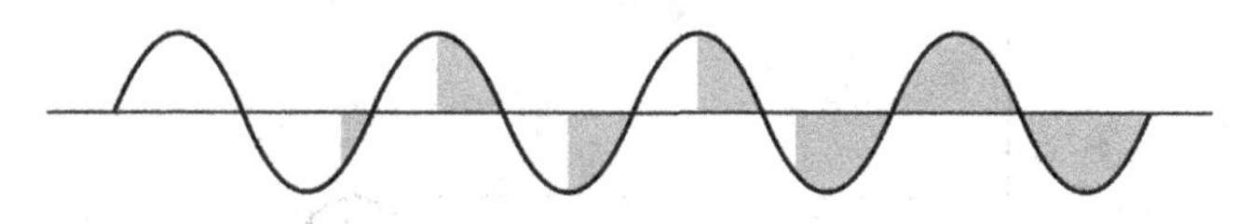

FIGURE 6.11
Voltage waveform applied to the motor in acceleration ramp.

As a result, the curves current by angular speed and current by time are shown in Figures 6.12 and 6.13, respectively.

Taking into account that the motor torque varies proportionally to the square of the voltage and that the current increases linearly, it is possible to control the motor starting torque as well as its starting current by the control of the effective voltage applied to the motor terminals.

The soft starters, by controlling the firing angle of the thyristor bridge, generate in the output a gradual and continuously increasing effective voltage until the rated voltage of the grid is reached. Thus, for a time, Ts, the control circuit, raises the voltage at the motor terminals, from the starting voltage ramp value, which, typically, can be adjusted from 15% to 100% of the grid voltage.

When setting a ramp time and starting voltage value, it does not mean that the motor accelerates from zero to its nominal speed at the time set in this adjustment. This also depends on the dynamic characteristics of the motor/load system, for example: coupling system or the moment of inertia

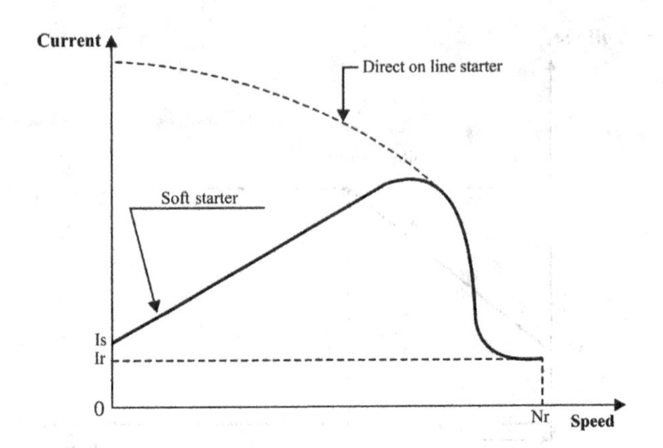

FIGURE 6.12
Current by angular speed curve in acceleration ramp.

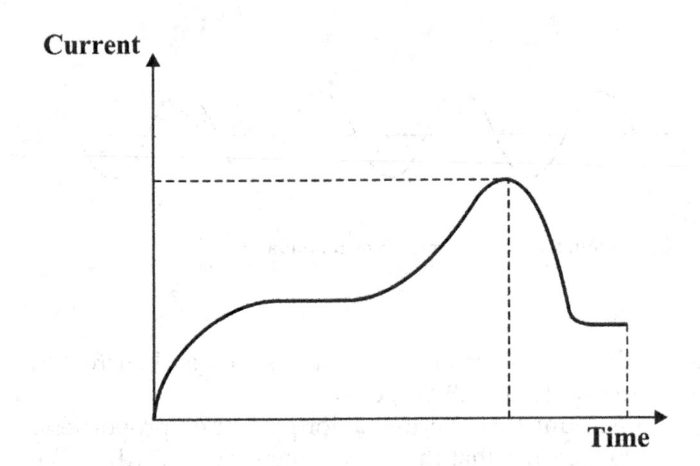

FIGURE 6.13
Current by time curve in acceleration ramp.

of the load reflected to the motor shaft. The voltage value and the ramp time are adjustable values within a range that may change according to manufacturer. There is no rule that can be applied to set the time value, as well as the initial voltage value, so that the motor can ensure the load acceleration. The best approach can be reached considering the acceleration time of the motor.

The starting voltage value, Vs, must be adjusted according to the type of load being driven. The following are two examples of applying the voltage ramp in acceleration.

Pumps: For this application, the starting voltage should not receive a high value setting in order to avoid the phenomenon known as water hammer, which is translated by the pressure wave of the liquid column during the start and stop processes. On the other hand, the voltage cannot receive a very low adjustment, otherwise, the starting process will not be performed. During motor acceleration, the motor torque must be at least 15% more than the maximum pump torque.

Fans: In the same way as pumps, the starting voltage setting value, Vs, must be low enough to allow proper motor torque to be applied to the load. The start time adjust, Ts, should not be too short. The starting current limiting can be used to extend the starting time, Ts, while the system inertia is exceeded. The motor starting torque must be at least 15% above the fan torque.

6.3.2 Voltage Ramp in Deceleration

A possible way for stopping the motor is by inertia, in which the soft starter takes the output voltage instantaneously to zero, so the motor gradually loses speed according to the kinetic energy of the load.

In the same way that soft starters allow smooth motor operation, they can also perform a smooth stop. In the controlled shutdown, the soft starter gradually reduces the output voltage to a minimum value in a predefined time, as shown in Figure 6.14.

On the deceleration ramp, the voltage, Vd, reduces its value in the form of a decreasing ramp to the final shut-off value in which the motor stops rotating, withdrawing the voltage from its terminals. The switch-off time, Td, can be

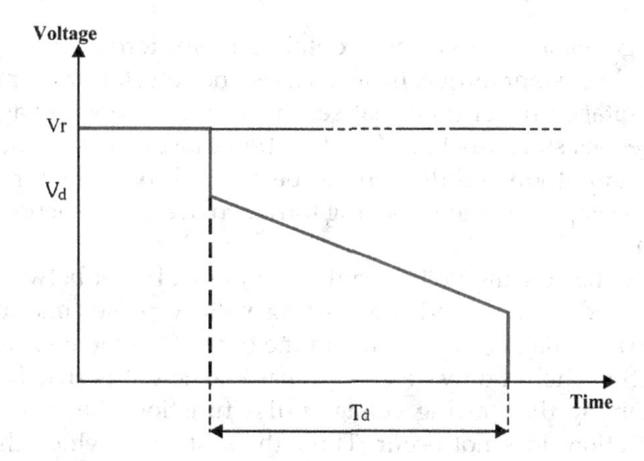

FIGURE 6.14
Deceleration ramp time.

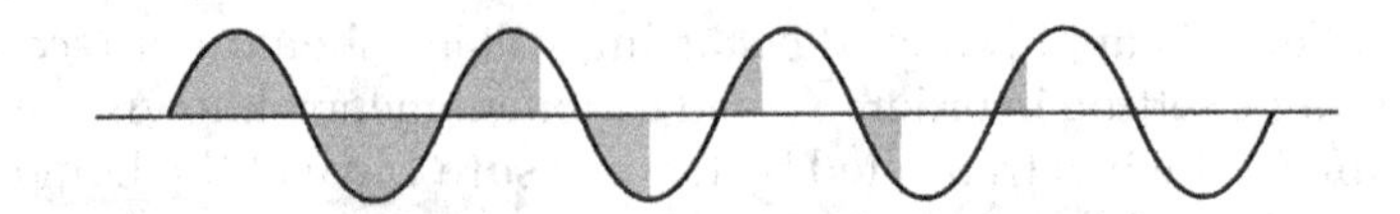

FIGURE 6.15
Voltage waveform applied to the motor during deceleration.

set from 1 to 20 seconds, depending on the manufacturer, and the switch-off voltage, Vd, can be parameterised up to 90% of the rated voltage. Figure 6.15 shows the voltage waveform applied to the motor during deceleration.

When the voltage applied to the motor is reduced, it loses torque and, consequently, loses speed; the driven load also loses. This feature is very common in applications requiring a soft stop from the mechanical point of view, for example, centrifugal pumps and conveyors.

In the case of centrifugal pumps, it is used to reduce water hammer, which can cause serious damage to the entire hydraulic system, by reducing the lifespan of components, such as valves and pipes, as well as the pump itself. It can also be used in situations of an industrial process, where a sudden stop can cause damage to the final product, such as conveyors for bottles of soft drinks.

6.3.3 Start Voltage Pulse (Kick Start)

Some soft starters are provided with a function called a kick start pulse with an adjustable value. It is applied to loads with high inertia which, at the moment of starting, require extra drive effort due to the high resistant torque.

This voltage must be sufficient to obtain a motor torque, which can overcome the load-resistant torque. In such cases, the soft starter normally needs to apply a voltage greater than that set in the acceleration voltage ramp by means of the kick start function. This function causes a voltage pulse of programmable amplitude and duration to be applied to the motor so that the motor can develop a sufficient starting torque to overcome friction and accelerate the load.

In practice, the starting voltage pulse (Vk) must be set between 75% and 90% of the rated voltage, and the starting voltage pulse time must be set between 100 and 300 ms, depending on the type of load to be driven.

This function should only be used in cases where it is strictly necessary, since by enabling the starting voltage pulse function, the current limiting current operation does not occur. Thus, the system in which the motor is inserted may suffer high voltage drops during the time set for the voltage pulse. Figure 6.16 shows the starting voltage pulse function (kick start).

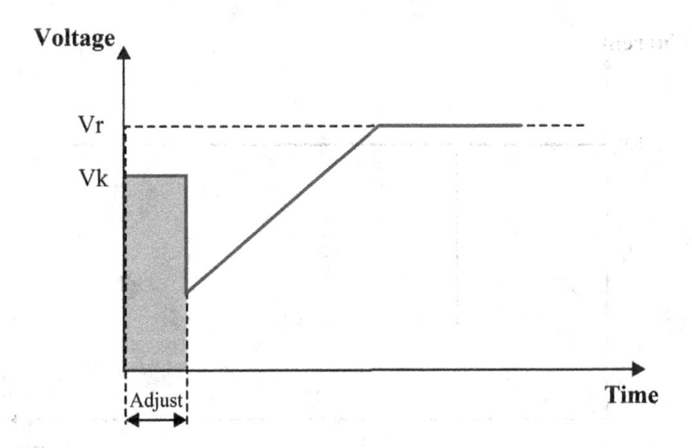

FIGURE 6.16
Kick start pulse function.

6.3.4 Current Limitation

In most cases where the load has a high inertia, this function is used, which causes the grid/soft starter system to supply the motor with only the current necessary to perform the load acceleration. This feature ensures a really smooth drive. Current limitation is also widely used in starting motors whose load has a higher moment of inertia.

6.4 Protections

The use of soft starters is not restricted to start motors, as they can also provide the motor the necessary protection. Thus, when a protection acts, a corresponding error message is issued to allow the user to visualise the occurrence in order to make the right decision. In the following are presented the main protections used in soft starters.

Immediate overcurrent: Adjusts the maximum current (Iov) value that the soft starter allows the motor for a defined period of time (see Figure 6.17).

Immediate undercurrent: Sets the minimum current value that the soft starter allows to the motor for a pre-set time period. This function is used to protect loads that cannot operate in a no-load condition, such as pumping systems (see Figure 6.18).

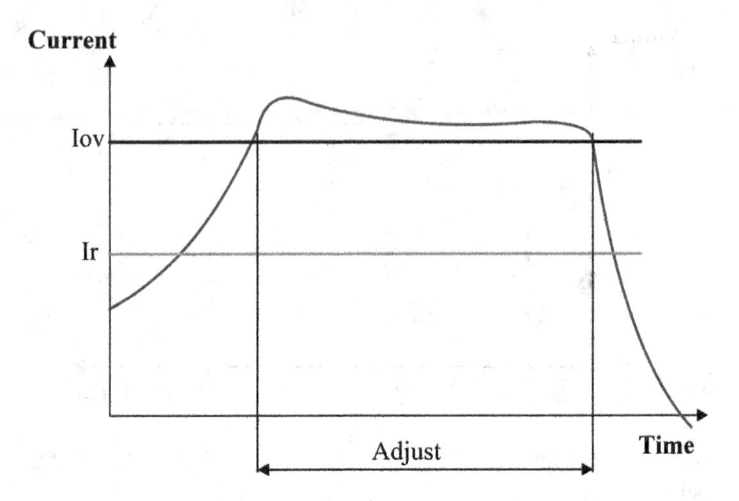

FIGURE 6.17
Immediate overcurrent protection.

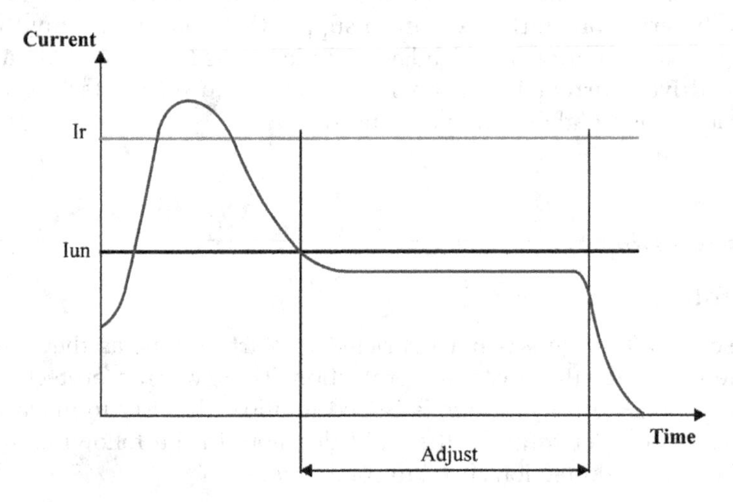

FIGURE 6.18
Immediate undercurrent protection.

NOTE: In addition to the above parameters, the soft starter has several protections, depending on the manufacturer of the equipment, e.g. thyristor over-temperature, reverse phase sequence, phase failure (grid), or motor phase loss.

6.5 Save Energy

When the motor operates under reduced load, it, therefore, operates at a low power factor. The soft starter has a function that optimizes the operating point of the motor, minimising the losses of reactive energy, providing only the active energy required by the load, which characterises an electrical energy saving procedure.

The save energy function is recommended in situations where the motor remains inoperative for a long period of time. This is done by reducing the voltage supplied at the motor terminals during the time the motor operates under reduced or no-load condition. Reducing the voltage reduces the current and, consequently, the iron losses, which are proportional to the square of the voltage.

According to the application, energy savings between 5% and 40% of the rated power can be obtained, considering that the motor operates under the same conditions, but at under the rated voltage, for a load on the shaft of only 10% of rated power. This function offers no advantage when applied in situations where the motor operates at reduced load for short periods of time.

In practice, this function only makes sense when the load is less than 50% of the rated load during an operation period of more than 50% of the motor run time. This function can be applied to loads such as sawmills, grinders, airport conveyors and similar loads. Thus, the voltage waveform applied to the motor is presented, as shown in Figure 6.19.

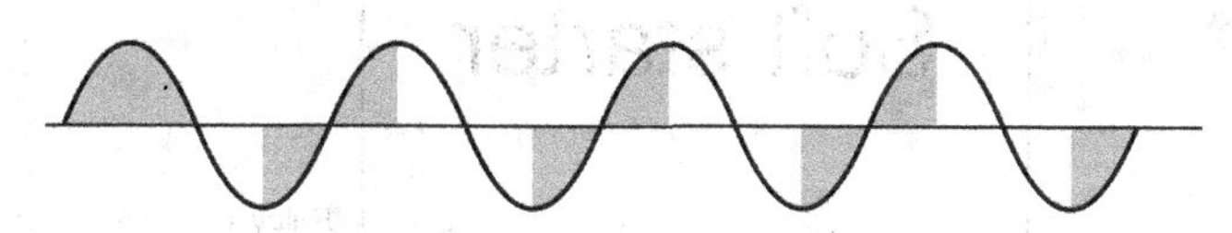

FIGURE 6.19
Save energy function voltage waveform.

6.6 Output Functions

Digital outputs: Typically, soft starters feature output relays to send information to external devices (plc, contactors, lamps, etc.). Figure 6.20 shows an output interface with two single pole single throw (SPST) relays and one Single Pole Double Throw (SPDT).

Different functions can be parameterised by the output relays. The following are some more important functions.

External bypass: Responsible for triggering the thyristor bypass contactor. It operates similar to "in full voltage," but should only be used when an external bypass contactor is required.

Direction of rotation 1 and 2: Actuators used to drive contactors responsible for determining the direction of rotation of the motor.

In operation: The relay is instantly switched on with when the soft starter receives a start command. It is switched off when the soft starter finishes the acceleration ramp.

DC braking: Output will be activated during DC braking.

Alarm: Output will be activated while soft starter has an alarm.

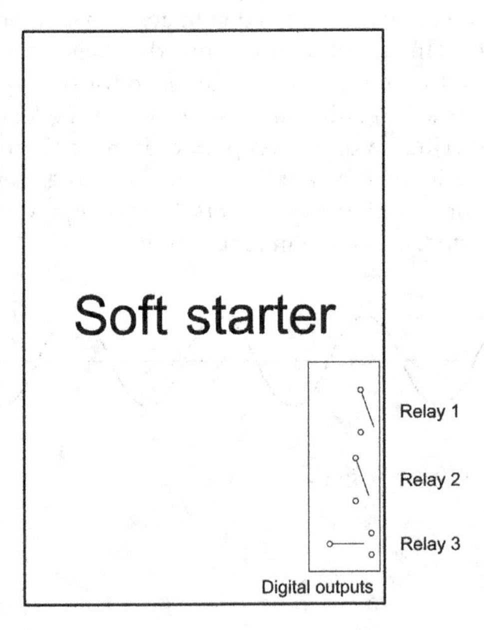

FIGURE 6.20
Output relay interface in soft starter.

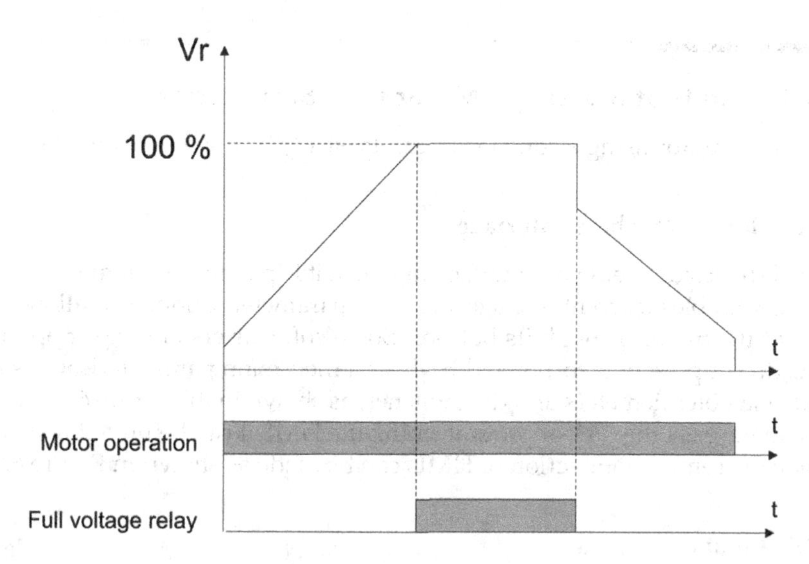

FIGURE 6.21
Full voltage relay operation.

Full voltage function: The relay is only switched on after the soft-starter reaches 100%. It is switched off when the soft starter receives a stop command, as shown in Figure 6.21.

Analogue: In the following are presented some typical output functions assigned to analogue outputs:

Current: Current in the soft starter as a percent of full load current (FLA).

Voltage: Proportional to the output voltage of the soft starter as a percent of rated voltage (Vr).

Power Factor: Load power factor considering harmonic currents.

6.7 Input Functions

In the following are presented some typical input functions assigned in a soft starter.

Digital: Start, stop, reset, error, alarm, jog and rotational direction.

Analog: Voltage ramp proportional to 0–10 V input.

6.8 Methods of Starting a Motor with Soft Starter

Starting a motor using a soft starter can typically be done as shown below:

6.8.1 Human Machine Interface

The HMI, besides being an interface in which it is possible to monitor the soft starter variables (current, voltage, etc.) and parameterisation, also allows the start of the motor through its buttons. Some soft starters of simpler applications do not present incorporated HMI, and their main parameterisations are made via potentiometers and dip switches, as shown in the Figure 6.22.

To make easy the task of programming the HMI, it can be placed in a door panel; the remote connection of HMIs can be made as shown in Figure 6.23.

6.8.2 Inputs

In an industrial application, when a large number of motors and equipment makes it not feasible to start them through the HMI, it is common to start the motors via digital or analogue inputs. Typically, there are several ways of connecting this type of input, most commonly, with voltages in the order

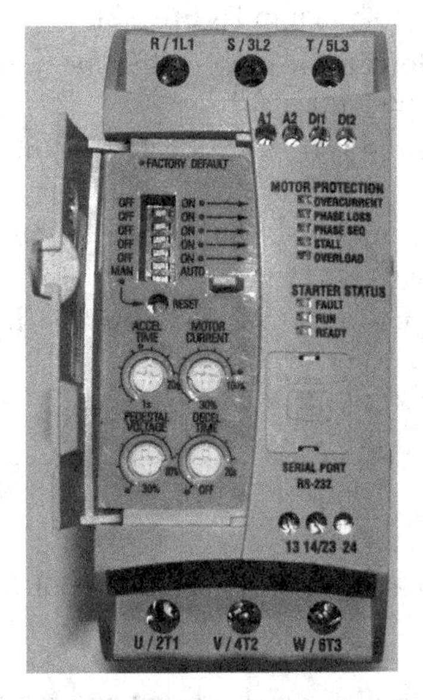

FIGURE 6.22
Simple soft starter without HMI.

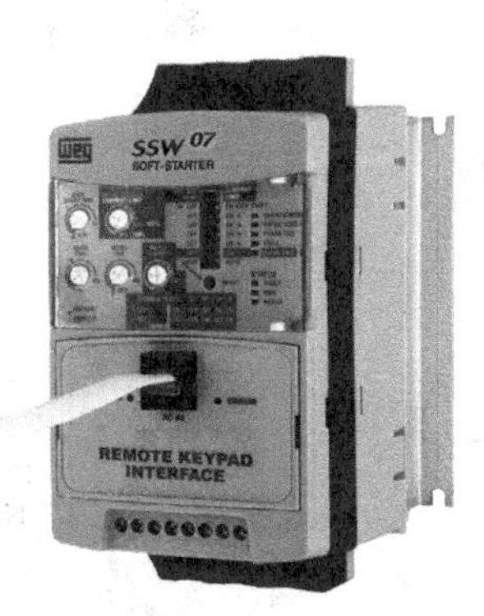

FIGURE 6.23
Remote HMI in a door panel.

1	DI1
2	DI2
3	DI3
4	DI4
5	DI5
6	24Vdc

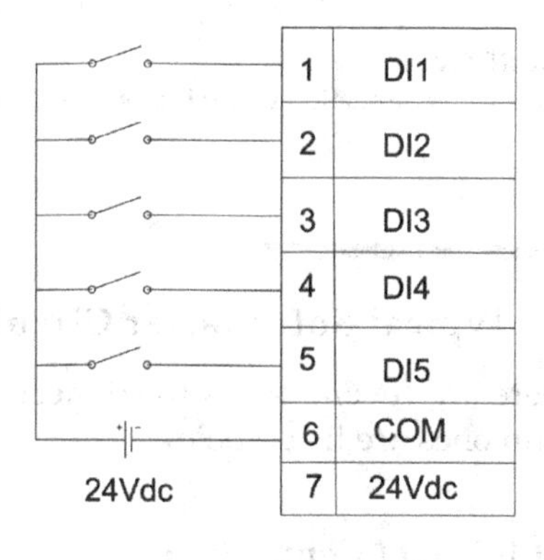

FIGURE 6.24
Inputs to motor start using soft starter.

of 24 Vdc by means of external sources or programmable logic controllers (PLC), or through a source already existing in the terminals, as shown in the Figure 6.24.

6.8.3 Industrial Networks

It is also possible to operate the soft starter through industrial networks in systems using PLCs and Supervisory Control and Data Acquisition (SCADA) supervision system, as shown in Figure 6.25. With industrial networks, in addition to the remote start, it is possible to monitor all variables and alarms of the soft starter using only one or two pairs of wires.

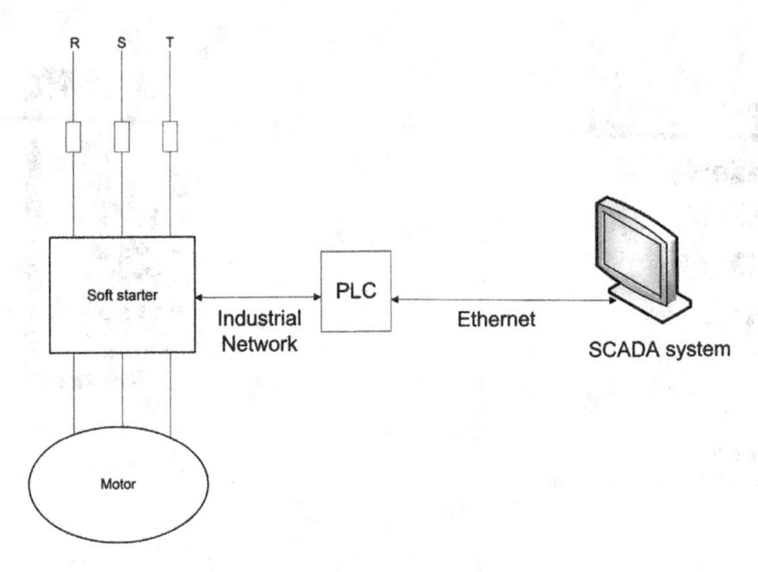

FIGURE 6.25
Soft starter connection using industrial networks.

6.9 Typical Soft Starter Circuits

There are several ways to connect and perform the soft starter control, the main ones are listed below:

6.9.1 Direct Connection

In this type of connection, the motor is connected directly to the soft starter. Depending on the soft starter model, it can be connected with fuses directly to the power supply (see Figure 6.26) or with contactors and overcurrent relays.

NOTE: Although soft starters can be installed with or without a contactor, it is recommended to use them for the following reasons:

- To meet local electrical regulations.
- For physical isolation when soft starter is not in operation.

6.9.2 Using Digital Inputs

In Figure 6.27, presented are some of the most used circuits employing digital inputs to drive soft starters: Two, three and four wires.

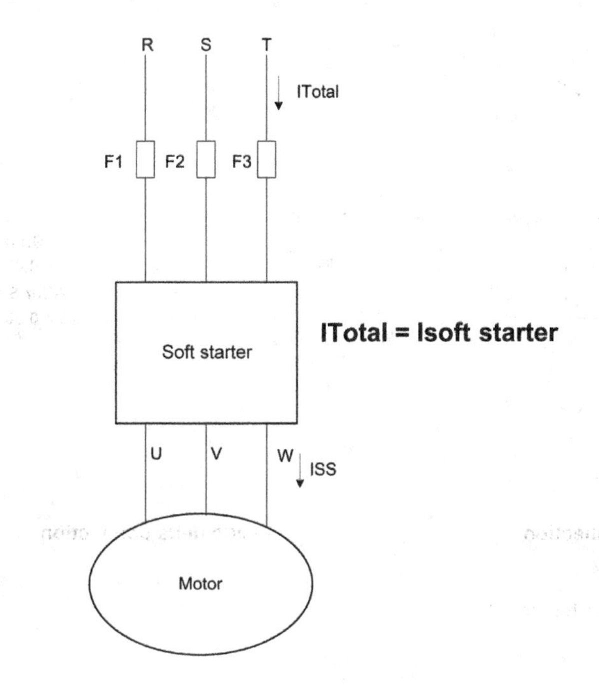

FIGURE 6.26
Soft starter direct connection.

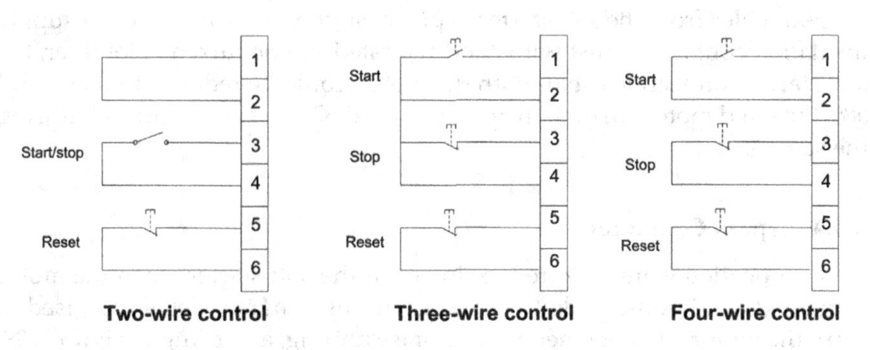

FIGURE 6.27
Typical soft starter wiring diagrams using digital inputs.

6.9.3 Inside Motor Delta Connection

In this type of connection at start, for the same motor power, the inside delta connection (6 leads motor) allows for a reduction of 33% of the soft starter current if compared to the conventional connection in star or delta (3 leads motor) connection. In the motor in steady state a reduction of 42% of the soft-starter current is achieved (see Figure 6.28).

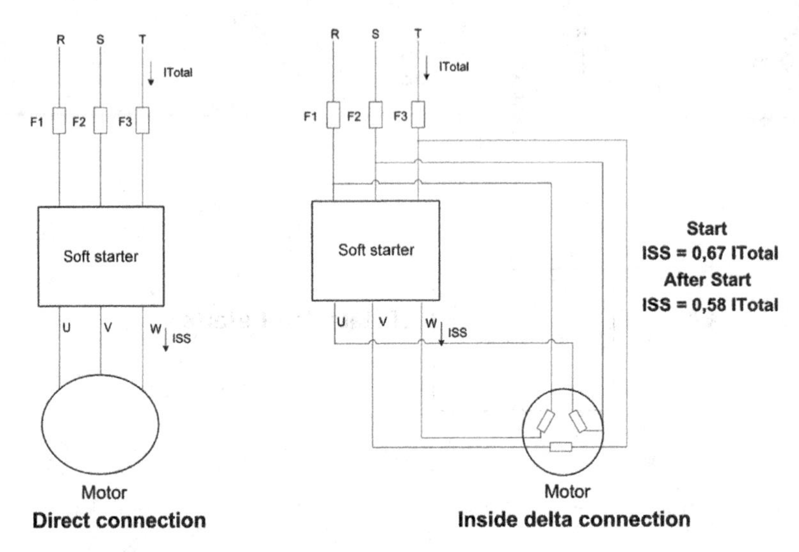

FIGURE 6.28
Three-wire soft starter circuit.

This type of connection provides a way of reducing cost and size. The connection cables from the soft starter to power supply, and/or the power supply insulation contactor, must withstand the rated motor current. However, the soft starter connection cables with the motor, could be reduced to withstand 58% the rated motor current (in operation) and 67% the motor current (during the start).

6.9.4 Bypass Contactor

This connection is made to reduce losses in the soft starter when the motor is in rated conditions. To do this, a contactor used in parallel is energised to carry the motor current when the motor is running, according to Figure 6.29.

NOTE: Some soft starters have a built-in bypass to reduce heating losses in the thyristors, providing size reduction and energy saving, as shown in Figure 6.30.

6.9.5 Multiples Motors Simultaneously

It is possible for one soft starter to be used to control the start of several motors. In order to make this connection, the soft starter's capacity must be greater than or equal to the sum of the powers of all motors.

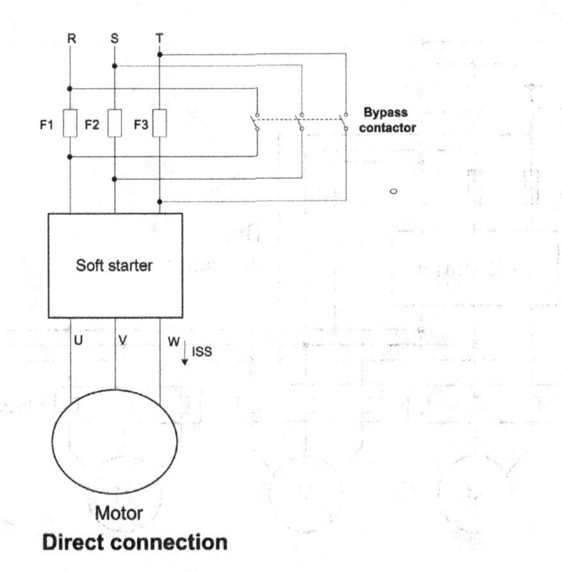

FIGURE 6.29
Bypass contactor circuit.

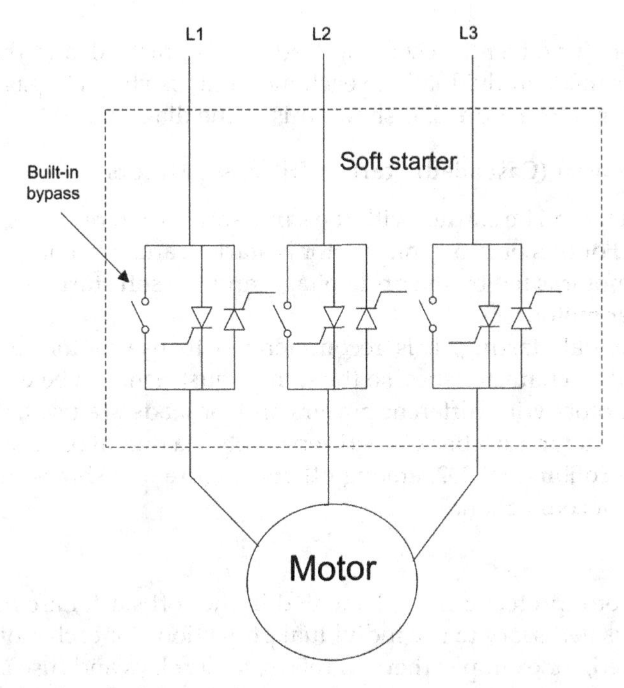

FIGURE 6.30
Bypass built in the soft starter.

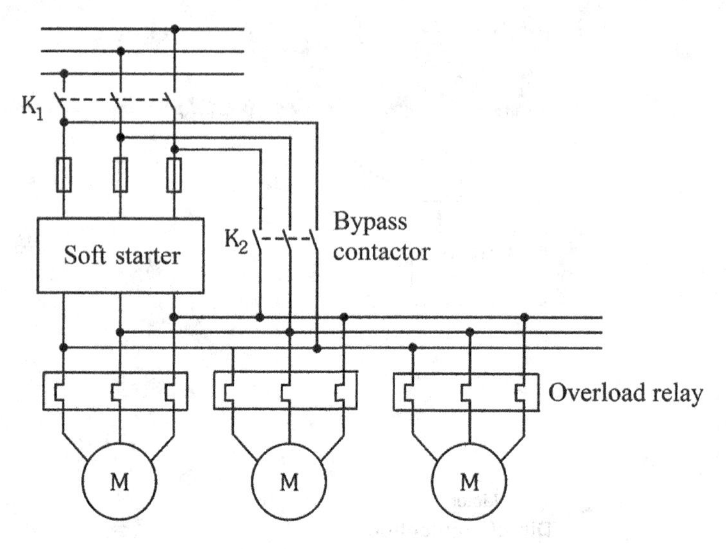

FIGURE 6.31
Simultaneous motor start using soft starter.

The motors have identical starting loads, as is practical, and the overload protection is made individually to each motor. It also has a bypass contactor that acts after start. Figure 6.31 shows this in the diagram.

6.9.6 Sequential (Cascaded) Start of Different Motors

Several motors can be started with the same soft starter, reducing the cost of the system. For this purpose, one motor is started, and when it is at the rated speed, the motor is fed by the grid voltage, and the soft starter is released to start another motor.

For sequential starting, it is recommended to use motors of the same power and load characteristics, so the same adjustment can be used for both motors. If motors with different powers and/or loads are used, the parameters of each motor must be adjusted separately via digital or network inputs (devicenet, profibus, RS 232, among others). Figure 6.32 shows the diagram of this type of connection.

NOTES:

- The motor protections implemented in the soft starter are not used, and it is necessary to use individual protections for each motor to be operated, for example: thermal relays, fault relays and fuses.
- It will result in a high complexity circuit, considering wires, contactors, timers, PLCs, etc.

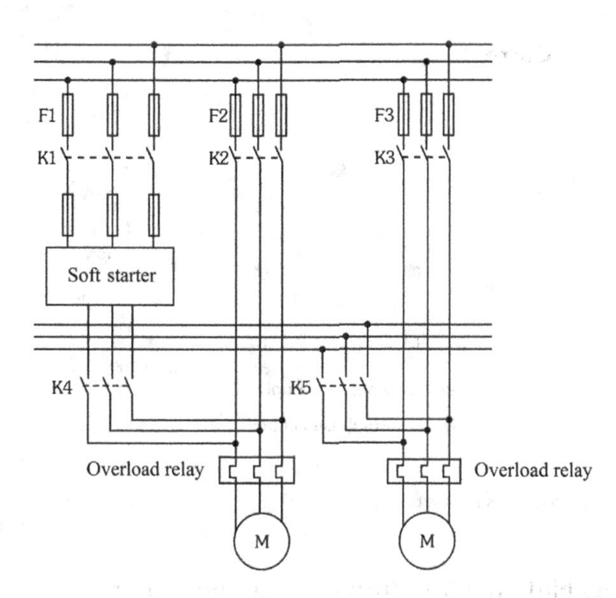

FIGURE 6.32
Diagram for cascaded start using soft starter.

6.10 Number of Phase Control

Typically, the soft starter uses 2 SCRs per phase in antiparallel connections to perform voltage control in the three phases of the motor. However, in order to reduce the number of components, and also the cost of the equipment, some manufacturers use the control of two of the three phases, with one of the phases passing through the soft starter without the voltage control.

This type of control is called a two-phase controlled soft starter. In typical applications, this will not cause any problems to the motor and start. However, some care must be taken.

6.10.1 Unbalance of Motor Currents in the Start

When using a soft starter with control in all three phases, the current will be balanced in the same way to each phase at the start. In the case of the soft starter that uses only two controlled phases, we will have a higher current in the uncontrolled phase and a smaller in the other two phases. This is represented in Figure 6.33.

NOTE: It is worth emphasising that this will happen only at the start and stop time, when the motor receives the rated voltage will the phases will be balanced.

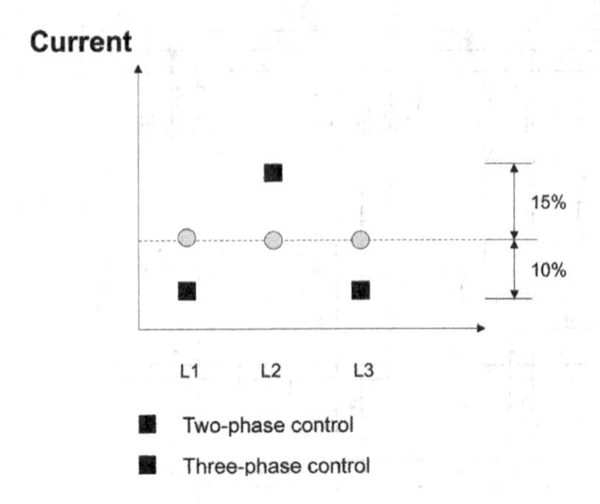

FIGURE 6.33
Unbalance of motor currents at start.

6.10.2 Impossibility to Make Inside Delta Connection

As one of the phases is not controlled by the soft starter, the rated phase current will reach the motor, causing heating in one of the windings, and may even damage the motor, as shown in the Figure 6.34.

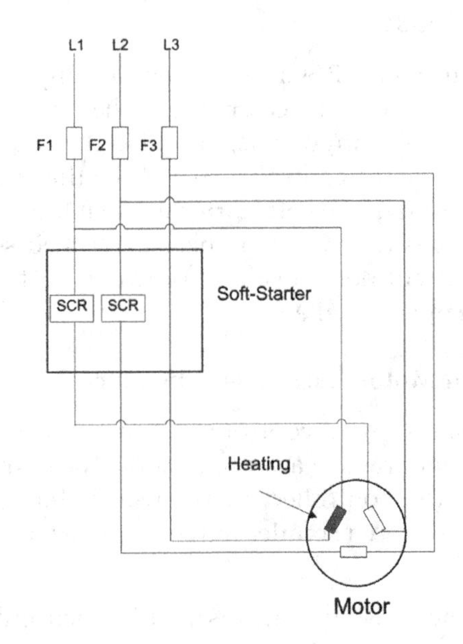

FIGURE 6.34
Impossibility to make connection inside delta with two-phase control soft starter.

6.11 Torque Control

In some softs starters, it is possible to control the torque of the motor instead of the voltage. This function is typically used to drive pumps and prevent water hammering on motors to avoid premature wear due to torque variations. It is possible to select between torque ramps or current ramps during start and stop.

The following Figure 6.35 shows a comparison of torque curves for a direct on line starter and a soft starter with a conventional voltage ramp control and a torque control.

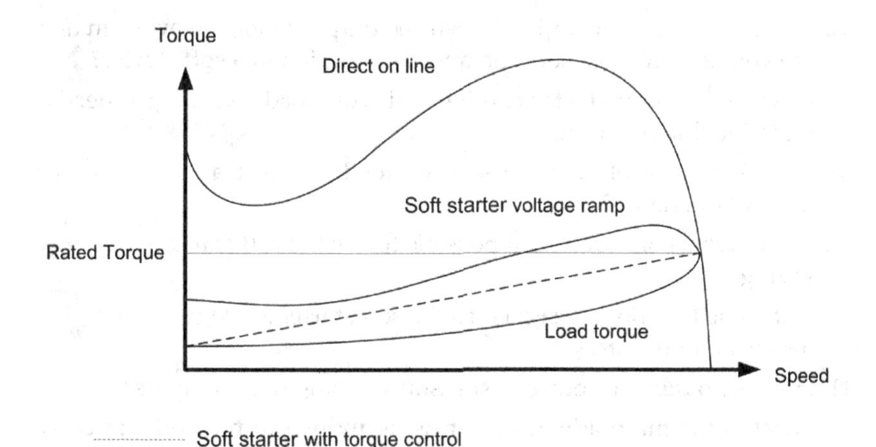

FIGURE 6.35
Comparison of torque curves for DOL starting and soft starter with a conventional voltage ramp control and a torque control.

Exercises

1. What is the purpose of using a soft starter switch for starting motors? Why is it better than star-triangle, primary resistance and autotransformer?
2. Compare the voltage, current and torque applied in motor start in a direct on line starter, star triangle and soft starter.
3. With the help of a diagram, describe the operating principles of a soft starter.
4. Describe the SCR operation in the soft starter power circuit.
5. List the main soft starter functions.
6. What is the relationship between the output voltage waveform and the voltage ramp acceleration and deceleration in a soft starter?
7. How can the soft starter can be applied to loads with high inertia? Describe this function.
8. How is it possible for the soft starter to protect a motor against no-load operation?
9. Under what conditions is it possible to use the soft starter for energy saving?
10. List some functions assigned to the soft starter analogue and digital inputs and outputs.
11. How is possible to control a soft starter using digital inputs?
12. What is the main advantage of using industrial networks to drive the soft starter?
13. Why is the inside delta soft starter connection used?
14. What is the purpose of using bypass contactors in soft starters?
15. Is it possible to start more than one motor with the same soft starter at the same time? If so, show the circuit.
16. How can motors be sequentially started using only one soft starter?
17. Describe a two-phase controlled soft starter. What precautions should be taken for this soft starter?
18. When is a soft starter with torque control used?

7

Variable Frequency Drives

There are numerous reasons for using speed control devices. In some applications, such as with pulp and paper industries, devices cannot operate without speed control, while others, such as centrifugal pumps, can benefit from energy reduction. The use of speed control devices in motors has an extensive range of applications in the industry.

A variable-frequency drive (VFD) is a type of adjustable-speed drive used to control induction motor speed and torque by changing motor input frequency and voltage.

Before we begin the study on the principle of operation of VFDs, some fundamental concepts to help in understanding this device operation will be presented.

7.1 Fundamental Concepts

7.1.1 Force (F)

Movement is the result of the application of one or more forces to a given object. Movement occurs in the direction in which the resulting force is applied. Force is a combination of intensity and direction. A force can be applied in the forward or reverse direction. The SI unit of force is Newton (N).

7.1.2 Speed (n)

Speed is the measure of the distance that an object can reach in a determinate unit of time. Usually, the unit used is meter per second (m/sec). For forward motion, speed is considered positive and is considered negative for reverse direction.

7.1.3 Angular Speed (ω)

Although force is directional and results in a linear motion, many industrial applications are based on rotational motions. The rotational force associated with rotating equipment is known as torque. Angular speed is the result of the torque application and angular rotation. It is usually measured as revolutions per minute (rpm).

7.1.4 Torque

Torque is the product of the tangential force (F) of the circumference of a wheel, and the radius of the centre of this wheel. The unit of torque most commonly used is newton-metre (N·m). The torque can be positive or negative, depending on the direction of rotation.

For a better understanding of these concepts, consider the example of Figure 7.1, in which a car is used to illustrate the relationships between direction, force, torque, linear velocity and angular velocity. In the car, a combustion engine develops rotational torque and transfers it through a transmission system to the wheels, which convert torque into a tangential force (F). The greater the magnitude of this force, the faster the car accelerates.

7.1.5 Linear Acceleration (la)

Linear acceleration (la) is the rate of change of linear velocity, usually shown in meters per second squared (m/sec^2).

$$la = \frac{\text{speed change}}{\text{time change}} \cdot (\text{m/sec}^2) \tag{7.1}$$

Linear acceleration is the increase of speed in any direction, with deceleration being the braking or reduction of the speed in any direction.

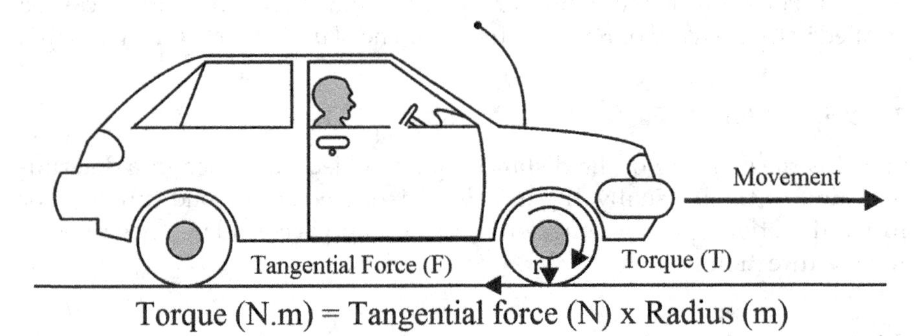

FIGURE 7.1
Example of relationships between force, radius and torque.

7.1.6 Rotational Acceleration (ra)

Rotational acceleration (ra) is the rate of change of angular speed, generally, in radians per second squared (rad/sec^2).

$$ra = \frac{\text{angular speed change}}{\text{time change}} \cdot (\text{rad/sec}^2) \tag{7.2}$$

7.1.7 Power

Power is the rate at which work is performed by a machine. The most commonly used power unit is watts (W). In rotary machines, the power can be calculated as the product of torque and speed. The power can be related to the torque in the following way:

$$\text{Power(kW)} = \frac{\text{torque}(\text{N} \cdot \text{m}) \times \text{speed(rpm)}}{9550} \tag{7.3}$$

7.1.8 Energy

Energy is the product of power over time and represents the amount of work done over a period of time. It is usually expressed in kilowatt-hours (kWh). In the example of the car in Figure 7.1, the fuel consumption over a period of time represents the energy consumed. The equation representing the energy is as follows:

$$\text{Energy}(\text{kWh}) = \text{Power}(\text{kW}) \times \text{Time}(\text{h}) \tag{7.4}$$

7.1.9 Moment of Inertia (J)

Moment of inertia (J) is the property that a rotating machine has to resist a change of rotation speed either by acceleration or deceleration. The unit for the moment of inertia in the SI is kg.m^2.

This means that to accelerate an object in a rotational motion from a velocity n_1 to a velocity n_2, an acceleration torque T_A (N·m) must be provided to promote acceleration. The time t required for speed change depends on the moment of inertia J (kg.m^2) of the drive system comprised by the VFD and the mechanical load. The acceleration torque will be:

$$T_A(\text{N} \cdot \text{m}) = \text{J}(\text{kg.m}^2) \times \frac{2\pi}{60} \times \frac{(n_2 - n_1)(\text{rpm})}{t(\text{sec})} \tag{7.5}$$

In applications where rotational motion is transformed into linear motion, for example, a conveyor belt, the speed (n) can be converted to linear velocity (V) using the rotating element diameter (d), as shown below:

$$V = (\text{m/sec}) = \pi \cdot d \cdot n(\text{rpm}) = \frac{\pi \cdot d \cdot n}{60} \tag{7.6}$$

So:

$$T_A(\text{N} \cdot \text{m}) = \text{J}(\text{kg.m}^2) \times \frac{2}{d} \times \frac{(v_2 - v_1)(\text{m/sec})}{t(\text{sec})} \tag{7.7}$$

7.2 Torque Relations in a Variable Frequency Driver

An induction motor can be modeled as a transformer in which the primary is the stator and the secondary is the rotor. By the equation of the asynchronous induction motor, the torque developed by the asynchronous motor is given by the following equation:

$$T = \varphi_m \cdot I_2 \tag{7.8}$$

And the voltage applied on the coil of a stator is given by:

$$U_1 = 4.44 \cdot F_1 \cdot N_1 \cdot \varphi_m \tag{7.9}$$

where:

- T is the motor torque (N.m)
- φ_m is the magnetic flux (Wb)
- I_2 is the rotor torque (A)
- U_1 is the stator voltage (V)
- F_1 is the grid frequency (Hz)
- N_1 is the number of coils

The alternating flux φ_1, resultant from the stator voltage U_1, induces a electromotive force (emf) in rotor (U_2), that produces a flux φ_2 proportional to U_2 and inversely proportional to the frequency. Therefore, we have:

$$\varphi_2 = U_2 / F_1 \tag{7.10}$$

In order to enable the operation of the motor with constant torque at different speeds, it is necessary to change the voltage U_2 proportionally with the change of the frequency F_1, thus maintaining a constant flow.

7.3 Variable Frequency Driver Components Blocks

In Figure 7.2, a block representation of variable frequency drives components and a commercial VFD are presented.

7.3.1 Central Processing Unit

The central processing unit (CPU) typically is composed of a digital signal processor (DSP) or a microcontroller, depending on the manufacturer. It is in this block that all the information (parameters and the system data) are stored, since a memory is also integrated into this set. The CPU not only stores the data and parameters related to the equipment but also performs the most vital function for the inverter operation: the generation of the trigger pulses, by means of a coherent control logic for the isolated gate bipolar transistors (IGBT).

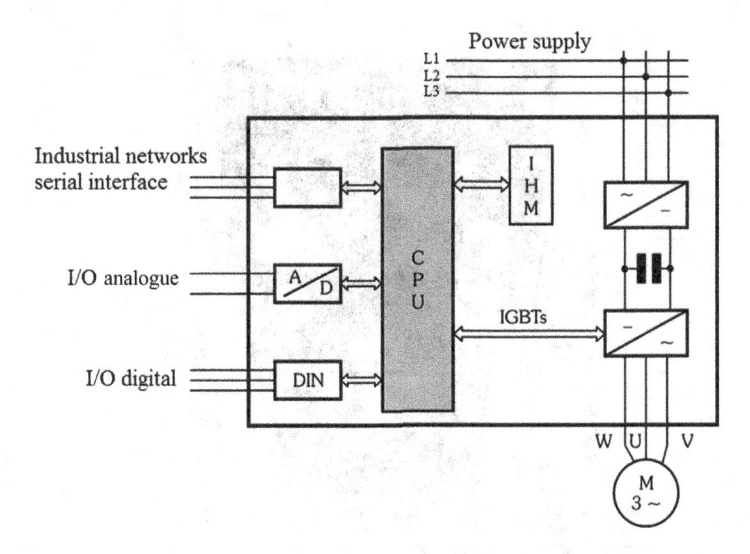

FIGURE 7.2
Block representation of the components of VFDs.

7.3.2 Human Machine Interface

The human machine interface (HMI) is responsible for monitoring function on the VFD (display) and parameterization according to the application (keys). In Figure 7.3, a commercial VFD HMI is presented.

With the HMI, it is possible to monitor different motor parameters, such as voltage, current, frequency, alarm status, among other functions. It is also allows monitor the direction of rotation, check of the operating mode (local or remote), switching of the VFD on or off, change of speed, change of parameters and other functions.

7.3.3 Input and Output Interfaces

Most VFDs can be controlled by two types of signals: analogue or digital. Normally, when it is desired to control an application in a close loop, motor speed uses an analogue command (voltage or current). Typically, the voltage is between 0 and 10 Vdc and the current 4 to 20 mA. The speed (rpm) is proportional to its value, for example:

$$1 \text{ Vdc} = 1000 \text{ rpm}, 2 \text{ Vdc} = 2000 \text{ rpm, and etc.}$$

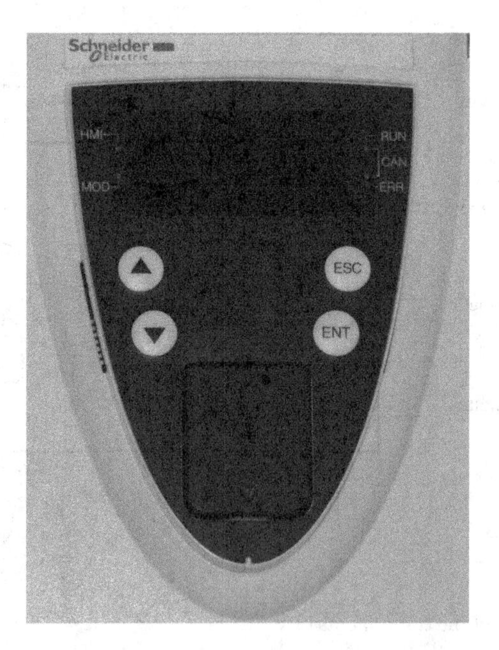

FIGURE 7.3
Commercial VFD HMI.

To reverse the direction of rotation, simply reverse the analogue signal polarity (from 0 to 10 Vdc clockwise and –10 to 0 Vdc counter clockwise). This is the most used system in automatic machines and tools, being the control analogue signal coming from the programmable logic controller (PLC).

In addition to the analogue interface, the VFD has digital inputs, which can also be used to drive it, as will be seen next. With a programming parameter, it is possible to select the valid input (analogue or digital).

7.3.4 Power Stage

The power stage consists of a rectifier circuit, which, by means of an intermediate circuit called "DC bus", feeds the inverter output circuit (IGBT module).

The technological advancements of power electronics allowed the development of frequency converters with solid state devices, initially, with thyristors and, currently, with transistors, more specifically, the IGBT.

Some definitions are important for the VFD operation, the symbols of which are used below to describe different types of VFD converters.

- **Rectifier:** A converter that transforms AC to DC (Figure 7.4).
- **Inverter:** A special type of converter that converts DC to AC (Figure 7.5).

In variable frequency drives, it is common to use an intermediate circuit, called a DC link with filter, to make the sine waveform (Figure 7.6).

With the union of these three modules, we have a variable frequency converter that can be applied in a three-phase induction motor to control the speed (Figure 7.7).

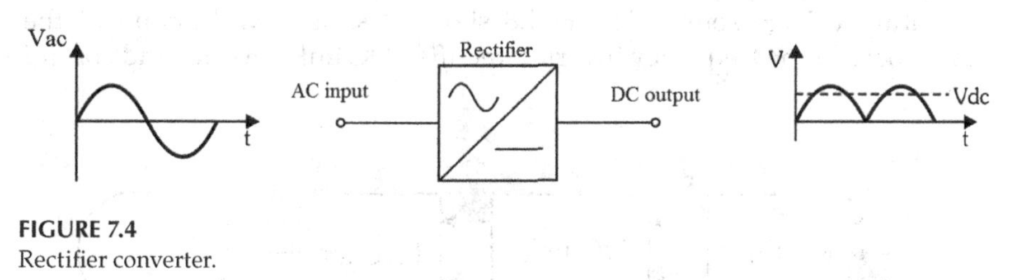

FIGURE 7.4
Rectifier converter.

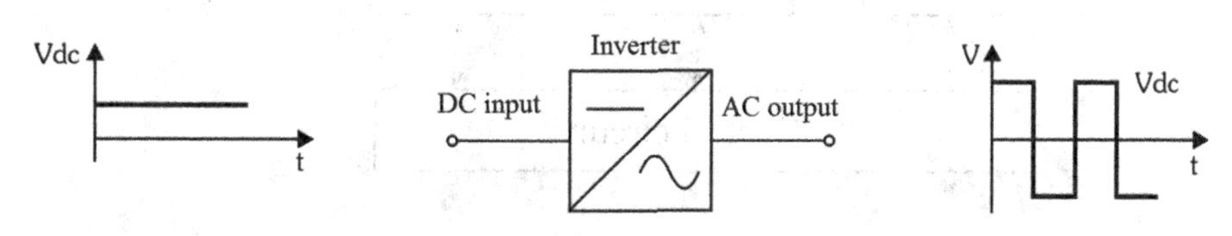

FIGURE 7.5
Inverter converter.

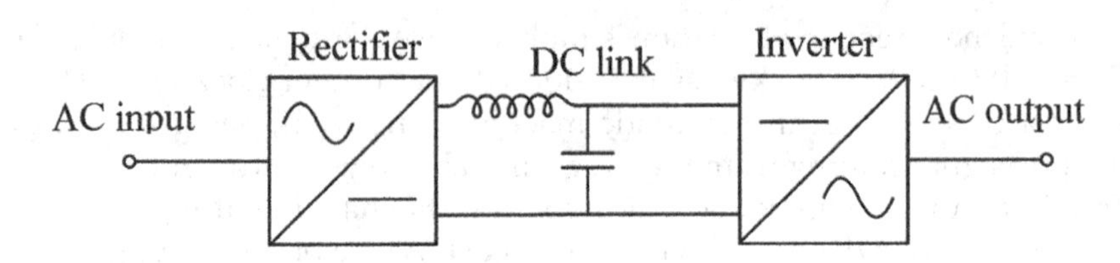

FIGURE 7.6
DC link in a VFD.

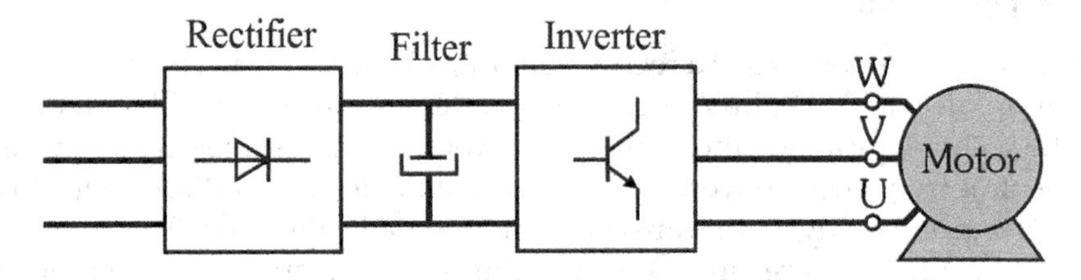

FIGURE 7.7
Rectifier, inverter a DC link forming a VFD.

Cycloconverters preceded, to some extent, the current VFDs. They were used to convert 60 Hz of the grid into a lower frequency, a AC-AC conversion. The VFDs use the AC-DC conversion and, finally, the DC-AC conversion again.

The VFDs can be classified by their topology, which is divided into three parts, the first for the type of input rectifier, the second for the type of control of the intermediate circuit and the third for the output.

As previously shown, the VFD can be considered as a variable frequency alternating voltage source. Figure 7.8 shows a simplified diagram of the main blocks of the frequency inverter: rectifier, DC link, inverter and control circuit.

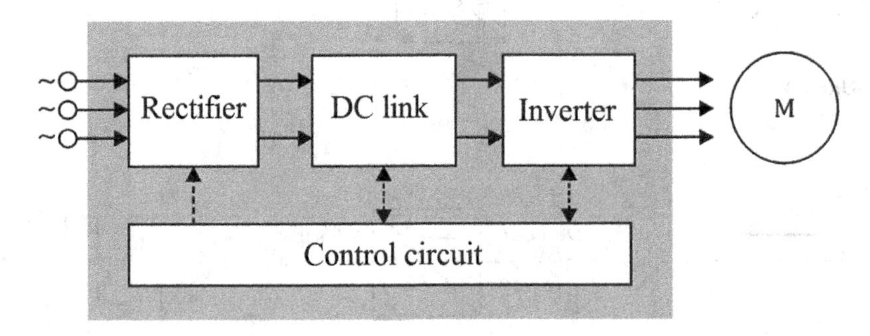

FIGURE 7.8
Simplified diagram of VFD blocks.

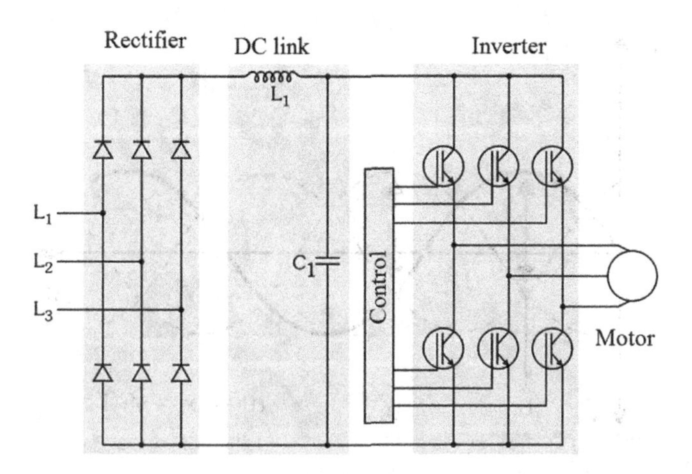

FIGURE 7.9
Rectifier, intermediate circuit, inverter and control circuit illustrated in details.

In Figure 7.9, the rectifier, DC link, inverter and control circuit are presented in more details.

The following are details of the blocks of the variable frequency driver:

- **Rectifier:** Responsible for converting the alternating signal coming from the grid in a constant voltage and frequency. In the input grid, the frequency is fixed at 50/60 Hz, being transformed by the rectifier into a continuous one (full wave rectifier). The filter transforms this voltage into a continuous value of approximately:

$$\mathrm{Vdc} = \sqrt{2} \cdot \mathrm{Vgrid} \tag{7.11}$$

Consider the three-phase alternating voltage with a fixed frequency of 60 Hz, as shown in Figure 7.10.

For the rectifying circuit, diodes are used, which are semiconductor components that allow the current to flow in only one direction: from the anode (*A*) to the cathode (*K*). It is not possible to control the current intensity. An alternating voltage on a diode is converted to a pulsating DC voltage. If a three-phase source is used in conjunction with an uncontrolled rectifier, the DC voltage continues to pulsate. Figure 7.11 shows the application of a three-phase voltage across phases L_1, L_2 and L_3 on the rectifier bridge containing the diodes D_1, D_2, D_3, D_4, D_5 and D_6.

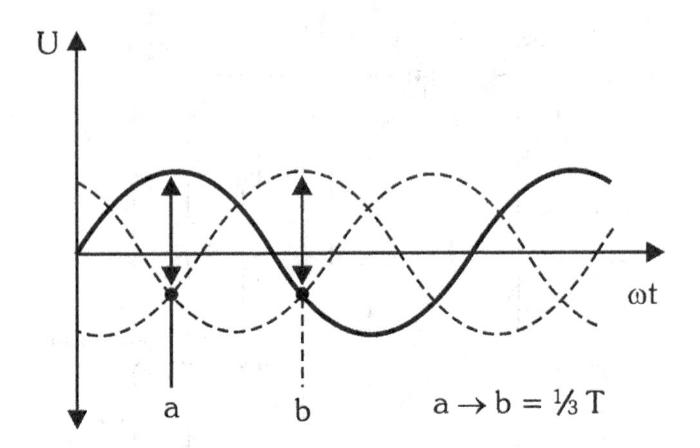

FIGURE 7.10
Power supply alternating three-phase voltage.

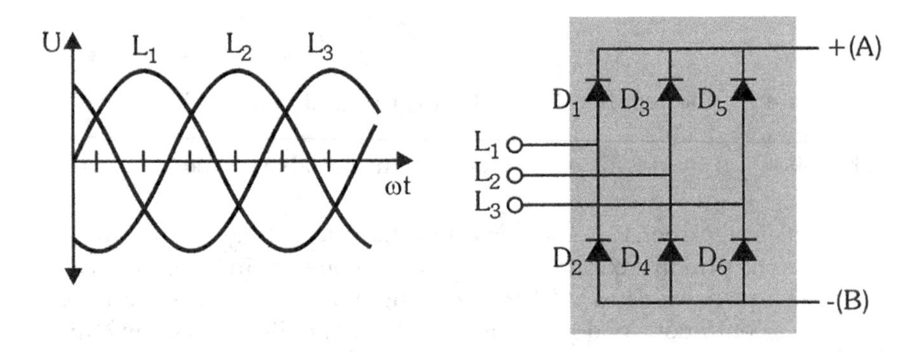

FIGURE 7.11
Application of a three-phase AC voltage source on a rectifier bridge.

As the result is obtained, a pulsed rectified waveform of the format is shown in Figure 7.12. This signal will have the oscillation (ripple) reduced by using of filter capacitors. This circuit forms a symmetric direct current source due to the existence of a ground point as a reference.

A continuous voltage +V/2 (positive) and a –V/2 (negative), with respect to earth, appears in a DC link, which feeds the inverter stage that is composed of six power transistors called IGBTs. With logic provided by the control circuit, the transistors act in order to change the current direction that circulates by the motor. Although it depends on the type of semiconductor used, the switching frequency is typically between 300 Hz and 20 kHz.

The filter, or DC link, has the function of controlling the voltage rectified with energy storage in the capacitors bank.

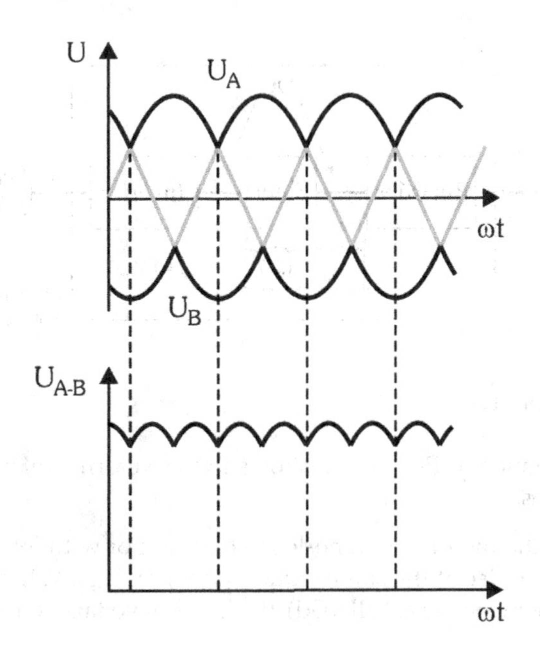

FIGURE 7.12
Rectified waveform.

7.3.4.1 Control System

The control system can be divided into four distinct parts:

- VFD control system
- Motor speed reading system
- Current reading system
- Interface systems that include:

Parameter setting by the user

Sending information to the operator and diagnostics of failure by HMI
Digital and analogue inputs to receive control signals (start, stop, etc.)
Digital and analogue outputs to send information (motor running, fault etc.)

7.3.4.2 Inverter

In this block are the IGBT's transistors, responsible for the DC voltage inversion coming from the DC link in an alternating signal, with variable

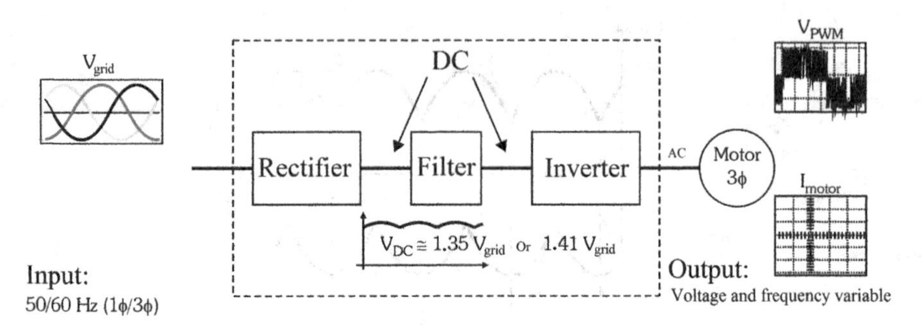

FIGURE 7.13
Waveforms in different stages of VFD.

voltage and frequency. Figure 7.13 shows the waveforms relative to the VFD operating stages.

NOTE: When the motor is in a no-load condition or with low loads, the DC link tends to stabilise at the same value $V_{DC} \approx 1.41.\ V_{grid}$. When, however, the motor has higher loads (e.g. full load), the DC link voltage tends to $V_{DC} \approx 1.35.\ V_{grid}$.

7.3.5 Switching Control

To understand how a frequency inverter switching control works, Figure 7.14 shows a VFD schematic diagram.

The voltage (DC) is connected to the output terminals by the thyristors T_1 to T_6, which operate in the cut-off or saturation modes of operation as a static switch. The operation of these circuits is done by the control circuit in order to obtain an alternating voltage system in which the frequencies are lagged

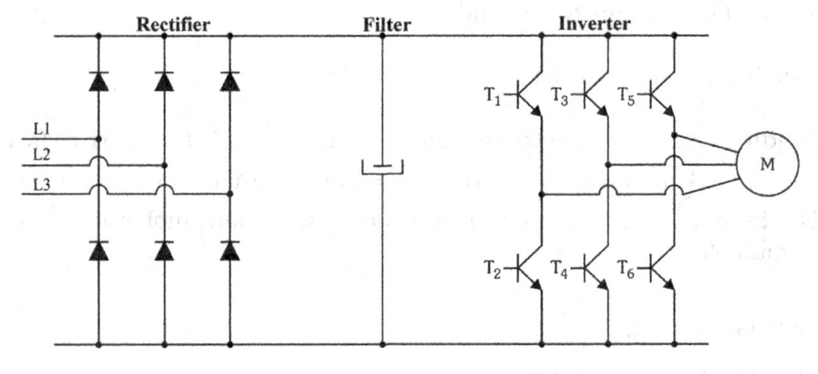

FIGURE 7.14
VFD diagram.

by 120°. The voltage and frequency must choose to allow the voltage U_2 to be proportional to the frequency f so that the flux φ and the torque will be kept constant.

The power transistors control circuit is responsible for generating the control pulses to the use of digital microcontrollers. This technique has become possible and extremely reliable. By controlling the transistor's base rate of change of switching, the frequency of the generated three-phase signal is controlled. Since the inverter receives a continuous current signal, the VFD frequency and output voltage to the motor are independent of the power supply, which allow the exceeding of the rated grid frequency.

To understand the operation of the inverter stage that transforms an AC current into a DC, we will start the study with a single-phase circuit. Figure 7.15 illustrates a single-phase inverter circuit feeding a single-phase AC motor.

To understand how the circuit works, let's check its operation in which the control logic that drives the transistors, always in pairs, is as follows:

- Transistors T_1 and T_4 are switched on, and T_3 and T_2 are switched off. At that time, the current flows in the direction of A to B, as shown in Figure 7.16.
- At the next instant of time, the transistors T_1 and T_4 will be switched off, and T_3 and T_2 will be switched on. The current flows from B to A, as shown in Figure 7.17.

According to this sequence, the inversion occurs in the direction of current, that is, the direct current becomes alternating, as the voltage that is applied to the motor. According to the change of the switching frequency of these

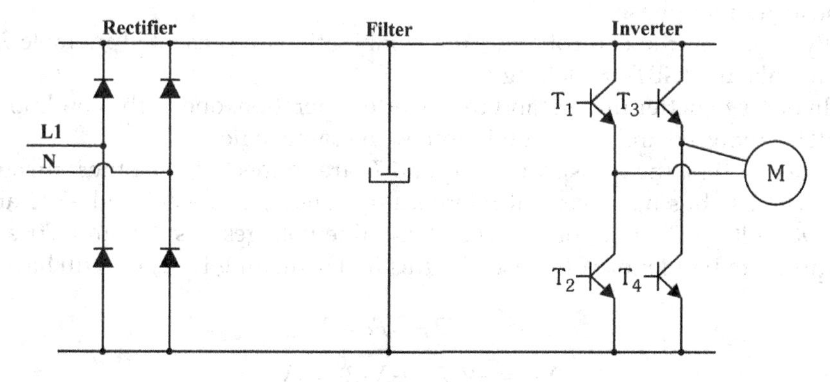

FIGURE 7.15
Diagram of a single-phase frequency inverter.

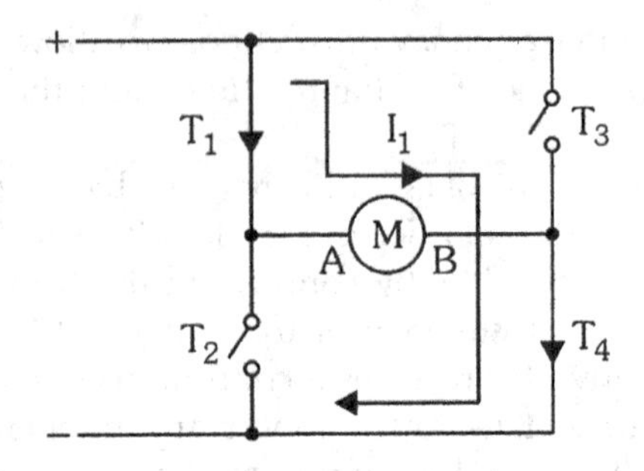

FIGURE 7.16
Single-phase inverter switching transistors T_1 and T_4.

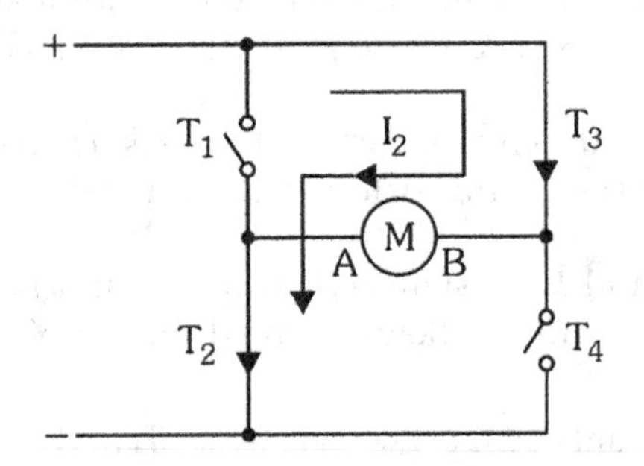

FIGURE 7.17
Single-phase inverter switching transistors T_2 and T_3.

transistors, the speed of the motor increases or decreases in direct proportion to the switching frequency, as shown in Figure 7.18.

Since most of the inverters used in the industry are three-phase, Figure 7.19 shows a three-phase inverter with a control logic for the pulses of six IGBTs, in order to generate an alternating and lagged output voltage of 120° from one to another phase.

By means of the control logic, the combinations represented in Table 7.1 will make the IGBTs switching.

In order to better understand the inverter operation, one of the conditions will be analysed and the rest will follow the same logic.

The first time the transistors T_1, T_2 and T_3 are connected, the others turned off. The DC bus has a central reference (ground), since +V/2 and –V/2 are as DC voltage. The motor is three-phase; line voltages Vrs, Vst and Vtr are required to be 120° out of phase. For this first switching time, we will have:

$$Vrs = +V/2 - V/2 = 0$$

$$Vst = +V/2 - (-V/2) = +V$$

$$Vtr = -V/2 - V/2 = -V$$

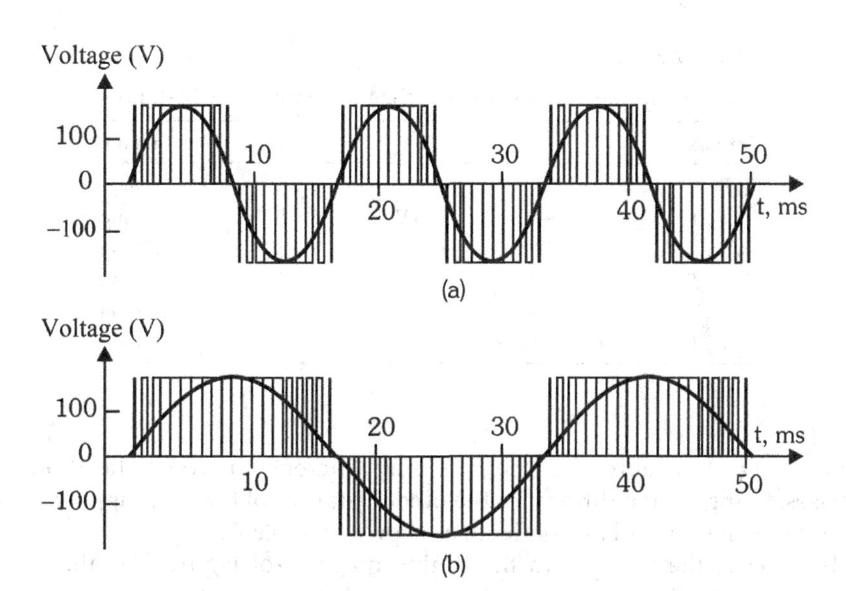

FIGURE 7.18
Variable control frequency: (a) 60 Hz waveform, 120 V; (b) 30 Hz waveform, 120 V.

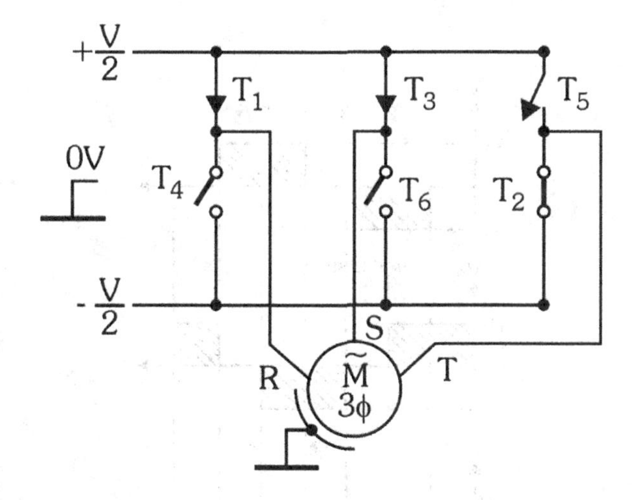

FIGURE 7.19
Representation of a frequency inverter for a three-phase circuit.

TABLE 7.1
Combinations for Activating the IGBTs in a Three-Phase VFD

1° Instant	2° Instant	3° Instant	4° Instant	5° Instant	6° Instant
T_1, T_2, T_3	T_2, T_3, T_4	T_3, T_4, T_5	T_4, T_5, T_6	T_5, T_6, T_1	T_6, T_1, T_2

TABLE 7.2

Voltages Applied to the Motor at the Respective Time Instances

Transistors	Vrs	Vst	Vtr	Instants
T_1, T_2, T_3	0	+V	–V	1° time
T_2, T_3, T_4	–V	+V	0	2° time
T_3, T_4, T_5	–V	0	+V	3° time
T_4, T_5, T_6	0	–V	+V	4° time
T_5, T_6, T_1	+V	–V	0	5° time
T_6, T_1, T_2	+V	0	–V	6° time

The voltage Vrs represents the potential difference between the R and S phases. Using the conditions for the other instances of time, the applied voltages to the motor will be obtained, as shown in Table 7.2.

By placing the voltages in the timing diagram of Figure 7.20, the three electrical phase lags of 120° are the expected voltage to drive a three-phase induction motor.

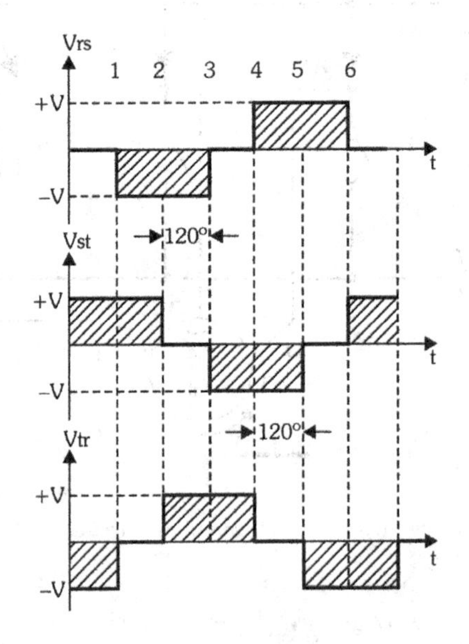

FIGURE 7.20
Three resulting phases at the output of the inverter.

7.4 Pulse Width Modulation

Because transistors work as "on or off" switches, the VFD output voltage waveform is always square. To obtain an output voltage closer to the sine wave, the transistors switch by modulating their pulse width through a technique called pulse-width modulation (PWM). There are several PWM modulation techniques, however, it is beyond the scope of this book to describe them in detail. However, we will present one of the most commonly used techniques for the control of frequency inverter switching.

With the use of the microprocessor, the PWM control functions are effectively performed by combining a triangular wave and a sinusoidal waveform that produces the output voltage waveform, as shown in Figure 7.21.

The triangular signal is the switching frequency of the inverter. The sinusoidal wave generator produces a signal that determines the pulse width and, hence, the rms inverter output voltage, as shown in Figure 7.22.

The IGBT is switched on for a short time, allowing only a small portion of current to reach the motor. In PWM, the IGBT is then switched on for longer periods of time, allowing larger currents in the motor until it reaches the rated motor current. After this, the IGBT is turned on for shorter periods

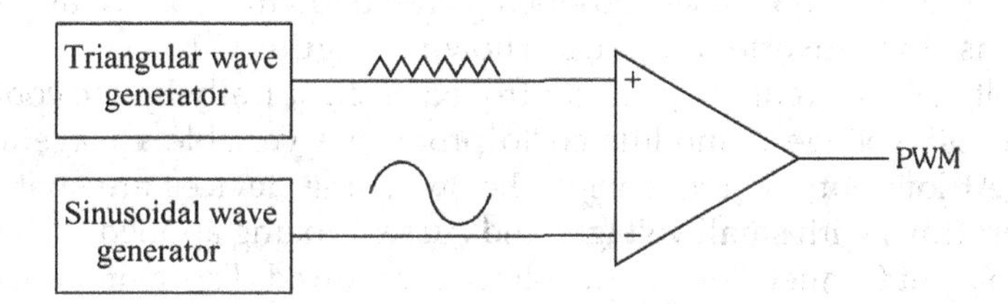

FIGURE 7.21
Generation of the PWM by the combination of a triangular wave and a sine wave.

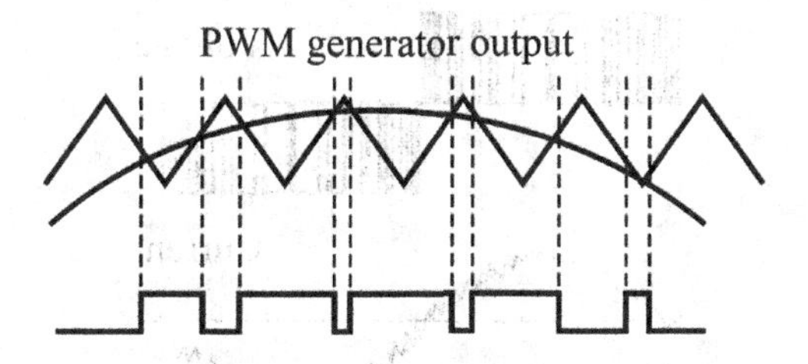

FIGURE 7.22
Output signal from PWM generator.

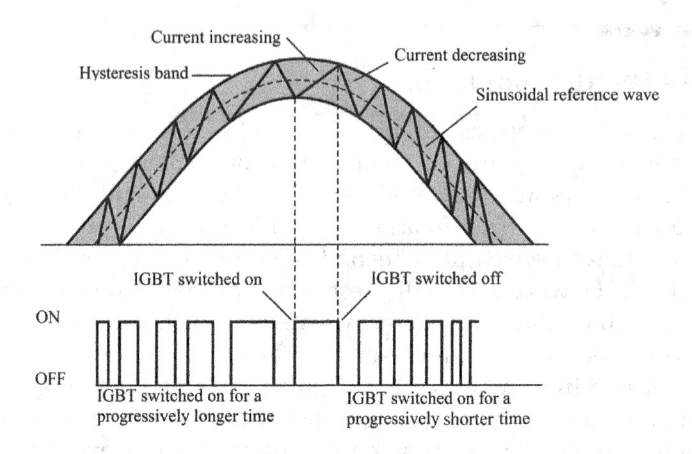

FIGURE 7.23
Switching the IGBTs in the inverter circuit.

progressively, reducing the current applied to the motor. Figure 7.23 shows how this process occurs.

The negative portion of the sine wave is generated by the switching of the IGBT connected to the negative value of the DC voltage.

The more sinusoidal the PWM output current, the greater the reduction of torque pulsations and losses, producing the following voltage and current waveforms at the inverter output, as shown in Figure 7.24.

The voltage and frequency are controlled electronically by the control circuit. The DC voltage is modulated to produce a variable voltage and frequency. At low output frequencies, the switching devices are switched on for a short time, with small voltages and currents being applied to the motor. At high output frequencies a high voltage is required. The switching devices are switched on for a long period of time, allowing higher currents and voltages to be applied to the motor. Figures 7.25 and 7.26, respectively, describe this behavior.

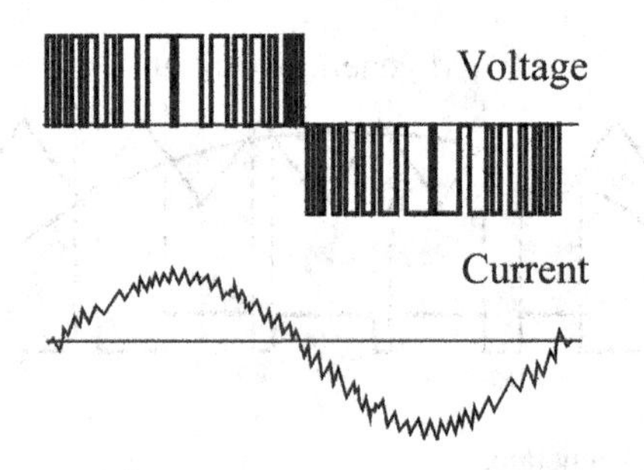

FIGURE 7.24
Voltage and current waveforms at the output of the frequency inverter.

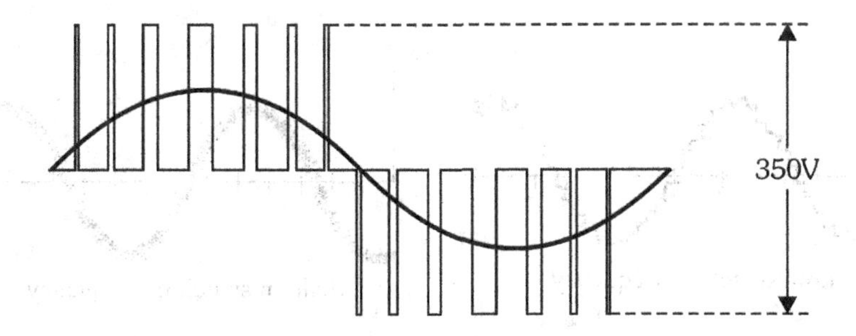

FIGURE 7.25
Switching devices connected in a short time, low voltage.

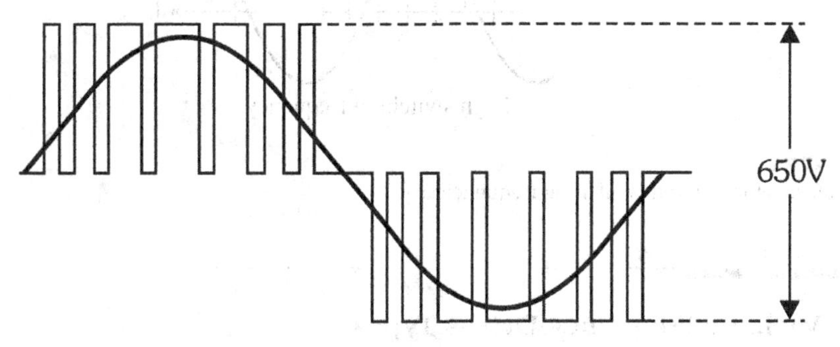

FIGURE 7.26
Switching devices connected over a long period of time, high voltage.

The higher the switching frequency, the more sinusoidal the resulting waveform will be, although the noise is significantly increased, since it is proportional to the switching frequency, as shown in Figure 7.27.

It can be stated that, regardless of the topology used, the principle of operation is based on a DC voltage in the DC link and must be converted to AC voltage to drive the motor. A VFD block circuit with the PWM topology has been shown previously, which is the most used in the actual VFDs. Since the voltage is fixed in the diagram, it is necessary to switch the output transistors by pulse width modulation to obtain a variable frequency AC voltage.

The inverters must ensure that the applied voltage change is proportional to the frequency, which is done by automatically adjusting the transistor triggers by microprocessor systems. For normal application motors, very precise adjustment of the speed or the torque control is not necessary. For these cases, a speed accuracy of 0.5% of the nominal speed, without load variation, and 3%–5% with load variation of up to 100% of the rated torque is quite reasonable. Usually, the frequency range is small, somewhere between 6 and 100 Hz.

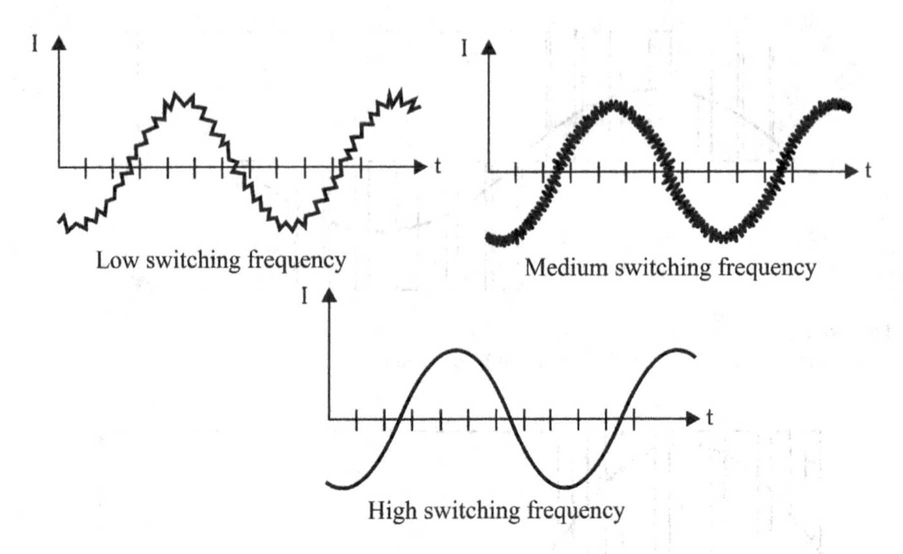

FIGURE 7.27
Waveforms for different switching frequencies.

7.5 Variable Frequency Drives Types

Starting devices that use PWM have different levels of performance which are based on control algorithms. There are four basic types of control widely used: scalar (Volts/Hertz), sensorless vector control, vector-flow control and field-oriented control.

Scalar control (V/Hz) is a basic method that provides a variable frequency for applications, such as fans and pumps. It provides reasonable control of torque speed at a low cost.

Sensorless vector control provides better speed regulation and has the ability to produce high starting torque.

Vector flow control enables more torque and velocity accuracy with dynamic response.

Field-oriented control allows the maintenance of speed and torque available for AC motors, providing performance of a DC motor for AC motors.

Field-oriented and vector-flow controls are the latest control techniques for frequency inverters and are not included in the scope of this book.

7.5.1 Scalar Control

Also called Volts/Hertz control, it is based on the original concept of the frequency converter. It imposes a certain voltage/frequency on the motor in

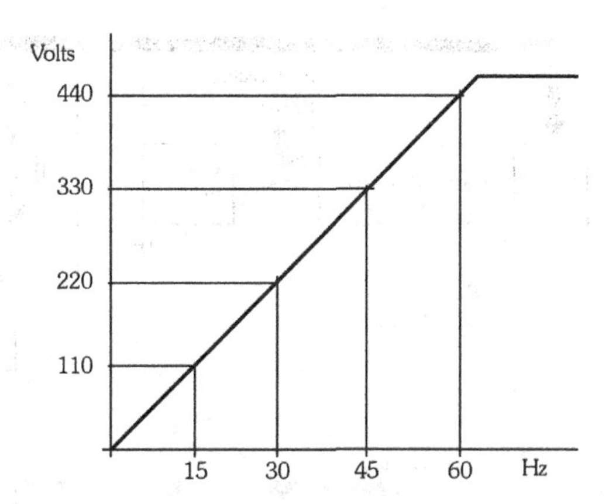

FIGURE 7.28
Voltage by frequency graph.

order to keep the V/f ratio constant—i.e., the motor works with an approximately constant flow. This type of control is applied when there is no need for quick responses to torque and speed commands. In Figure 7.28, a graph illustrates this behavior.

NOTE: The voltage will be increased in the same ratio up to the rated motor voltage (440 V in this case), because, if the voltage continues to increase with increasing frequency, it may cause damage to the motor windings.

This VFD family is composed of systems whose requirement is restricted to the control of the motor speed, without torque control and without knowledge of the dynamics of the process under control. They are systems that provide a certain speed error that, given the application, can be easily assimilated by the controlled system. The motors driven by this family of converters have, or must meet, normal requirements and the control is done in open loop (no feedback), i.e., there is usually no speed sensor installed on the motor shaft to feed the drive controller structure. The frequencies operated usually range from 10 to 60 Hz.

For scalar control, the simplest form is volt/hertz, which uses the speed reference through an external source, such as an analogue input, and the voltage and frequency applied to the motor changes. The acceleration and deceleration times can be adjusted by the user. By keeping a constant ratio of V/Hz, it is possible to control the motor speed. Figure 7.29 shows a block diagram for the VFD scalar control.

Typically, the current limit block monitors the motor current, changing it when it exceeds a predetermined value. The current reading from a current transducer is to protect and indicate the current value, and there is no

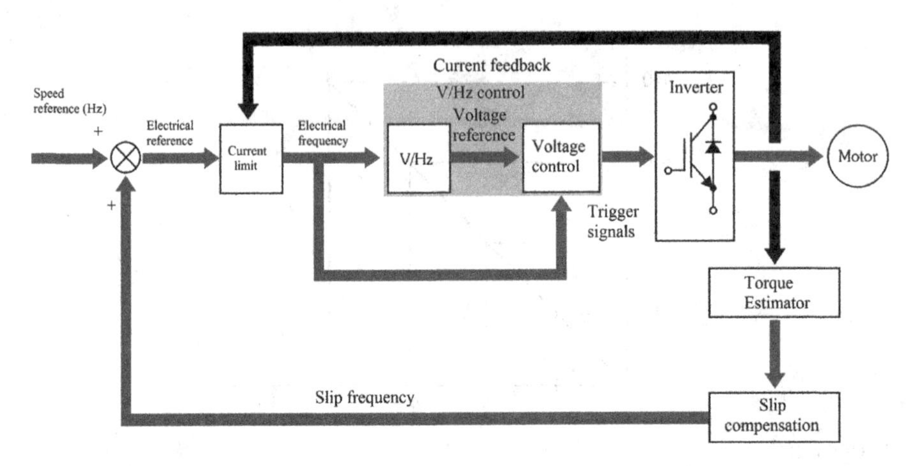

FIGURE 7.29
Block diagram for scalar control.

control strategy for the circuit. This current measurement cannot distinguish between torque current and motor current magnetisation.

The current measuring circuit is intended to perform the following tasks:

- To protect the motor against overload.
- Provide protection for electronic components.
- Provide current limit. The control system reduces the command frequency when the current exceeds a predetermined value. Generally, the current limit is set to 150% of the rated motor current.

The V/Hz block converts the current command to a V/Hz ratio. The ratio between voltage and frequency is kept constant at all control steps. Another important point is the adjustment of the rates of change of these two values that determine the motor acceleration. The voltage and base frequency for this ratio are taken from the motor nameplate data. This block supplies the voltage amplitude for the voltage control block, made through the thyristors firing angle. In this way, the current flow to the motor is determined. If this angle is incorrect, the motor can operate in an unstable mode.

An additional feature in new devices is the slip compensation block to improve speed control. This block changes the frequency reference when the load changes to maintain the speed close to the desired value. This type of control is not recommended for applications requiring high performance, where the motor runs at very low speeds, or for those that require direct torque control.

7.5.1.1 Scalar Control Characteristics

By increasing the signal frequency imposed on the motor armature and keeping the voltage value, the magnetising current of the machine falls proportionally and, with it, the magnetic flux establishes in the air gap. Consequently, by dropping the magnetic flux, the torque made available by it decreases. It is the operation with field weakening. The electromagnetic motor weakens, and so it is possible to determine an area above the nominal frequency (60 Hz), called the field weakening region, where the flux begins to decrease, so the torque begins to decrease, too.

The torque by speed motor curve driven by VFD can be represented as shown in Figure 7.30.

It is possible to note that the torque remains constant up to the nominal frequency. Above this point, the torque begins to decrease.

Particular care must be taken in the application of VFDs to drive motors in low rotation, because the motors are self-ventilated. At low rotations, typically below 50% of the rated speed, airflow through the frame is poor. The heat withdrawal presents difficulty, and the power supplied by the motor is reduced to not damage the insulating materials of its armature winding.

The manufacturers propose an operating curve, as shown in Figure 7.31, to avoid machine damage.

One solution would be to specify the motor with a higher service factor, or to increase the insulation class for the coils to withstand the highest temperature, or to specify a motor with a larger frame for greater area for thermal exchange.

In three-phase induction motors with independent ventilation, the exchange is independent of the speed in the shaft. Thus, the torque requested to it can be optimized. Within a range from 10 Hz to the rated frequency, it is possible to have a torque of 90% of the rated torque, as shown in Figure 7.32.

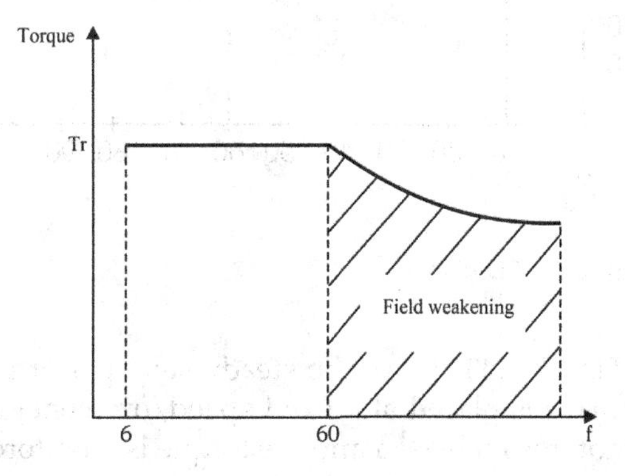

FIGURE 7.30
Field weakening region.

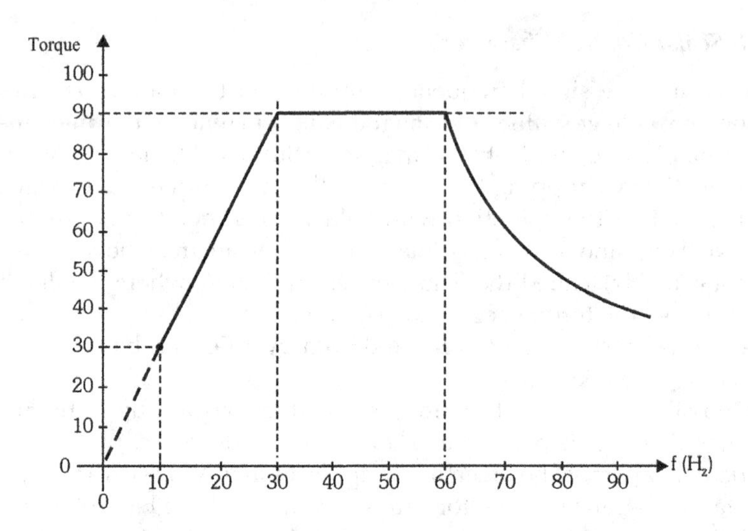

FIGURE 7.31
Operating curve for induction motor.

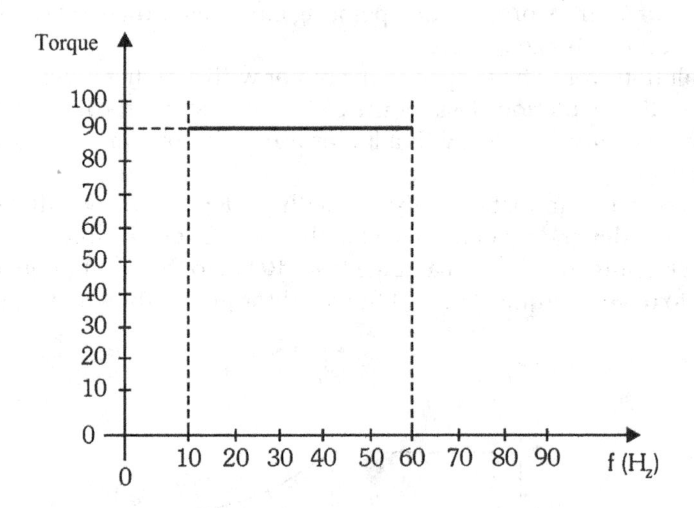

FIGURE 7.32
Torque frequency curve in VFD.

The graph of Figure 7.33 shows the steady-state performance of a volts/hertz VFD. The drive is placed at a fixed speed/frequency reference. Then, the load on the motor is increased and the torque is monitored. Note that the device's ability to maintain high output torque at low speeds falls significantly below 3 Hz. This is a normal feature of a VFD scalar.

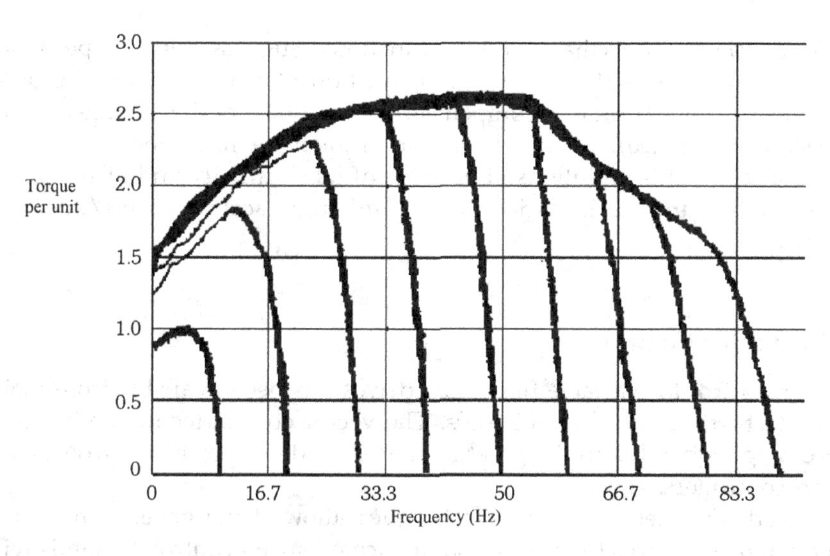

FIGURE 7.33
Graph torque by velocity for scalar VFD.

7.5.1.2 Final Considerations on Scalar VFDs

With respect to the scalar control applied to a VFD, it is possible to make the following statements:

- It presents a lower cost in relation to vector control.
- Scalar control is used in normal applications that do not require high dynamics (large accelerations and braking), high accuracy or torque control.
- Accuracy up to 0.5% of rated speed for systems with no-load variation, and 3%–5% with load variation from 0% to 100% of the maximum torque.
- This type of control is performed in open loop, i.e., without reading the motor speed through a sensor, and the speed accuracy is a function of the motor slip, which changes depending on the load.

This control is not suitable for motors running at low speeds (below 5 Hz), because the torque at low speeds is generally small because the voltage drop significantly affects the magnitude of the flow producing current. Many VFDs include an extra starting torque, which allows the V/F ratio to be increased at start up to increase the flux and, consequently, the starting torque.

Some VFD models have special functions, such as slip compensation (which reduces speed variation as a function of load) and voltage boost (increasing the V/f ratio to compensate for the effect of the voltage drop on the stator resistance), causing the motor torque to be maintained.

Scalar control is widely used because of its simplicity and also because most of the applications do not require high accuracy and/or speed control.

7.5.2 Vector Control

Vector control for variable frequency drives has been available from major manufacturers since the mid-1980s. The vector control technique has only become possible due to the great advances made in power electronics and microcontrollers.

The advancement of control techniques allowed the generation of new command structures to meet the sophisticated speed control demands with fast and high precision responses.

Vector control is recommended in applications where high dynamic performance, fast responses and high speed control accuracy are required. It is recommended that the motor provides accurate torque control for an extended range of operating conditions. DC machines with closed-loop control systems already met these demands. They were often used due to the proportionality of the armature current, the flux and the torque which allow a direct way for its control.

With the increasing evolution of power electronics, the DC motors application has been drastically reduced. Three-phase induction motors are used with VFDs with vector control with almost the same results.

The main idea of vector control is to use modeling and control of an induction motor in alternating current as if it were a direct current motor.

7.5.2.1 Principles of Direct Current Motor

To understand vector control, it is necessary to know the operating principle of the DC motor in which the torque and speed controls are performed independently. To accomplish these controls, the stator (field) and rotor (armature) windings are powered by two independent voltage sources.

The principle of operation of a shunt DC motor is relatively easy to understand by means of a set of simple equations. Figure 7.34 shows the DC motor elements.

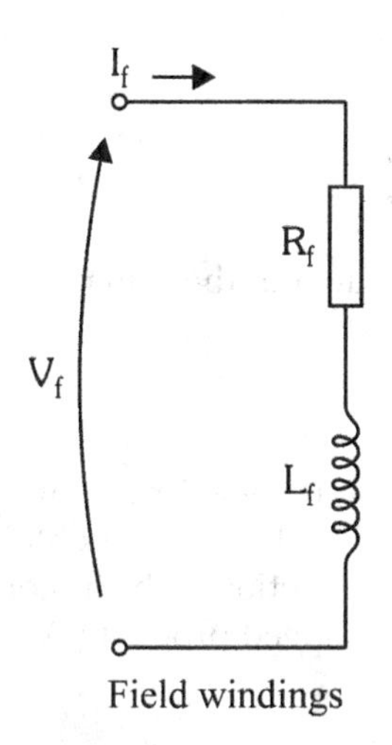

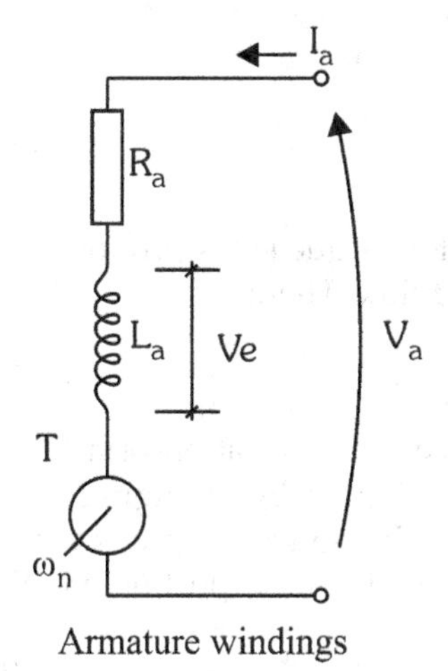

FIGURE 7.34
DC motor description.

where:

- R_f is the field resistance
- R_a is the armature resistance
- T is the torque
- ω_n is the speed
- L_f is the field inductance
- I_f is the field current
- L_a is the armature inductance
- V_a is the armature voltage
- I_f is the armature current
- V_f is the field voltage

The armature voltage (rotor) V_a is the sum of the internal voltage of the armature V_e and the voltage drop due to the resistance of the armature ($I_a \cdot R_a$). In this way, we will have:

$$V_a = V_e + I_a \cdot R_a \tag{7.12}$$

The DC motor speed is directly proportional to the armature voltage and indirectly proportional to the field flux φ, which depends on the excitation current of the stator (I_f). The motor speed can be controlled by adjusting the voltage of the armature that controls V_e, or the field current that controls the flow φ.

Thus:

$$\text{Motor speed} \propto \frac{V_E}{\varphi} \tag{7.13}$$

The motor torque (T) is proportional to the product of the armature current and field flux. Then:

$$T \propto I_A\, \varphi \tag{7.14}$$

The direction of rotation of a DC motor can be reversed by changing the polarity of φ, or by changing the polarity of I_A. This can be obtained by inverting the power supply or magnetic field connections. The motor output power (P) will be proportional to the torque and speed product (n).

$$P \propto T \cdot n \tag{7.15}$$

From these equations, the following conclusions can be drawn for the speed control of a DC motor:

- The DC motor speed can be controlled by adjusting the armature voltage, field flow or both. Generally, the field flow is kept constant, with the motor speed being increased by increasing the armature voltage.
- When the armature voltage V_a has reached its maximum output value, a further increase in speed can be obtained by reducing the field flow.
- The motor can develop its maximum torque outside its rated speed range. This is possible because the torque does not depend on V_a. The maximum torque is possible at speeds higher than rated or even with the motor stopped.

7.5.2.2 Vector Control Principles

To understand the vector control operation, the induction motor modeling is done as if it were a transformer. The motor speed is directly related to the voltage applied in the armature, and the torque is a function of the flux in the air gap.

To describe the motor operation, it will be represented through an equivalent circuit. By this, circuit it becomes clear what happens in the motor when the stator voltage and frequency are changed or when the torque required by the load increases or the slip changes. There are many different versions of the equivalent circuit, which depend on details and complexity. The stator current I_S is circulating in the stator winding from the supply voltage V. Figure 7.35 shows an induction motor equivalent circuit.

where:

V is the stator voltage supply
R_S is the stator resistance
E_S is the induced stator voltage
X_S is the stator reactance (60 Hz)
E_R is the induced rotor voltage
R_R is the rotor resistance
N_S is the number of turns in stator
X_R is the rotor reactance
N_R is the number of turns in rotor
X_M is the magnetising inductance
I_S is the stator current
I_M is the magnetising current
R_C is the core losses, bearings friction, ventilation losses, etc.
I_R is the rotor current
S is the slip

The main components of the induction motor equivalent circuit are:

Resistances: Represent the resistive losses in an induction motor and are composed of:

- Losses in stator winding resistance (R_S)
- Losses in the resistance of the rotor windings (R_R)
- Iron losses that depend on the value and density of flux in the iron core
- Friction and ventilation losses (R_C)

Inductors: Losses associated with the fact that not all the flux produced by the stator windings crosses the air gap into the winding of the rotor, nor does any flow reaching the air gap produce torque. They are:

- Stator reactance (X_S)
- Rotor reactance (X_R)
- Magnetising inductance (X_M, whose function is to produce the flux of the magnetic field)

Unlike the DC motor, the AC induction motor does not have separate field windings. As shown in the equivalent circuit of Figure 7.35, the current in the stator has a dual function:

- Provide the current (I_M) that produces a rotating magnetic field.
- Supply the current (I_R) that is transferred to the rotor to produce the torque on the shaft.

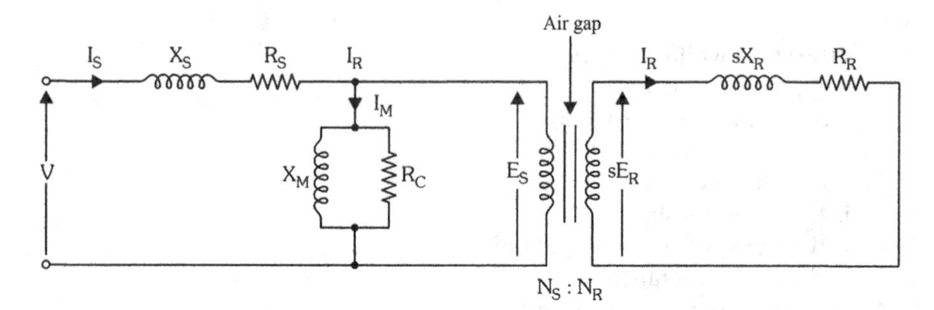

FIGURE 7.35
Equivalent circuit of induction motor.

The voltage (E_S) in the stator is different from the supply voltage V by the voltage drop in X_S and R_S. X_M represents the magnetising inductance of the core, and R_C represents the energy dissipated in the core by means of friction in the bearings and losses of ventilation. The rotor part in the equivalent circuit consists of the induced voltage ($S.E_R$), which is proportional to the frequency and, consequently, also to the slip.

This equivalent circuit is somewhat complex to analyse due to the existence of a transformer between the stator and the rotor, which has a transformer ratio that changes when the slip is changed. Fortunately, the circuit can be simplified mathematically by adjusting the rotor resistance and reactance values by the ratio $N_2 = (N_S/N_R)^2$, i.e., by transferring the stator side of the transformer. Once these components have been transferred, the transformer is no longer relevant and can be removed from the circuit. This mathematical manipulation must also adjust the rotor voltage that depends on the slip. The equivalent circuit can be rearranged and simplified, as shown in Figure 7.36.

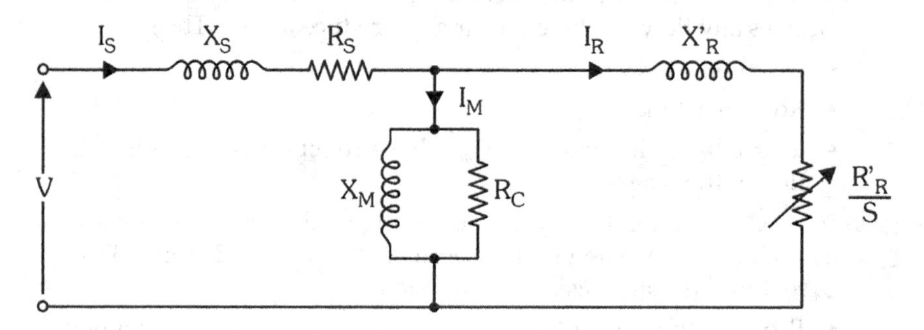

FIGURE 7.36
Simplified equivalent circuit of induction motor.

where:

$$X'_R = N^2.X_R \tag{7.16}$$

$$R'_R = N^2.R_R \tag{7.17}$$

$N = N_S/N_R$ is the turn ratio between stator and rotor.

In this modified equivalent circuit, the rotor resistance is represented by an element dependent on the slip S. In this way, the voltage that is induced in the rotor can be seen and, consequently, the current that depends on the slip. When the induction motor is powered by a voltage source and constant frequency, the current depends, first and foremost, on the slip.

The equivalent circuit can be modified and simplified to represent only the most significant components that are:

Magnetising inductance (X_M)

Variable rotor resistance (R'_R/S)

All other components, which are of small magnitude, have been disregarded. Figure 7.37 shows the new simplification of the AC induction motor circuit.

The magnetising reactive current I_M is independent of the load, being responsible for the rotating magnetic field creation. This current is lagged by approximately 90°, and its value depends on the stator voltage and its frequency. To keep a constant flow in the motor, the V/f ratio must be kept

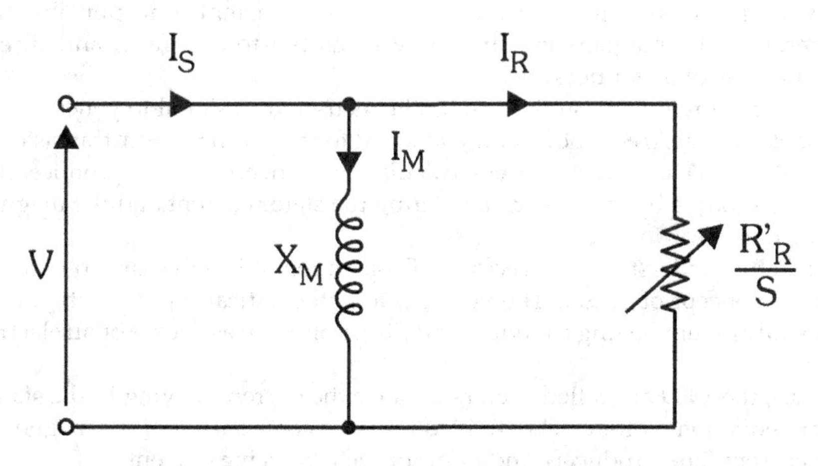

FIGURE 7.37
More simplified equivalent circuit of induction motor.

constant. In this way, the magnetising inductance presented is obtained as follows:

$$X_M = j\omega L_M = j(2\pi f)L_M \tag{7.18}$$

The magnetising current is:

$$I_M = \frac{V}{j(2\pi f)L_M} \tag{7.19}$$

$$I_M = k\left(\frac{V}{f}\right) \tag{7.20}$$

and

$$k = \frac{1}{j(2\pi)L_M} \text{Constant}$$

The active current I_R produces the rotor torque, depends on the load characteristics and is proportional to the slip. It is in phase with the stator voltage.

In an AC induction motor, the current for the flux production (I_M) and the current for the production of the torque (I_R) are in the "interior" of the motor and cannot be measured externally or controlled separately. These two currents are lagged approximately 90°, and their vector sum forms the current in the stator, which can be measured.

When induction motors are used, the stator current is responsible for generating the magnetising flux, as well as the torque flow, and direct torque control is not possible.

The speed reference on the scalar VFD is used as a signal to generate the variable voltage/frequency parameters and to trigger the power transistors.

In the VFD vector, a current calculation is necessary to produce the required torque by the motor, calculating the stator current and the magnetising current.

To better understand the vector VFD operation, it is necessary to remember the concept of vector. The vector is a mathematical representation of a physical amount having a module and direction, such as a force or an electric current.

Thus, the VFD are called vectors because the current flowing in the stator of an induction motor can be divided into two components: I_M, or magnetising current (flux producer), and I_R, or torque producing current.

Thus, the total current is the vector sum of these two components, in which the torque produced in the motor is proportional to the vector product of the two current components.

This makes the vector control in an AC motor harder than in a DC. This is the challenge to control: distinguish and control these two current vectors, without separating them into two distinct circuits.

The vector control strategy is to calculate the current of each of the vectors and to allow the separation of the control of the flow current and/or the torque current in all speed and torque conditions. Its purpose is to keep the current constant in the motor.

The calculation of these current vectors involves the measurement of the available variables: stator current (I_S), stator voltage (V_S), phase relation, frequency, shaft speed, etc. and to apply to them a model that includes the motor constants.

It is necessary to know or calculate the following motor parameters:

- Stator resistance
- Rotor resistance
- Stator inductance
- Rotor inductance
- Magnetising inductance
- Saturation curve

Variables, such as stator and rotor resistances and inductances, magnetising inductance, number of poles, etc. must be taken into account. Due to the fact that there are many variables, there are many possible applications of motor models, from a simple estimation of the conditions to those that need all the parameters for a more complete and precise control. The more detailed the motor model, the more processing power is needed.

In a motor without load, almost all the current in the stator (I_S) is composed of the magnetising current. Torque production current is only required to compensate for losses through ventilation and friction in the motor. The slip is approximately zero, and the current in the stator is delayed in relation to the voltage by approximately 90°, so the power factor is approximately zero ($\cos\varphi = 0$).

In the motor under load conditions, the stator current I_S is the vector sum of the magnetising current I_M (which remains unchanged in relation to the no-load condition), with a large increase in the torque production current. The stator current lags in relation to the voltage by a wide angle φ, so the power factor is low ($\cos\varphi \ll 1$). The slip remains small for this loading situation.

In a motor coupled to a high load, the current in the I_S stator is the vector sum of the magnetising current, with a large increase in the torque producing current, which increases the load torque in the same proportion. The stator

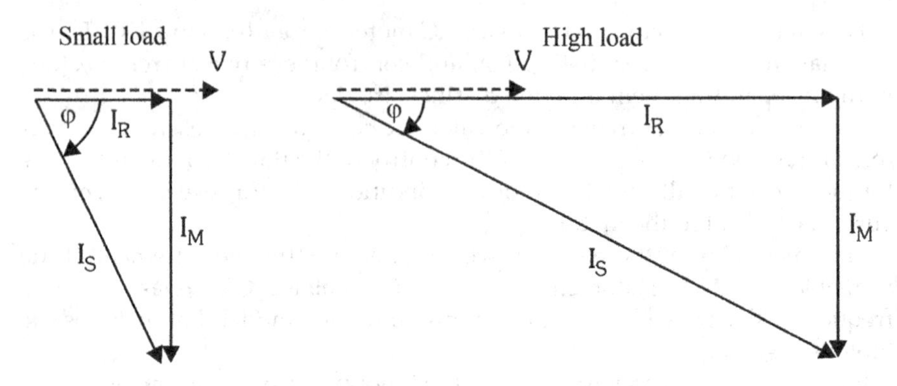

FIGURE 7.38
Different load conditions on the AC induction motor.

current delay for the applied voltage is given by the angle φ. Then, the power factor becomes larger, close to the nominal load power factor (cosφ = 0.85).

Figure 7.38 illustrates these load conditions on the AC induction motor.

The main part of a vector controlled system is the active motor model that continually models the internal motor conditions and performs the following operations:

- Calculates continuously the torque production current in real time by the following processes.
- Stores the motor constants in the memory to be used as part of the calculation.
- Measures the stator current and voltage at each phase.
- Measures speed (encoder) or calculates speed (without encoder).
- Calculates continuously the current to flux in real time.
- Implements a speed control loop by comparing the measured speed with a desired speed value, providing an output to act on the motor torque control.
- Implements a control loop by comparing the current torque, calculated from current and speed measurement, by sending an output signal to the PWM logic control circuit.
- Constantly updates this information and keeps robust control over the process.

For decomposition and calculation of these currents, it is necessary to represent the induction motor behavior by means of a mathematical model. In order to allow the application of this control in a three-phase induction motor, microprocessors with high processing capability are required, and digital signal processors (DSP) are used more frequently.

Some VFD models have these pre-programmed values for some types of motors, and most use self-tuning routines to perform the calculation of these parameters. In the auto-tuning routine, the evaluation of the internal variables of the motor, in a dynamic process, is passed to the controller system. The regulation of the three-phase induction machine has become more accurate and closer to the control achieved with the DC machine.

Therefore, by having this amount of knowledge, it knows the energy flows that the machine needs by analysing the current of the armature. The signal coming from the motor shaft, collected by a pulse tachogenerator, provides a closed control loop, which allows:

- High dynamic performance.
- Smooth operation in the speed range specified for the drive.
- Small oscillations in the motor torque when variations occur in the load.
- High speed accuracy.

For proper dynamic responses to the VFD, the calculation model needs to be done at least more than 2,000 times per second, which will give an update time of less than 0.5 ms.

While this is easily accomplished with high-speed microprocessors, the ability to continuously model the motor at this speed has only become feasible in the last 10 years or more with the development of 16-bit microprocessors. This control initially had a very high cost. Luckily, the cost of processors dropped, and the processing speed increased significantly.

If an analysis of the power circuit of a vector and scalar VFD is made, there will be no difference. The main difference between traditional scalar and vector control is almost the entire control system, where control of the active motor model is implemented to control the switching of the inverter IGBTs.

7.5.2.3 Open-Loop Vector Control (Sensorless)

This type is simpler than with control with a sensor, but presents torque limitations, mainly, in very low rotations. At higher speeds, it is practically as good as vector feedback control.

This VFD has a lower performance rating, but is superior to the V/F VFD. The sensorless vector control, as well as V/F, continues to operate as a frequency-control device with slip compensation keeping the motor speed close to the desired one. The current torque estimator block determines the percentage of current that is in phase with the voltage, providing

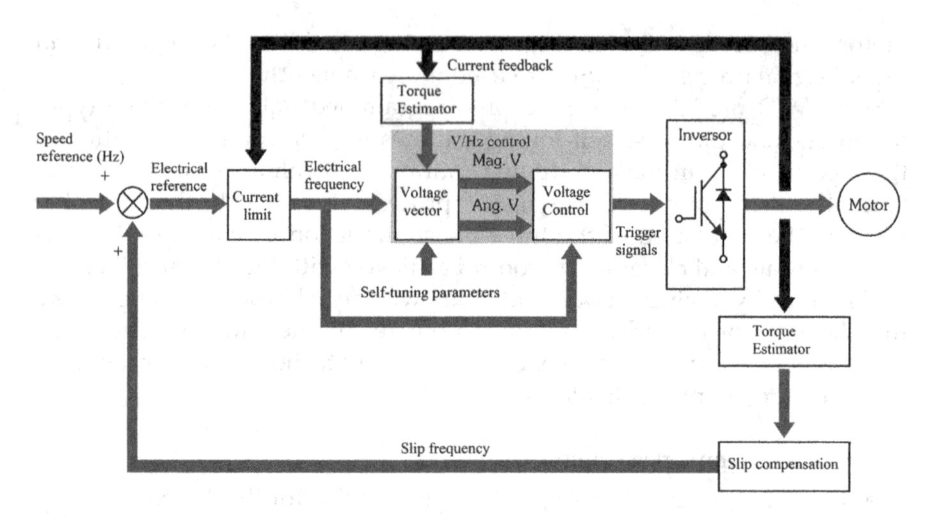

FIGURE 7.39
Block diagram for vector control in open loop (sensorless).

an approximate torque of current. It is used to estimate the amount of slip, providing better speed control over the load. Figure 7.39 illustrates a block diagram with the open-loop control stages.

This type of control presents improvements compared to the V/Hz technique by providing both magnitude and angle between voltage and current control instead of V/Hz, which controlled only the magnitude. The motor voltage angle controls the current amount going to the motor flux enabled by the current torque estimator. By controlling this angle, operation at low speeds and torque control is improved relative to V/Hz.

In the following, it is possible to summarise the features of these VFDs:

- Speed regulation: 0.1%
- Torque adjustment: no
- Starting torque: 250%
- Maximum (non-continuous) torque: 250%
- Open loop control
- The VFD already knows the motor parameters by the auto-tuning
- It has better performance when compared to scalar control (V/f)

7.5.2.4 Feedback Control

In this type of control, the feedback is made by current sensors and position sensors (encoders), and the control is done by decoupling the stator current

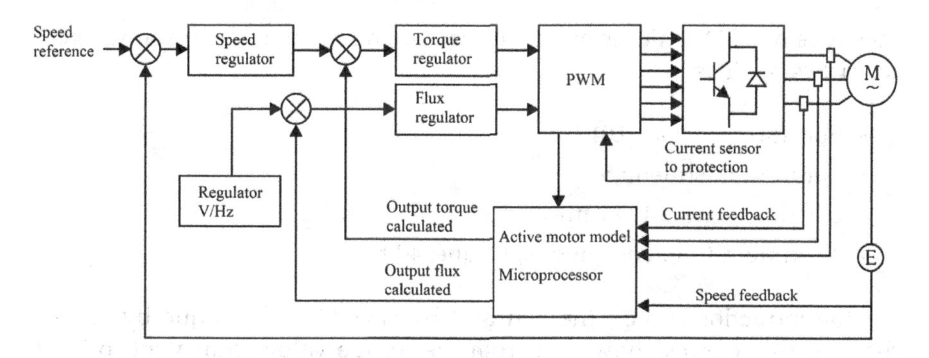

FIGURE 7.40
Block diagram for a VFD with vector control.

into two components, one that produces torque and another that produces flux in the air gap.

The control block diagram shown in Figure 7.40 is, essentially, a closed-loop cascade control with speed and torque control loops, where:

- The speed control set the output frequency that is proportional to the speed.
- The torque loop controls the input current in the motor, which is proportional to the torque.
- The speed reference command is given from the user, the value being entered in a comparator that provides the control for the speed regulator.
- The difference signal between the desired and current speed becomes the setpoint for the controller. This signal is compared to the current value of the motor and determines whether the motor needs to be accelerated or decelerated.
- There is a separate control loop for the current flow (V/f regulator).
- Finally, the signal is sent to the PWM control section, which controls the IGBTs so that the desired voltage and frequency are generated for the output according to the PWM algorithm.

This control provides the following advantages:

- High speed control accuracy.
- High dynamic performance.
- Linear torque control for position or traction applications.
- Smooth operation at low speed and without torque fluctuations, even with load variation.

This type of VFD achieves excellent regulation and dynamic response characteristics, such as:

- Speed regulation: 0.01%
- Torque adjustment: 5%
- Starting torque: 400% max.
- Maximum (non-continuous) torque: 400%

For the induction motor, the "current by speed" and "torque by speed" characteristic curves show that from the torque value equivalent to 150% of rated torque, the two curves represent a similar behavior. Thus, it can be concluded that torque and speed have a linear behavior with current, as shown in Figure 7.41.

The torque by speed curve illustrates the relationship between the torque developed by the motor and its rotation. At the time of starting, when the motor is connected directly to the grid, the torque will be approximately 2–2.5 times the rated torque and will be reduced as the speed increases until it reaches a value of 1.5–1.7 of the rated torque at approximately 30% of rated speed.

As the speed increases, the torque increases again until it reaches its maximum value (80% of rated speed), reaching its rated value at the rated speed. This behavior is represented by the full line in Figure 7.41.

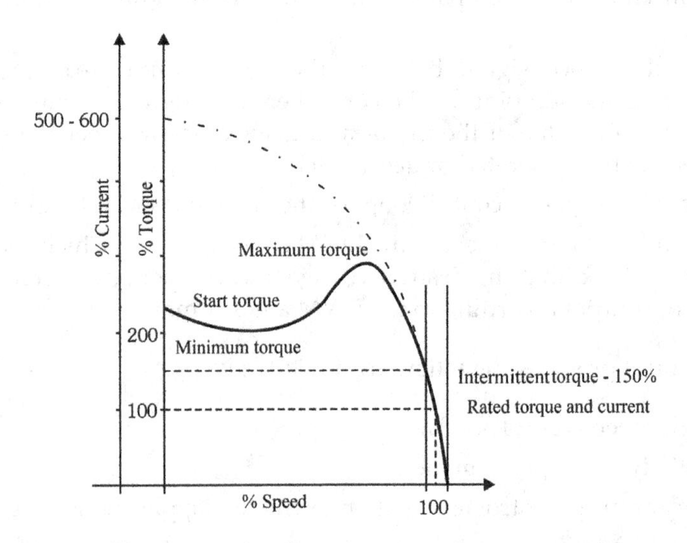

FIGURE 7.41
Characteristics curves current by speed and torque by speed.

The current by speed curve is the dashed line shown in Figure 7.41, which illustrates the relationship between the current consumed by the motor as a function of its speed. At start up, when the motor is connected directly to the grid, the current flowing through it will be five to six times greater than the rated current, being reduced as the speed increases until reaching a value determined by the load coupled to the motor. If the rated load is applied to the motor, the current absorbed by it will also be the rated.

Figure 7.42 shows the torque by speed curve behaviour when the motor is powered by a VFD.

The motor in the example is four-pole 60 Hz, i.e., its synchronous speed will be 1800 rpm, and the speed with rated load will be 1750 rpm. It is possible to observe that, with the motor with rated load, there is a difference of 50 rpm due to the slip. Likewise, for a frequency of 30 Hz, the synchronous speed will be 900 rpm for the nominal torque, with the slip being 50 rpm, and the motor speed being 850 rpm.

When driving motors with VFDs, the starting current is almost the same as the rated current, and when the motor is supplied from 3 or 4 Hz, a torque of 150%, it is sufficient to drive any load coupled to the motor.

Finally, the main differences between the vector and scalar controls are that the scalar control only considers the amplitudes of the instantaneous

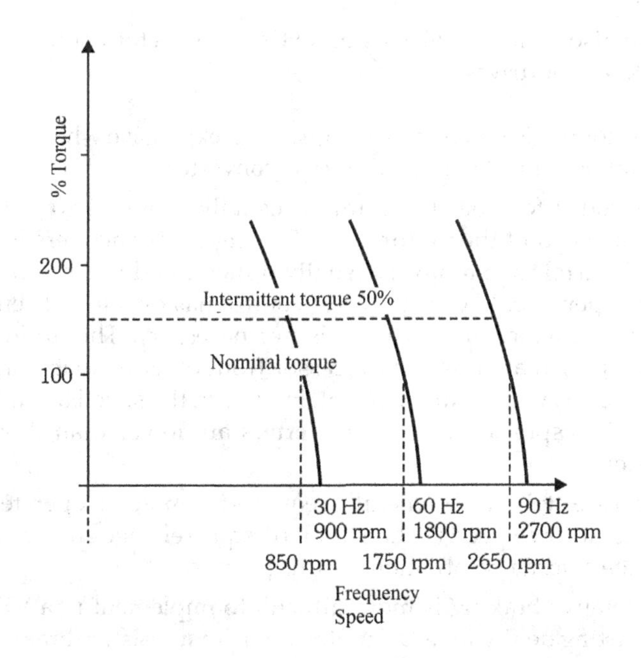

FIGURE 7.42
Torque by speed curves in a motor to different speeds.

TABLE 7.3

Comparison between Types of Control in a VFD

Control Type	Scalar Control	Vector Control	
Feedback	–	Open loop	Closed loop
Zero speed torque	Impossible	Possible	Impossible
Speed regulation	Depends on the slip	0.01%	0.2%
Torque	To low	high	Only in the limit

Note: Zero Speed Torque is the motor's ability to "hold a load" by torque control, with a crane as an example of typical application.

electrical values (flows, currents and voltages), referring them to the stator, and its equation is based on the equivalent circuit of the motor, that is, are permanent regime equations. Vector control, however, allows the representation of the instantaneous electric amounts by vectors, in the dynamic spatial equations of the motor, with the amounts referred to the flux submitted to the rotor, i.e., the induction motor is seen by the vector control as a DC motor, with independent regulation for torque and flow. The Table 7.3 presents a comparison of the possible control types and their characteristics for the VFD.

There are also some drawbacks of VFDs with vector control when compared to DC motor drives:

- Vector control is much more complex and expensive when compared to a simple control using an AC/DC converter.
- An encoder for speed reading is usually required to obtain the current speed of the motor shaft. Placing these encoders in a standard squirrel cage motor is usually difficult and makes the system more expensive. Recently, vector control has developed sensorless control, in which the encoder is not necessary. The approximate speed is calculated by the processor from other available information, such as voltage and current. However, the speed accuracy and dynamic response of these converters are lower than those with encoders.
- The nature of the VFD generally requires the motor to operate at high torques at low speeds. The standard squirrel cage motor requires separate external ventilation for this purpose.
- Regenerative braking is more difficult to implement in a VFD than in a starting device for a DC motor, being that resistive braking is the most used alternative.

Exercises

1. Describe the following concepts used to better understand variable frequency drives:
   a. Force
   b. Speed
   c. Angular speed
   d. Torque
   e. Linear acceleration
   f. Rotational acceleration
   g. Power
   h. Energy
   i. Moment of inertia
2. Explain, with the help of equations, why to keep the operation of the motor with constant torque at different speeds, it is necessary to change the motor voltage and frequency proportionally.
3. Describe the main VFD components.
4. List the converters used in the VFD power stage.
5. Show the waveform in the rectifier, filter and inverter VFD stages.
6. Explain the inverter stage operation.
7. What is PWM, and what is its relation to the VFD output waveform?
8. What is the relationship between noise and switching frequency?
9. Why in the VFD scalar voltage/frequency ratio kept constant only up to the rated motor voltage?
10. With the block diagram, describe the scalar VFD principle of operation.
11. What happens when the inverter operates at low and above rated speed?
12. List the main features of scalar VFD.
13. When is the vector control VFD recommended?
14. Describe the control strategy used in vector VFDs.
15. With the help of a block diagram, explain the sensorless and feedback vector control VFD.
16. Although vector control has several advantages in relation to DC motor control, it has some disadvantages. List them.

8

Parameters Description of VFD

VFD functions are performed according to predefined parameters allocated in the CPU. The parameters are grouped according to their characteristics and particularities, as shown below:

Reading parameters: Variables that can be monitored on the display, but cannot be changed by the user, such as voltage percent, current percent, active power, etc.

Regulation parameters: These are the adjustable values to be used by the VFD functions, such as initial voltage, acceleration ramp time, deceleration ramp time, etc.

Configuration parameters: These are defined as the VFD characteristics, the functions to be carried out, as well as the input and output functions.

Motor parameters: Indicate the rated motor characteristics, such as motor current adjustment, service factor, voltage, etc.

NOTE: There is a parameter that loads the factory default settings. The overall parametrization is chosen to suit the largest number of applications, minimising the need for reprogramming during start up.

8.1 Data Input and Output Systems

The data input and output system is responsible for the interconnection between the user and the VFD. They are responsible for entering information into the VFD or for the VFD to send information to the user. The main ones are listed below:

Human machine interface (HMI): It is a data input/output device, where the operator can input the VFD operating parameters, such as speed adjustment, acceleration/deceleration time, etc. It is also possible to access the operation data, such as motor speed, current, error indication, etc.

Analogue inputs and outputs: Responsible for controlling/monitoring the VFD via analogue electronic signals, i.e., signals in voltage (0... 10 Vdc) or current (0... 20 mA, 4... 20 mA) and which, basically, allow speed control (input) and current or speed (output) readings.

Digital inputs and outputs: Provide the VFD controlling/monitoring via discrete digital signals, such as on/off switches. This type of control basically allows access to simple functions, such as rotation direction selection, locking, speed selection, etc.

Communication interface: Allows the VFD to be controlled/monitored remotely. This communication is performed by pairs of wires, and several converters can be connected to a central computer or operated by PLC, fieldbus, RS232 or RS485 networks, among others.

8.2 Speed Control Forms in a VFD

The main VFD function is the speed change in an electric motor. The main forms to perform this speed change are presented below.

8.2.1 HMI

One of the most used forms to control the speed of a VFD is by using HMI keys. It is necessary to put it in local mode, and, with the keyboard, to increase and decrease the motor speed locally, as well as to reverse the direction of the rotation of the motor.

8.2.2 Digital Inputs

In an industrial application, it is unfeaseble to operate a VFD locally via the HMI keys. Thus, most of applications are performed by means of remote commands. To do this, the VFD must be placed in the remote control mode and, with external buttons, the motor must be started or stopped, and its direction of rotation reversed.

8.2.2.1 Multispeed Function

Multispeed function is used when pre-programmed fixed speeds are desired. It allows the control of the output speed relating the values defined by the parameters according, to the logical combination of the digital inputs programmed for multispeed.

To activate the multispeed function, it is first necessary to have the speed reference source given by the multispeed function, to set the VFD in remote mode and to program one or more digital inputs for multi-speed, according to Table 8.1.

The multispeed function has the advantage of pre-programmed fixed reference stability and also guarantees immunity against electrical noise. Figure 8.1 shows a speed change in a commercial VFD.

TABLE 8.1

Speed Change according to the Multispeed Function

8 Speeds			
	4 Speeds		
		2 Speeds	
D12	**D13**	**D14**	**Frequency Reference**
Open	Open	Open	PA
Open	Open	0 V	PB
Open	0 V	Open	PC
Open	0 V	0 V	PD
0 V	Open	Open	PE
0 V	Open	0 V	PF
0 V	0 V	Open	PG
0 V	0 V	0 V	PH

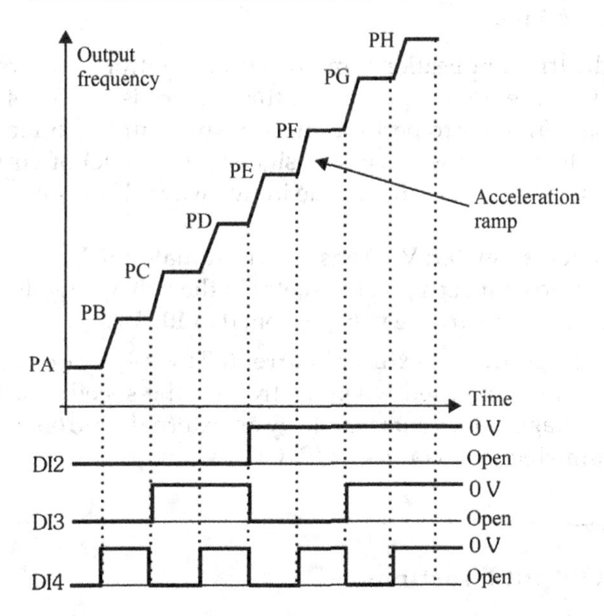

FIGURE 8.1
Multispeed ramp in a VFD.

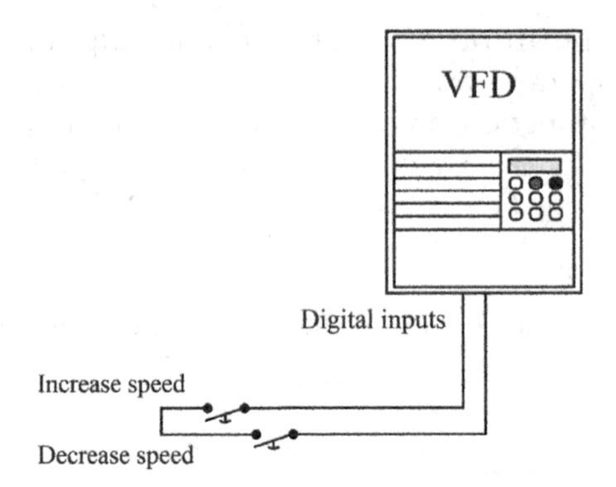

FIGURE 8.2
Speed control with two digital inputs.

8.2.2.2 Speed Control with Two Digital Inputs

In this control type, the VFD speed can be increased or decreased by pulses at the digital inputs, i.e., the speed can be increased by one digital input and decremented by another digital input. This function is very useful when it is desired to control speed without the use of analogue inputs or via buttons on a local panel, as shown in Figure 8.2.

8.2.3 Analogue Inputs

In many industrial applications, motor speed control is desired from 0% to 100%. As was previously presented, this control is not feasible if digital inputs are used. In order to perform this type of control, working with VFD analogue inputs, by means of voltage signals (0–10 Vdc) or current signals (4–20 mA), is needed. This can be done in two ways. They are:

- **By the potentiometer:** VFD has at its terminals a 10 Vdc source, so, it is possible to connect a potentiometer in the voltage divider configuration to apply a variable voltage from 0 to 10 Vdc.
- **By voltage source or external current:** This type of configuration is one of the most used when controlling the speed remotely. The supply voltage or current is made by an external controller, such as a programmable logic controller (PLC).

8.3 Relay Output Functions

With VFD digital outputs (relays), it is possible to remotely check the VFD status, as well as some alarm conditions, as if the motor is running, whether the acceleration ramp has been completed, or whether the speed or current has reached a certain limit.

Most of VFD manufacturers have this digital relay output function. Table 8.2 shows a typical operation of a two-output VFD relay (RL1 and RL2).

Figure 8.3 shows the digital output functions operation as described in Table 8.2.

TABLE 8.2

Typical Relay Output Functions

Output Function	Parameter X RL1	Parameter Y RL2
Fo > *Fx*	0	0
Fe > *Fx*	1	1
Fs = *Fe*	2	2
Io > *Ix*	3	3
No function	4 e 6	4 e 6
Run	5	5
No error	7	7

where:
Fo is the output frequency (motor)
Fe is the reference frequency
Fx is the frequency *Fx* (adjustable)
Io is the motor current
Ix is the current *Ix* (adjustable)

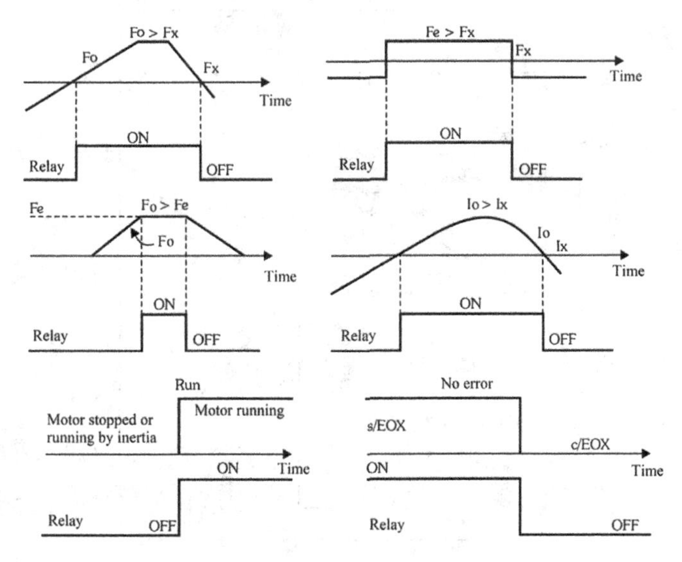

FIGURE 8.3
Operation of the digital output functions described in Table 8.2.

NOTE: In the same way that the VFD digital outputs (relay) have been selected, it is possible to use analogue outputs for reading motor variables, such as current, voltage, torque, etc.

8.4 VFD Input and Output Connections

The signal connections (analogue and digital inputs and outputs) are made in the electronic control card connector. In Figure 8.4, connection terminals for the power and ground circuit in a commercial VFD are presented.

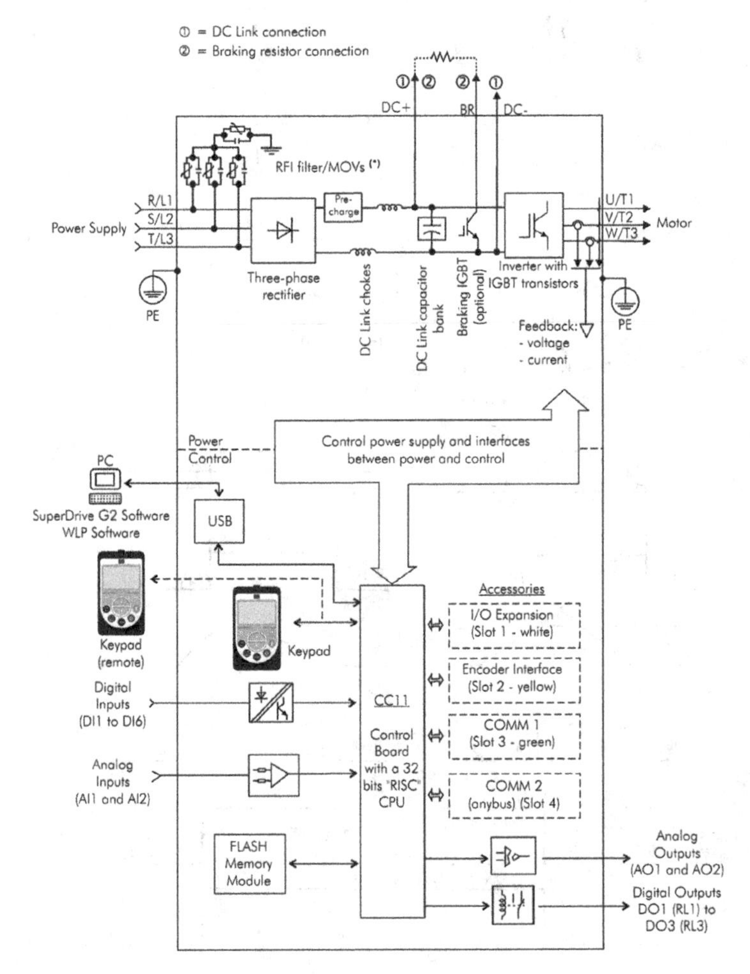

FIGURE 8.4
Connections diagram in a commercial VFD.

8.5 Typical VFD Wiring Diagrams

For a better understanding, in Figure 8.5, a typical VFD configuration with digital and analogue inputs and outputs, with some function parameterised, such as speed control, start, reverse, and current reading is presented.

With the use of automated systems with programmable logic controllers, these VFD signals are connected directly to the PLC, as shown in Figure 8.6.

Is important to remember that all analogue inputs and outputs require two cables per function, plus one connection for the shielded cable.

Manual or automatic control systems have been operated with relative success for a long time with electrical connections of digital and analogue

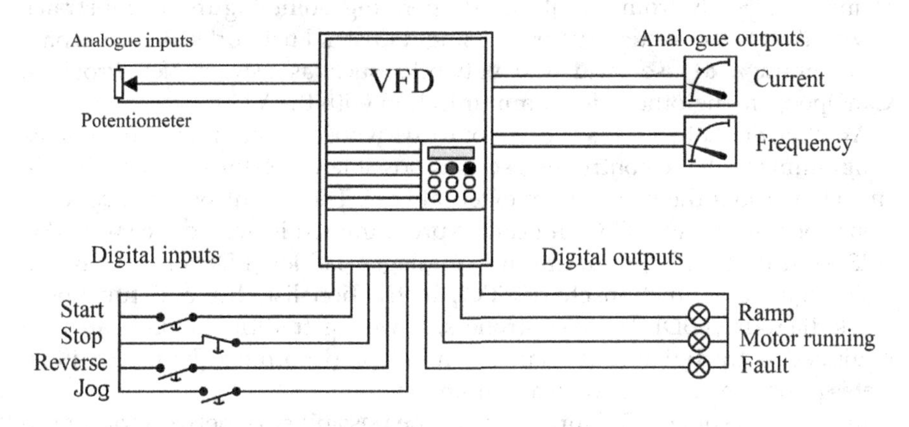

FIGURE 8.5
Typical connection of digital and analogue inputs and outputs in a VFD.

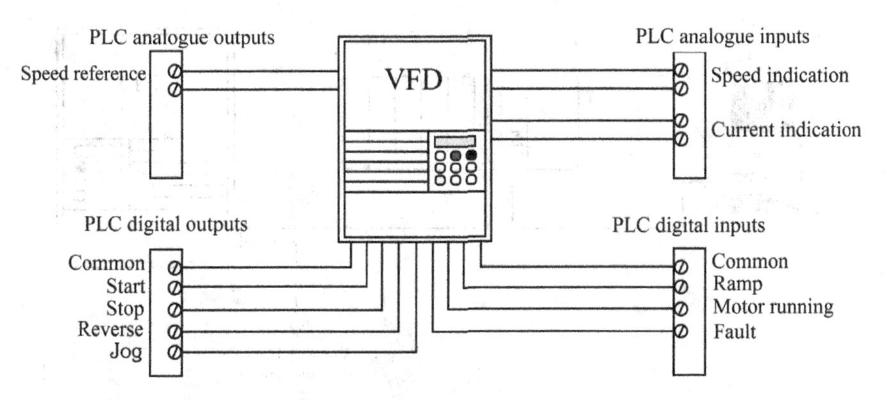

FIGURE 8.6
VFD connection with PLCs.

inputs. The major disadvantage of these systems is that all inputs and outputs require one cable per function, plus one common cable.

As control systems have grown in complexity, and the amount of information required by field sensors has also increased, the number of conductors needed to implement control systems has increased in the same proportion, becoming a problem from the point of view of cost and complexity. As more field devices are integrated into the control system, increasing the number of cables becomes a critical installation problem.

Communication using industrial networks eliminates this problem and allows the VFD connection to automated systems, such as PLC, with a minimum of cabling. VFD variables and status can be transferred over the networks between the PLC and the VFD. Configuration parameters can also be changed remotely from a centralised operating point. Figure 8.7 illustrates a typical VFD connection diagram using industrial networks. It has become very common to use industrial networks such as Device Net, Profibus, CanOpen, among others, for communication with the VFD.

Another market trend is the use of cards whose function is to operate as programmable logic controllers, so it is possible to perform logics directly in VFD without the need for an external PLC. These controllers have functions very similar to a PLC and can be programmed in accordance with IEC 61131-3 in up to five programming languages: ladder (LD), structured text (ST), sequential function chart (SFC), instruction list (IL), and functional block diagram (FBD). These controllers, as well as traditional PLCs, are also equipped with digital and analogue inputs and outputs. Figure 8.8 illustrates a commercial VFD with a built-in PLC.

In addition, some VFDs already have the possibility of network programming with dedicated software, for example, VFD parameterisation using a Bluetooth® connection.

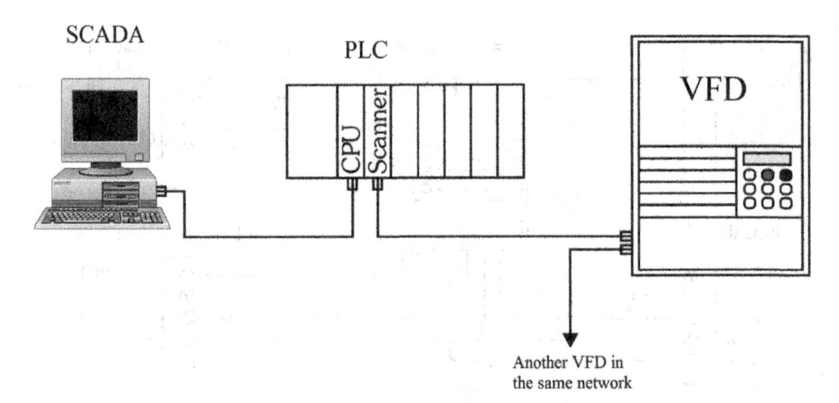

FIGURE 8.7
Typical connection diagram of VFD with industrial networks.

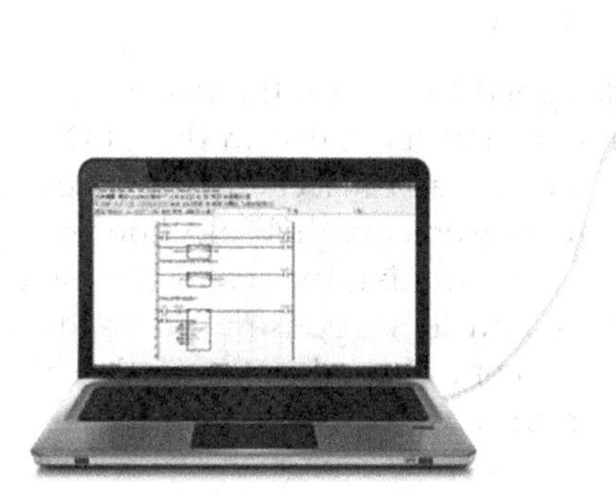

FIGURE 8.8
VFD with built-in PLC. (Courtesy of Weg.)

8.6 Transferring Configuration Using Human Machine Interface

In many models, it is possible to transfer the values of the parameters from one VFD to another, allowing the configuration with greater agility. This function can only be used when the VFDs have the same model (voltage and current) and compatible software versions.

Figure 8.9 shows the HMI withdrawal and connection to transfer configuration.

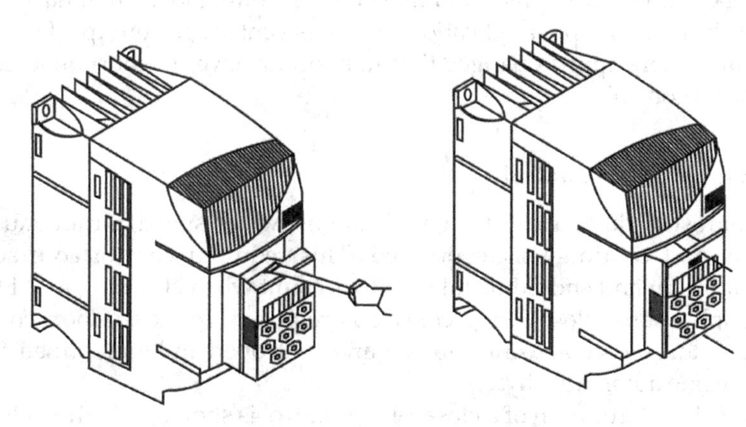

FIGURE 8.9
Removing and connecting the HMI VFD.

8.7 Application in Process Control

A variable frequency driver is designed to control the frequency and voltage applied to the motor. The digital control applied in the VFD automates this process. For example, when the operator selects motor speed through a potentiometer, the control system implements this selection by adjusting the output frequency and voltage to ensure that the motor runs at a certain speed. The control system accuracy and its response to the operator command are determined by the type of control used in the VFD. This type of control may be the one described in the following topics.

8.7.1 Open-Loop Control

In many applications, the VFD is used to control the load speed based on a setpoint (SP) provided by an operator or controller of a process, such as a PLC.

Conventional VFDs are devices that have the function of controlling the amplitude and frequency of the output voltage. The current flowing depends on the motor and load conditions, which are not controlled by the VFD, although they are the result of the application of a voltage with variable frequency. The only current control that exists is the current limit whose value reaches a maximum level, for example 160% of a total load current. There is no actual speed information to check if the motor is running at the desired speed. This control is important, because if the load on the motor shaft changes, the motor slip will vary by changing the speed, and the drive will not adjust its output to compensate for the process changes. This method is called open-loop control and is suitable for controlling loads that have small changes in more simple applications, such as centrifugal pumps, fans, and conveyors where speed changes that may occur have no consequences for the process control.

8.7.2 Closed-Loop Control

In the industry, there are critical applications whose speed/torque must be precisely and continuously controlled. This required control accuracy is very important and should be taken into account when choosing a VFD. For these applications, closed-loop control is required. In closed-loop control, the speed/torque is read continuously, and controlled and maintained at the desired value automatically.

A typical configuration of a closed-loop control is shown in Figure 8.10 and consists of the following elements:

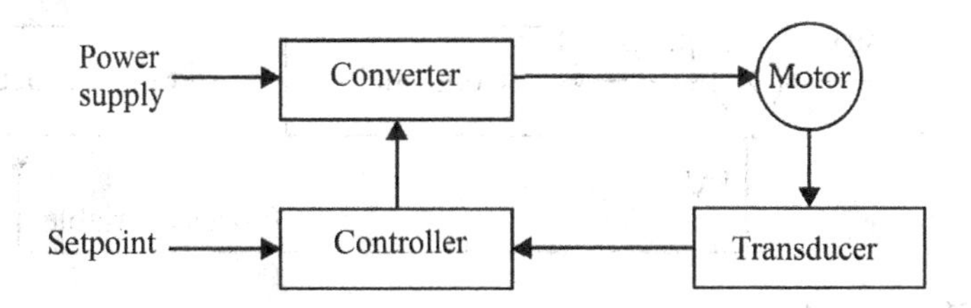

FIGURE 8.10
Typical configuration of a closed-control loop with VFD.

Motor: whose function is to convert electrical energy from the power supply into mechanical energy required to move the load.

Transducer: to measure the amount of load to be controlled. It uses a feedback signal to control the system. When the speed is the variable to be controlled, the transducer can be a tachometer (analogue system) or an encoder (digital system). When the position is the variable to be controlled, the transducer will be a resolver (analogue systems) or an absolute encoder (digital systems), and there may be other lower cost ways for measuring speed and position, depending on the accuracy required. If the variable to be measured is current, the transducer is a current transformer (CT).

A **converter** controls the amount of electric power to the motor. In this case the converter is the VFD.

A **controller** compares the desired speed or position value, called the set-point, to the measured value, called the process variable (PV), and provides a control output that adjusts speed and torque to reduce error (SP–PV) to zero. This control can be done in an external controller, such as a PLC, or in more a modern VFD as a built-in function.

The desired load speed can be manually adjusted by the operator via a potentiometer (analogue system) or by a keyboard (digital system). If the VFD is part of a more complex control system, the desired value will pass through a PLC in analogue values (4–20 mA or 0–10 V). This controller compares this set-point value with the current value of the variable and calculates the error by sending a correction signal to the VFD.

The closed-loop control was redesigned in Figure 8.11 to emphasise the most important control aspects. The term closed-loop emphasises the control system nature in which feedback is provided from an output variable of the motor entering the control system.

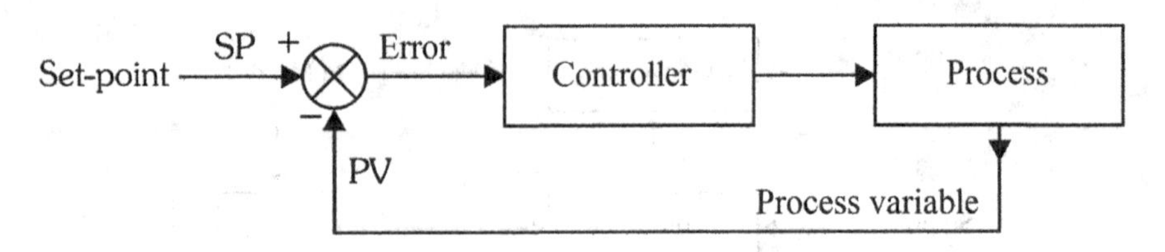

FIGURE 8.11
Closed-loop control representation.

Most VFDs have the PID function that can be used to control a closed-loop process. This function plays the role of a proportional, integral and derivative regulator applied to speed control. The speed is changed in order to keep the process variable (the one to be controlled, for example, water level of a reservoir) at the desired value (set-point).

In a closed-loop speed control, we have the following elements:

- Measurement of the process variable (speed) by means of an encoder.
- Comparison of the process variable (speed measured) with the set-point (desired speed value) to provide the SP–PV error signal.
- The error signal is then processed by the controller to adjust the output signal for the process, in this case, the motor speed.

An example of such control is a VFD that drives a motor pump that circulates a fluid in a given pipe. The VFD itself can control the flow in this pipe using the PID regulator, without the need for an external controller. In this case, for example, the set-point can be given by an analogue input or via a digital set-point, and the feedback signal of the flow reaches the analogue input. Other application examples: level control, temperature, etc.

Figure 8.12 represents an application of a VFD with PID control.

8.8 VFD Functions

In addition to the previously mentioned features, the VFD also features programmable functions that allow configuration of the drive system according to the user's needs. The following are some functions available in most VFDs.

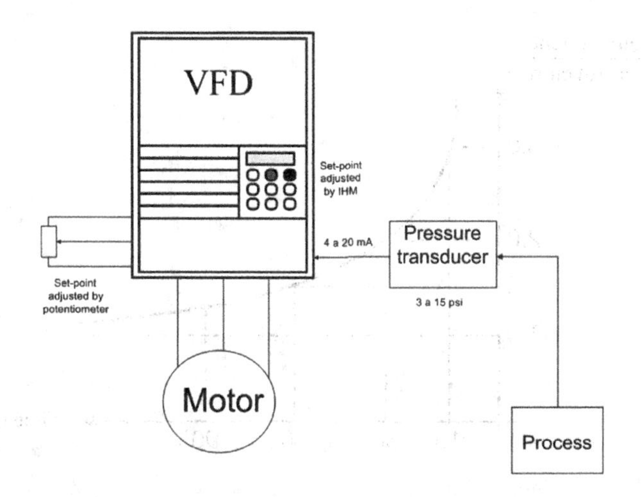

FIGURE 8.12
Application of a commercial VFD with PID control.

8.8.1 Types of Acceleration and Deceleration Ramps

It is possible to choose the type of acceleration and deceleration ramp for the VFD. Linear is the most commonly used, with the type *S* ramp recommended to reduce mechanical shocks to the motors during acceleration and deceleration ramps, as shown in Figure 8.13.

8.8.2 Motor Overload Current

This parameter is used for protection against motor overload. The motor overload current is the current value from which the VFD checks if the motor

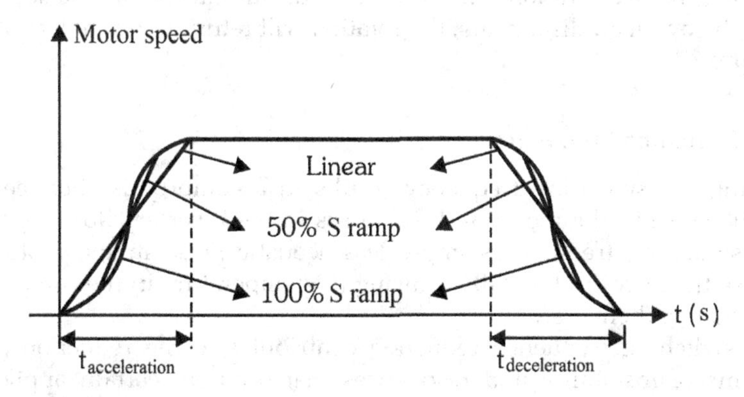

FIGURE 8.13
Types of acceleration and deceleration ramps in a VFD.

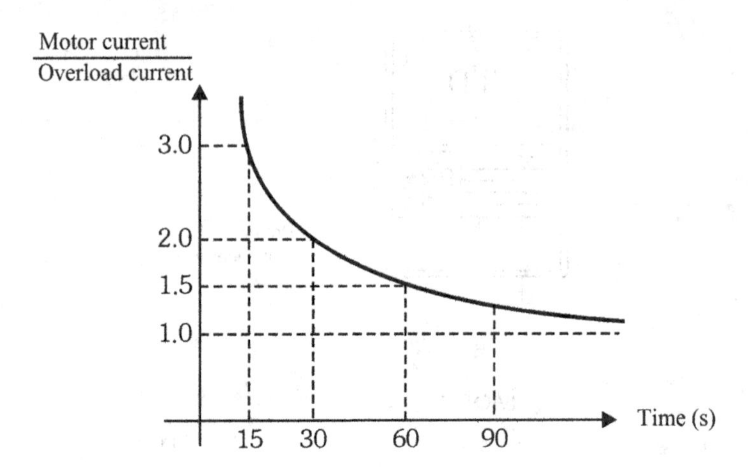

FIGURE 8.14
Overload detection function.

is operating at overload. The higher the difference between the motor current and the overload current, the faster the motor protection will act, as shown in Figure 8.14.

8.8.3 Maximum Output Current Limiting

This function is used to prevent the motor from stopping for instantaneous overloads. As the load on the motor increases, its current also increases. Depending on the value of the current rise, the overcurrent protection can be triggered. With the maximum output current function, if the current attempts to exceed the value set in this parameter, the motor speed will be reduced by following the deceleration ramp until the current falls below the set value. When the overload disappears, the rotation will return to normal, as shown in Figure 8.15.

8.8.4 Switching Frequency

Selecting the switching frequency results in a compromise between the acoustic noise in the motor and the losses in the inverter IGBTs (heating). High switching frequencies imply less acoustic noise in the motor, but increase the losses in the IGBTs, raising the temperature in the components and reducing their useful life.

The switching frequency reduction contributes to the reduction of the problems of instability and resonances that occur in certain application conditions, as well as the emission of electromagnetic energy. However, the reduction in switching frequency reduces the quality of the output waveform.

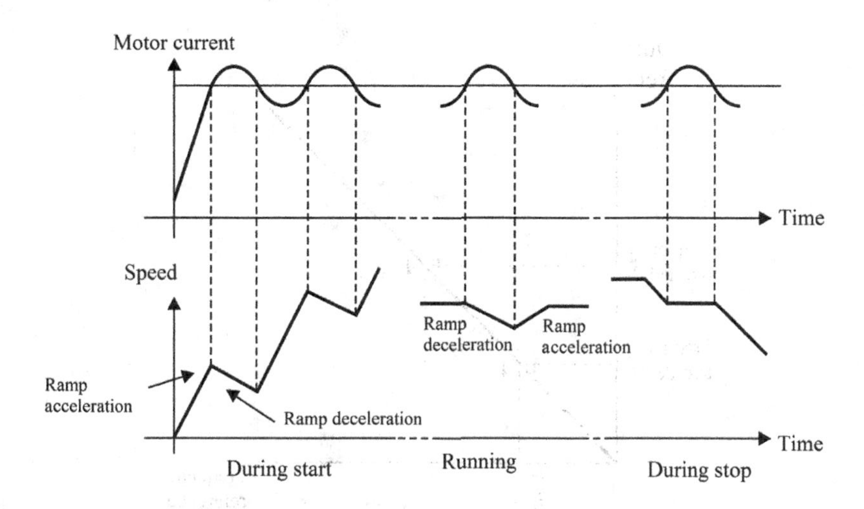

FIGURE 8.15
Maximum output current limiting actuation.

Typical frequency values for switching frequency drives range from 2.5 to 15 kHz.

8.8.5 Avoided Frequencies

This function prevents the motor from operating at output frequencies (speed), where the mechanical system goes into resonance causing vibration. Figure 8.16 illustrates the behaviour of this parameter.

8.8.6 Automatic Cycle

The automatic cycle has the purpose of activating a motor in a certain sequence of operation that will be repeated in defined time instants. In Figure 8.17, the frequency of each step is presented, as well as its duration, which can be configured independently.

This type of application does not require an external command to change speeds. It presents precise and stable operating times, does not present external influence, but great repeatability, and also has immunity to electrical noise.

8.8.7 Manual Boost Torque (Compensation IxR)

For frequencies below 30 Hz, the term corresponding to the motor resistance (R) starts to have an influence on the current calculation.

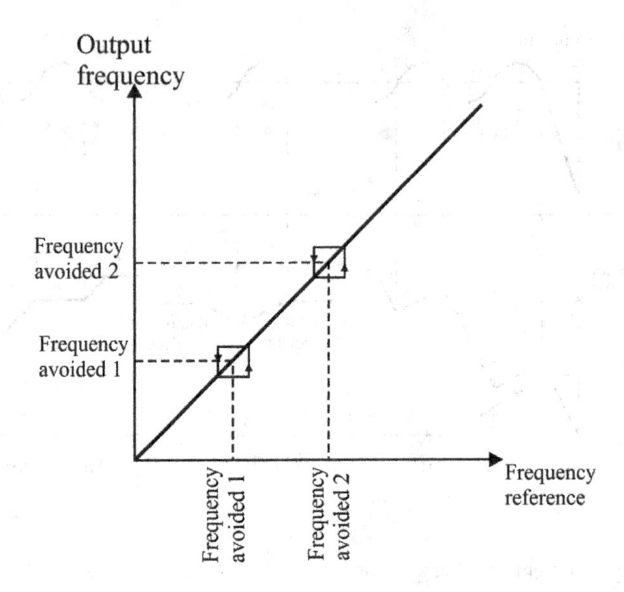

FIGURE 8.16
Frequencies avoided in a VFD.

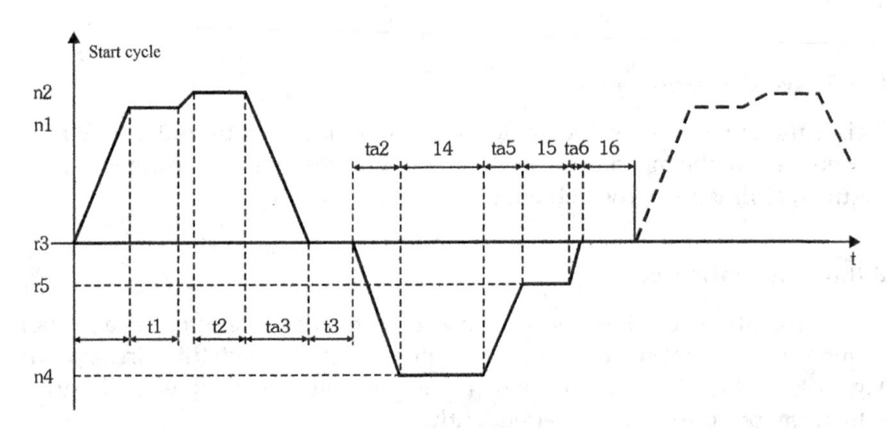

FIGURE 8.17
Automatic cycle function.

In order to avoid loss of torque, the stator voltage at low frequencies must be increased.

The manual torque boost has the function of compensating for the voltage drop in the stator resistance of the motor. It operates at low speeds, increasing the VFD output voltage to keep a constant torque in V/F operation. The best setting for this parameter is the lowest value

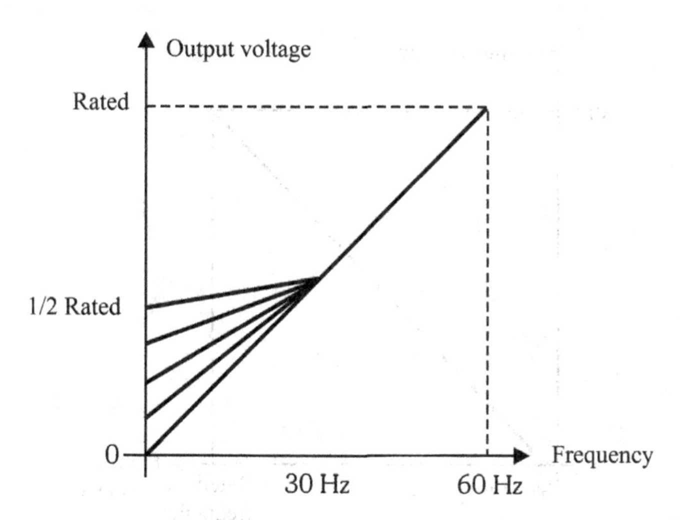

FIGURE 8.18
Manual torque boost.

that allows the motor to start and operate satisfactorily. Higher values than necessary greatly increase the motor current at low speeds, which can force the VFD to an overcurrent condition. Figure 8.18 illustrates this parameter.

Some VFDs have an automatic compensation parameter, depending on the active current of the motor.

8.8.8 V/F Curve Adjust

Defines the V/F curve used in scalar control. This parameter can be used in special applications in which the motors used need voltage and/or frequency different from the power supply.

Figure 8.19 shows this parameter with a 440 V Voltage grid applied in a 380 V motor.

8.8.9 Braking

In applications where the induction motor is employed in processes requiring fast shutdowns, a very short deceleration time is required, and the electrical or mechanical braking feature must be employed. During braking, the rotor frequency is greater than the stator frequency, which causes a reverse energy flow from the rotor to the stator. The motor then operates as a generator, injecting this energy into the VFD DC bus, which can lead to overvoltage.

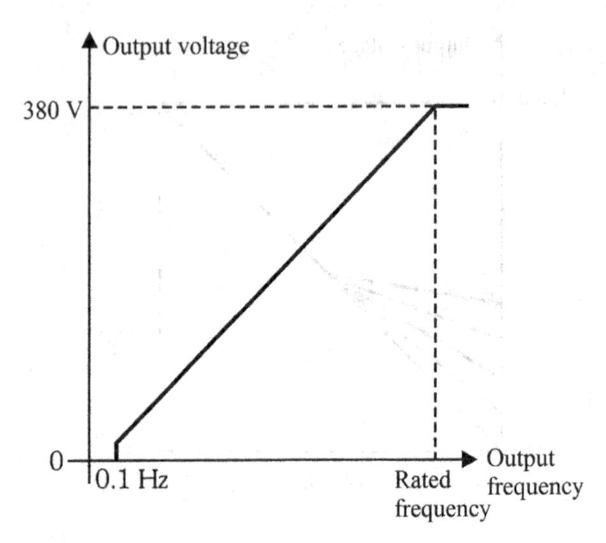

FIGURE 8.19
V/F curve adjust.

When the VFD is switched off, a braking action is required to reduce the speed of the mechanical load. To reduce this kinetic energy, it is necessary to remove the kinetic energy from the AC motor and mechanical load, and turn it into another form of energy so that the system can stop. This stop becomes more difficult as the load size increases.

Most starting devices for motors and VFDs stop the motor by removing power from the motor. The rotary assembly comprising the motor and the load stops after a natural deceleration time.

This type of stop is suitable for most mechanical loads, such as conveyor belts, fans, and others. The stopping time depends on the load inertia, the friction of the load and the type of process. For other processes, braking is required to provide a shorter deceleration time, as shown in Figure 8.20.

Old systems use mechanical braking, where kinetic energy is converted to heat by friction. Although this system is relatively efficient, the cost of maintenance is high.

With VFD use, electric braking is an appropriate method for braking the motor. Electric braking is based on a system in which the motor is used at the instant of deceleration as a load-driven generator. During electric braking, the conversion of mechanical energy into electric can be done in four ways:

- Dissipating heat in the motor rotor—DC injection braking.
- Dissipating heat in the motor stator—flux braking.

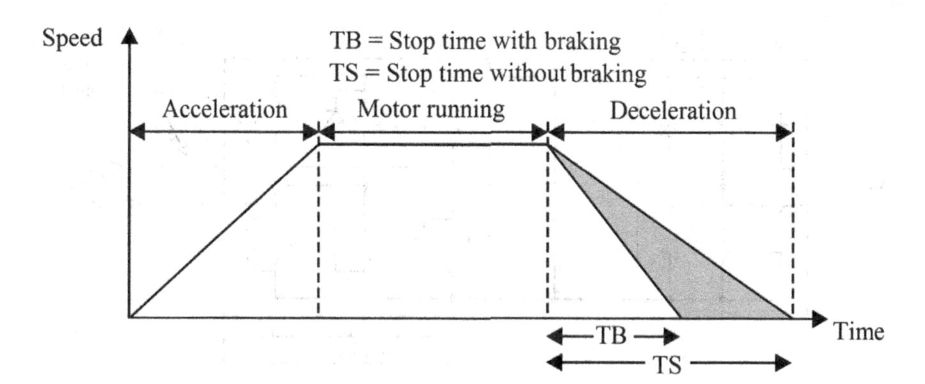

FIGURE 8.20
Deceleration times with braking and without braking.

- Dissipating the heat in an external resistor—rheostatic braking.
- Returning electric power to the power supply—regenerative braking.

Electric braking has several advantages over mechanics:

- Reduction of maintenance of braking elements.
- Speed can be more precisely controlled during the braking process.
- The energy dissipated in the braking can be collected and returned to the power supply.

Electric braking can be done by one of the procedures or a combination of them.

8.8.9.1 DC Injection Braking

Allows the motor to stop by applying direct current. The magnitude of the direct current that defines the braking torque and the period during which it is applied are parameters that can be specified by the user. This mode is generally used with low inertia loads, and can cause excessive motor overheating when stop cycles are too repetitive.

The basic DC braking principle is to inject DC current into the motor stator to cause a stationary magnetic field in the motor. This is achieved by connecting two phases of the induction motor to a DC source. The injected current must be at least equal to the motor excitation current or the current of the motor in a no-load condition.

This type of brake is done as follows: the VFD control sequence is modified so that the IGBT in one phase is switched off, while the other two phases

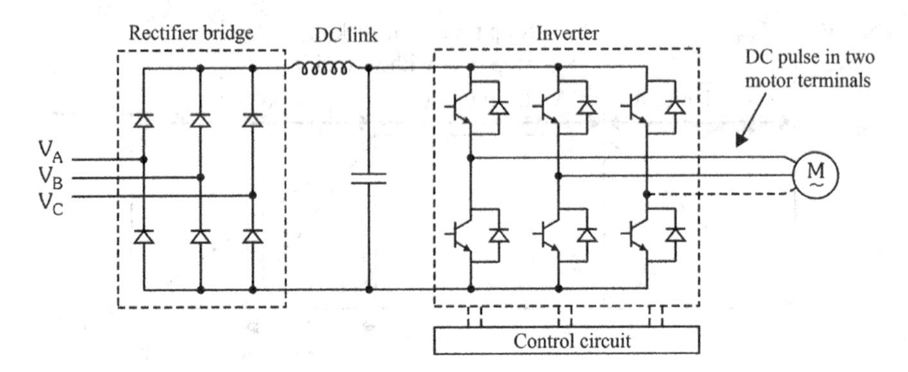

FIGURE 8.21
Braking a motor with DC injecting direct.

provide an output with a DC signal with a control magnitude and duration of this signal (Figure 8.21).

As the rotor bars cut off the magnetic field generated by the DC injection, a current will be developed in the rotor with a proportional amplitude and frequency at the speed, resulting in a braking torque that is proportional to the speed. The braking energy is dissipated as loss in the rotor windings that will dissipate heat and is limited by the maximum temperature rise allowed by the insulation of the motor.

DC braking allows a fast motor stop in which direct current is applied. The current applied to the DC braking, which is proportional to the braking torque, can be set as a percentage (%) of the rated VFD current. Figure 8.22 shows the DC braking operation.

In a form of ramp, DC braking is applied after the motor is switched off and when the motor reaches a certain speed, which is established by a VFD configuration parameter.

NOTE: Before starting DC braking, there is a dead time, when the motor runs freely, necessary for demagnetising the motor. This time is a function of the motor speed at which DC braking occurs.

8.8.9.2 Rheostatic Braking

Typically, the braking is used to lower the speed to a defined value, from which DC current is applied to the motor, achieving rapid braking and preserving the VFD.

Several VFD models offer the option of using dynamic braking modules, which are banks of electrically controlled resistors connected to the DC circuit, which provide a braking torque close to the rated motor

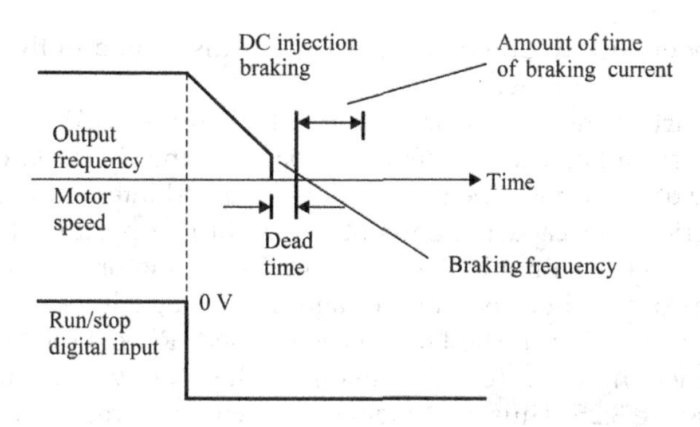

FIGURE 8.22
DC braking in a form of ramp.

torque, ensuring the dissipation of the braking energy in these external resistors.

To understand the operation of motor dynamic braking, consider the following figures.

Figure 8.23 shows a VFD driving the motor to achieve a rotation that is transmitted to the load. Due to the inertia of the load, when it is necessary to stop this load or reduce its rotation, the VFD will energise the motor with a lower frequency and amplitude voltage, causing the electromagnetic field inside the motor to rotate with a slower speed.

In this situation, the motor behaves as a generator, as shown in Figure 8.24, where the voltage induced in the rotor has a greater amplitude

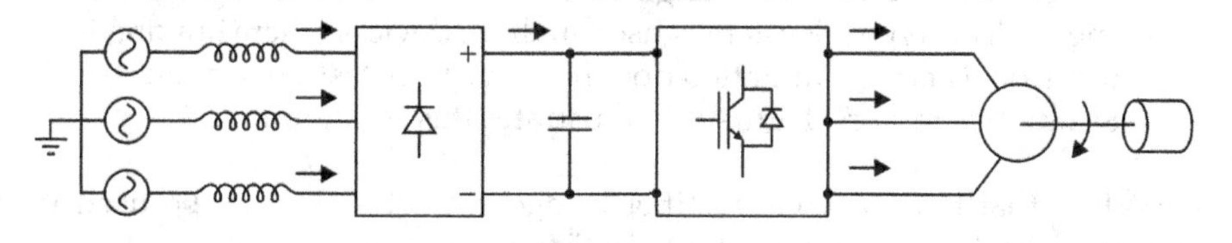

FIGURE 8.23
Typical VFD operation.

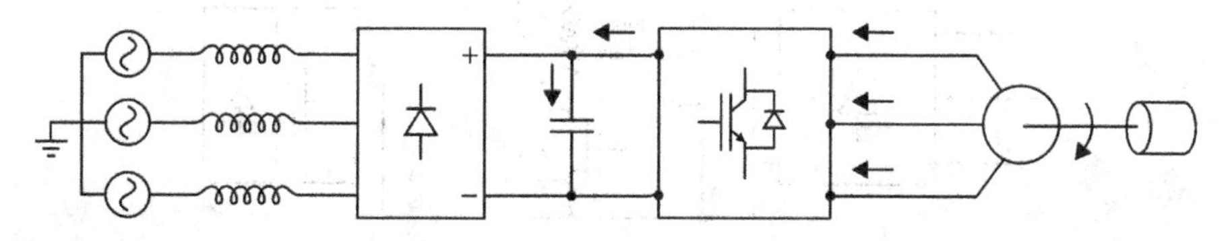

FIGURE 8.24
VFD operation at motor stop.

than the power supply. Some of the energy is dissipated in the motor and some in the IGBT's bridge.

Another part of the energy will be rectified in the IGBT bridge diodes, which are accumulated in the capacitor bank of the DC link, causing the diode bridge of the input to be reverse polarised and interrupting the power flow of the grid to the capacitor bank of the DC link. A portion of this energy returns through the output IGBTs to magnetise the motor.

A problem occurs because the voltage in the capacitors rises until the overvoltage protection in the DC link is tripped. Rheostatic braking consists of connecting an external resistor to the VFD via the DC link, as shown in Figure 8.25. Thus, the energy that would be returned to the link is dissipated in the form of heat, as a simple solution to the overvoltage problem in the DC link.

For cases where the power dissipated is quite large, a solution may be the return of the energy to the grid called regenerative braking.

8.8.9.3 Regenerative Braking

From the VFD point of view, regenerative braking is seen in a similar way to DC braking. When braking occurs, the motor power flow path to the diodes of the rectifier bridge through the DC links is required. To perform regenerative braking, there are two methods:

- If a thyristor-controlled rectifier bridge is used instead of a diode bridge, regeneration is only possible by changing the DC link polarity. This can be achieved through a switch between the rectifier and the capacitor that will change direction according to the flow of energy. This type of system is used in drive devices where braking is occasional and commutation does not have to be fast, such as electric locomotive motors. Figure 8.26 illustrates this configuration.

For VFDs that have a diode rectifier bridge, thyristor sets can be used to extract the braking power, as shown in Figure 8.27.

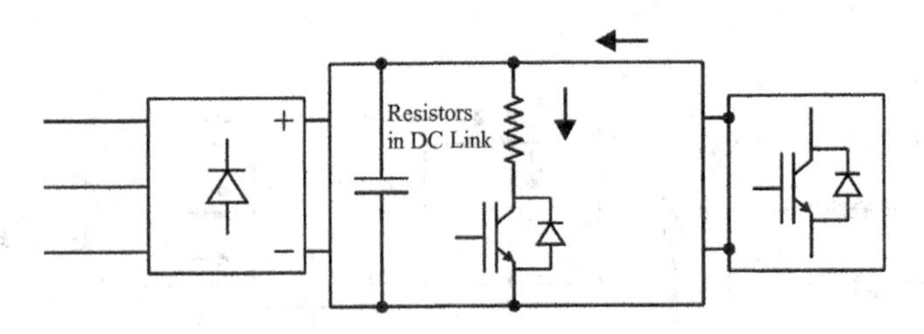

FIGURE 8.25
Resistors placed in the DC bus.

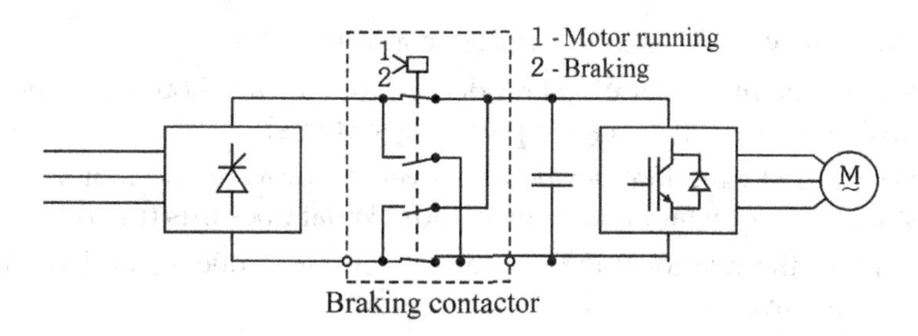

FIGURE 8.26
Regenerative braking with rectifier bridge (thyristors).

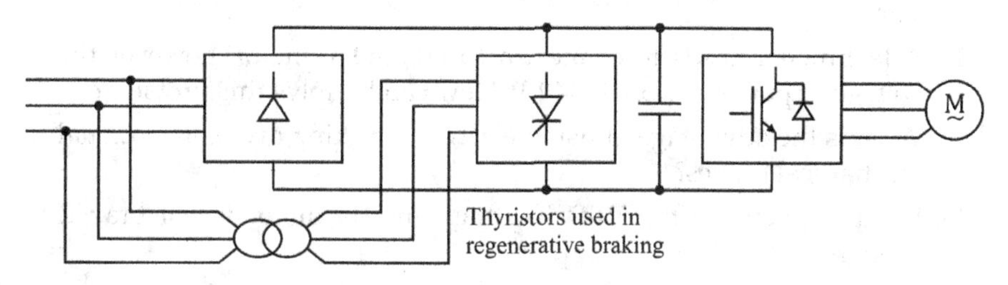

FIGURE 8.27
Regenerative braking with rectifier bridge (diodes).

The main disadvantage of regenerative braking is the initial cost compared to the return of energy savings. This return depends on the type of application, the inertia of the load and the braking duration required.

Exercises

1. List the types of parameters of a VFD.
2. Describe the following forms of VFD control:
   a. HMI
   b. Digital inputs
   c. Multispeed function
   d. Speed control with two digital inputs
   e. Analogue inputs
3. How is possible to control the motor speed using analogue inputs?
4. What is the function of the VFD output relays?

5. How can VFD be used for process control?
6. What type of acceleration and deceleration ramp should be used to avoid mechanical shocks in pumping systems?
7. What VFD function can be employed to prevent the motor from stopping for instantaneous overloads? What does this function do?
8. Explain the relationship between switching frequency and voltage output waveform.
9. How can VFD use reduce resonance in an electric motor?
10. Is it possible to use VFD to reduce the loss of torque during motor start?
11. A technician needs to connect a 15 HP/460 V motor, however the voltage of power supply is 440 V. How can he solve this problem?
12. What is the advantage of using electrical braking over conventional mechanical brakes?
13. What happens to the VFD if a high inertia motor is not braked properly?
14. Describe the three most used motor braking systems: DC injection braking, rheostatic braking and regenerative braking.

9

VFD Protection and Installation

VFDs are provided with thermal protections, overvoltage, under-voltage, overcurrent for the motor, as well as protections to itself. A microprocessor reads the measured current and motor speed data to promote these protections, sending alarm signals or de-energising the VFD. The most common protections are:

- Short circuits between phases and between phase and ground
- Overvoltages and under-voltages
- Phase imbalance
- Phase loss

This chapter presents protection elements built in to VFD, together with the design and application characteristics.

9.1 VFD Electric Protection

As with other electrical equipment, protection devices against short circuit and overload must be provided to prevent damage to the power supply, VFD or motor.

The most commonly used devices are high speed fuses or circuit breakers. In spite of this, the VFD itself must be sized to withstand the short circuit at the point where it will be installed.

The most commonly used protection is type 2 coordination, according to IEC 60947, which should have the following characteristics:

- No risk to persons and installations, i.e., safe shutdown of the short-circuit current.
- After a fault, there should be no loss of equipment settings.
- The insulation must not be damaged in the event of a fault.

- The elements (circuit breaker, VFD and contactor) must be able to operate as soon as the cause of the short circuit is eliminated.
- Surface welding of the contactor contacts is allowed, and it is possible to separate them manually.
- The following will present some of these VFD protections.

9.2 Built-In VFD Protections

The protection is divided into two main groups: protection for the VFD and for the motor.

In modern devices, most of the protection functions are implemented electronically within the VFD control system. However, external sensors are required to monitor the current or temperature directly. The increase in temperature in a motor and VFD is the main cause of system shutdowns. The temperature rise is usually the result of a high current flow; current measurement is a common method used for overload and short circuit protections.

VFDs generally have a considerable number of circuits for the protection itself, the cables and the motor. However, the protection inserted does not protect the input side of the VFD, which is comprised of the power supply cable and the rectifier stage. Short-circuit and ground faults must be provided in the control panel and can be achieved by:

- Moulded case circuit breakers.
- High speed fuses.

The protections described below are usually available for most VFDs.

9.2.1 AC and DC Under-Voltage

Under-voltage protection monitors the three phases as well as the DC bus voltage. If there is a fault, it will act to avoid damage to the VFD.

If the supply voltage drops below the rated value due to a power supply failure, the VFD will have difficultly being damaged. The rectifier circuit can safely operate at any voltage between zero and the maximum expected operating voltage. In this way, a power supply under-voltage is not characterized as a problem for the VFD.

Under-voltage protection is primarily required to ensure the power supply for the VFD in order to operate within its specifications. If a power source has a voltage below rated voltage, the following problems may occur:

- The DC bus load relays can be switched off.
- Microcontrollers or DSPs may enter an undetermined state.
- Tripping circuits will not have enough voltage and current to turn switching devices on and off.
- If there is insufficient current to switch on, a power device may become saturated, increasing losses.
- If the reverse polarity is insufficient, the power device will turn off very slowly, or not turn off, i.e., the switching devices will fail.

9.2.1.1 Grid Under-Voltage Fault

The grid under-voltage fault can be detected by monitoring the three input phases and comparing these to a desired value. Power supply under-voltage can cause a complete power failure or a short voltage drop. Since the power to the control circuits is obtained directly from the DC bus, it is not necessary to stop the VFD immediately after a fault in the power supply. If necessary, the VFD can continue to operate, initially by obtaining power from the large capacitors located on the DC bus.

As the voltage in the DC bus begins to drop, the output frequency can be reduced to allow the motor still work through the mechanical load inertia. This situation can only be maintained for a short period of time. Otherwise, the control circuit can be programmed to stop immediately after the power supply goes into fault.

9.2.1.2 DC Bus Fault

The DC voltage bus can be monitored by a comparator circuit (hardware or software) that compares the bus voltage with a minimum working value. When the voltage value drops below this minimum value, the VFD must shut down. This minimum working value is usually set to the lowest possible voltage value minus 15%. For example, if a VFD is designed to operate at 380–460 Vac ± 10%, the lower value specified for operation should be 342 Vac with an equivalent voltage of 485 Vdc. The protection tripping point must be set to 485 Vdc –15%, which will be 411 Vdc.

In addition to this, some modules must be individually protected. For example, each driver module must have its own under-voltage protection circuitry to ensure voltage for switching. If this protection is triggered, a signal must be sent to the main processor indicating that device has failed. Local under-voltage protections are usually only placed in critical modules, such as transistor driver circuits.

9.2.1.3 AC and DC Overvoltage Protections

It is known that all electrical components fail whenever they are exposed to a voltage above the rated. In VFDs, overvoltages can occur due to the following reasons:

- High voltage in the power supply.
- High voltage generated by the connection of a motor when it behaves like a generator when trying to reduce the speed of a load with a high inertia.

In a VFD, the capacitor bank, the DC bus connected to the power supply module and the switching devices have the smallest tolerances to high voltages.

The capacitor bank usually consists of individual capacitors in series or in parallel. When the capacitors are connected in series, the voltage division will not be perfect, and the maximum voltage will be less than the individual capacitors' sum. For example, if two 400 Vdc capacitors are connected in series, the individual voltage rating of the association would be 800 Vdc. However, the safety operation voltage would be only 750 Vdc due to unequal voltage division characteristics. This value is due to the capacitor leakage current and the value of the discharge resistor in parallel with each capacitor.

The peak voltage on the DC bus is $\sqrt{2}$ times the supply voltage. If the maximum voltage of the capacitor is 750 Vdc, it allows a further 10% change in the supply voltage.

The power circuit semiconductors in the rectifier and inverter are generally designed for a maximum voltage of 1200 Vdc. Although this appears well above the maximum capacity of the capacitor, the voltage on the devices will be much higher during the shutdown of the devices, particularly during fault conditions, due to characteristic circuit inductances. The voltage peaks can reach about 400 V, so the bus voltage before failure should be limited to approximately 800 Vdc, depending on the drive and switching devices design.

In analogue converters, overvoltage protection is usually hardware through a simple comparator operating at a fixed value. In digital converters, the overvoltage protection is generally guaranteed by the microprocessor. This is possible because the DC bus voltage changes relatively slowly due to the filtering effect of the capacitors.

In digital converters, the processor can also provide some overvoltage control. Most bus overvoltages are caused by incorrect deceleration adjustments in high inertia loads. If the deceleration time is set too short in comparison with the natural time of the load stop, the motor behaves as a generator and the energy is transferred from the motor to the DC bus.

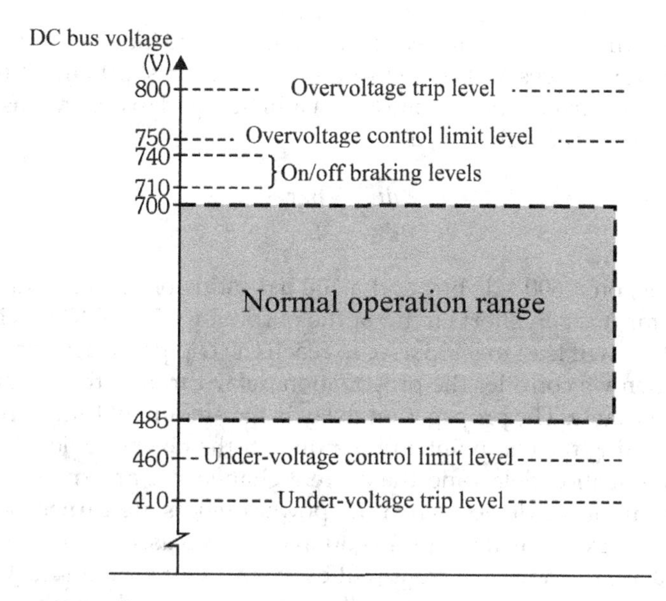

FIGURE 9.1
Typical under-voltage and overvoltage protection values in a VFD.

The DC bus can increase its voltage until the overvoltage point is reached. Some VFDs have an adjustable feature in which the controller increases the deceleration time to prevent overvoltage shutdown on the bus. The voltage bus allows the increasing of the voltage to a safe value, typically 750 Vdc, and the deceleration rate is controlled to keep the voltage below the safety value, which is 800 Vdc.

Under-voltage and overvoltage protections are normally monitored by the DC bus, because this is the DC source for the inverter and for the control circuits. Figure 9.1 illustrates the typical under-voltage and overvoltage protection values in a VFD.

9.2.2 Overcurrent Protection

The overcurrent protection purpose is to avoid failures in power semiconductor devices (IGBTs, BJTs, MOSFETs, GTOs, etc.) during short circuits between phases on the motor side. The most effective method of protection is to disconnect all switching devices from the VFD when the current exceeds a certain limit. The protection level is dependent on the operating characteristics of the VFD.

Usually, the trip current is around 200% of the rated VFD, with current limits at 150%, or sometimes 180%. To maximise the operation, it is possible to work close to the trip current value if the current growth rate (*di*/*dt*) is controlled. This can be achieved by introducing an inductor between the semiconductor devices and the output terminals. If a short circuit occurs at the output, the current change rate (di/dt) will be equal to the DC bus voltage, Vbar, divided by the inductance L:

$$\frac{di}{dt} = \frac{\text{Vbar}}{L} \tag{9.1}$$

For example, on a 600 Vdc bus and a 100 μH inductor, the current increase will be 6 amp/μsec. A short circuit at the output of a 50 kW VFD with a trip current 200 A will lead to 33.33 μsec to reach the trip point. This value is significant when we consider the propagation delay through the feedback and protection circuits. The propagation delay is the amount of time between the current reaching the trip point and turning off the power devices.

If the propagation delay and the current change rate are known, then the actual current of the device, when the power devices are turned off, can be estimated. For example, if the propagation delay is 3 μsec and the di/dt equal to 6 amp/μsec, the device current will be 18 amps larger than the trip point current when the device switches off. This current is called current overshoot.

While large inductors reduce this overshoot and offer other advantages, they will also introduce losses, being bulky and expensive. For these reasons, it is important to minimise the propagation delay in the overcurrent protection circuit. To minimise propagation delays in microprocessors, it is common for overcurrent protection to be carried out entirely by hardware, even in a fully digital VFD.

Overcurrent events can also cause a sudden increase in load torque in the motor. The current increases are relatively slow, allowing it to be monitored and controlled by the microprocessor. The current rise can be limited by up to 150% of the rated VFD current. The system current limit control regulates the output frequency in a way that reduces motor torque. If the overcurrent is due to a temporary motor overload, the motor speed can be reduced. Typical limits for overcurrent protection and current limits are summarised in Figure 9.2.

9.2.3 Earth Leakage Protection

Earth leakage protection is designed to detect phase-to-ground short circuits at the VFD output and immediately disconnect it. This protection is not intended for human beings against electric shock. The trip points are set at values much higher than those supported without causing harm to the human being, its main characteristic being the protection of the VFD only.

Earth leakage protection is generally implemented through a differential toroidal transformer, constructed from a magnetic toroidal core through which the bus cables or motor power cables pass. A low current in the secondary

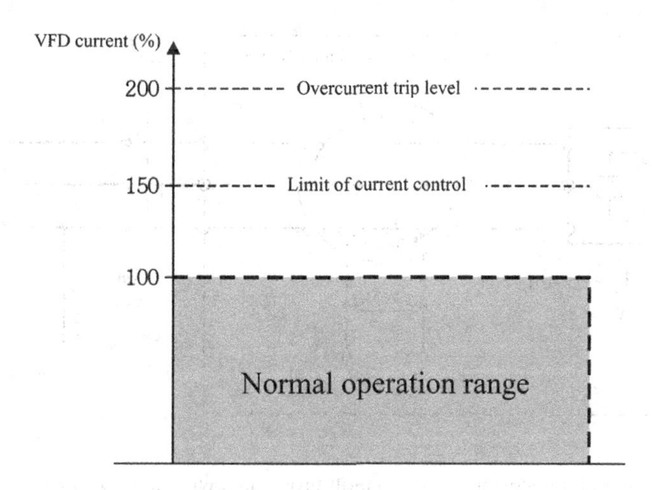

FIGURE 9.2
Typical limits for overcurrent protection and current limits.

winding obtained from the toroid is connected to the protection circuit. If the vector sum of all the currents passing in the core is zero, the flux in the core will be zero. If the value zero is registered, the VFD is in normal operation.

Figure 9.3 illustrates a toroidal differential transformer for earth leakage protection in normal operation (without earth leakage).

If a ground leak occurs, there will be a current path to the earth, the sum of the currents through the transformer core no longer being zero, and there will be a flux in the core, as shown in Figure 9.4.

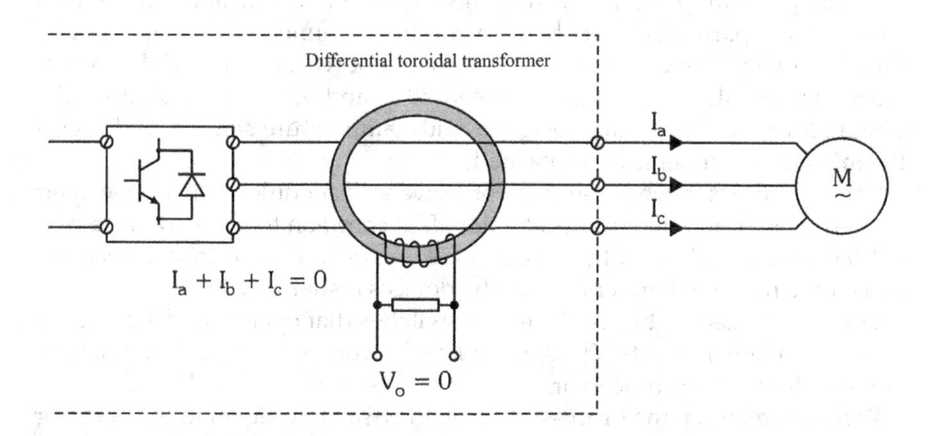

FIGURE 9.3
Toroidal differential transformer for protection ground fault. Normal operation without earth leakage.

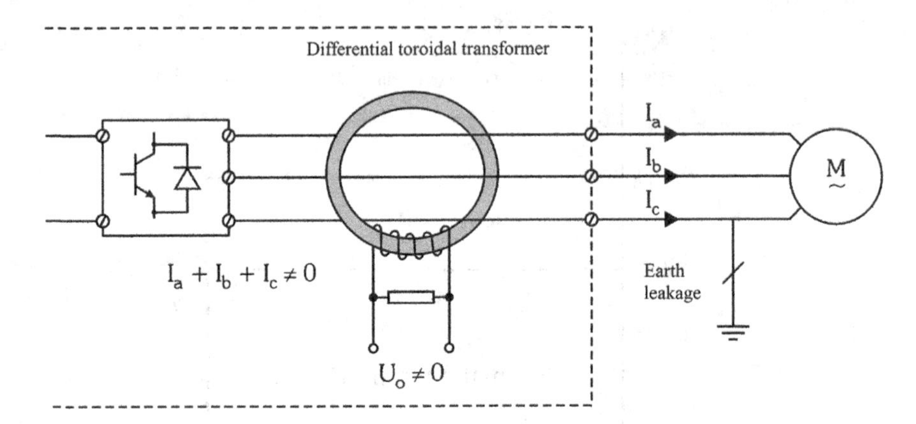

FIGURE 9.4
Toroidal differential transformer for earth fault protection, with ground leakage.

This flow results in a current generated in the secondary winding of the transformer that will be converted into voltage by means of a resistor in parallel. A comparator circuit detects fault and shuts off all VFD power devices. Typically, the trip level is 5 A. Care should be taken when establishing the set-point for the earth leakage protection trip. In every VFD, some leakage current always appears due to the high switching frequency and the capacitance of the cables with respect to earth.

9.2.4 Heat Sink Over-Temperature Protection

Over-temperature protection is designed to prevent overheating in the VFD components, particularly at the semiconductors junction that is typically limited to 150°C. To ensure that this limit is not reached, heat sink temperatures are usually maintained between 80°C and 90°C, depending on the design. Most heat sinks are designed with temperature sensors to detect if the maximum temperature is reached.

Other modules, such as sources or drivers of modules, must have their individual over-temperature protection. It is common to measure the ambient temperature close to the electronic controls to ensure that it does not exceed the nominal temperature of the devices (usually ±70°C).

Low cost VFDs use bimetallic micro switches that operate at defined temperatures. Higher quality VFDs use silicon junction sensors to send the temperature to the microprocessor.

With this method, the processor can send a warning signal to the operator before the VFD switches off. In VFDs that use more advanced technology, some corrective action can be taken automatically, such as reducing the motor speed or PWM switching frequency, to prevent motor shutdown.

9.2.5 Motor Thermal Overload Protection

Almost all actual VFDs include some protection function against thermal motor overload. The simplest way of protection is the use of a VFD digital input that shuts off when an external device, such as a relay or thermistor, is activated. Many manufacturers make a specific input to a thermistor, which needs to be placed in the motor windings, eliminating a temperature relay. The inputs are usually placed with a resistor connected to the terminals, which must be removed when the VFD is started, and this resistor is a common cause in starting difficulties in motors with inexperienced users.

The most common method used for motor thermal overload protection is indirect measurement through the motor current. A motor protection model is made as part of a control program on a microprocessor. Motor current measurement is required for other purposes. The current also serves to provide a thermal motor model, which can continuously estimate the thermal conditions in the motor and shut down the VFD if the temperature limit is exceeded.

The simplest model of the motor is to simulate an overload by integrating the motor current over time. This simple method does not provide reliable protection to the motor because the cooling and heating time constants of the motor change at different speeds.

Currently, the motor protection features have become much more sophisticated using the motor frequency as input. The cooling system performance at different speeds can also be modelled. In order to have a more complete model, motor parameters, such as rated speed, current, voltage, power factor and power must be used to enable a thermal motor model to be implemented in software, providing protection against overheating without the need for direct temperature measuring devices.

For these models to be accurate, the motor conditions must be stored in VFD non-volatile memories to avoid loss of information when power is interrupted. In this case, non-volatile memory as electrically erasable programmable read-only memorys (EEPROMs) in association with microprocessors is generally used.

Direct measurement can also be employed through specific sensors placed on the motor, such as thermostats, thermistors, thermocouples, and thermoresistors, which measure the actual temperature in the motor windings in real time and trigger the overload protection when the rated temperature value is exceeded.

9.2.6 VFD Protections Overview

The block diagram in Figure 9.5 is a summary of the protections commonly used in VFDs.

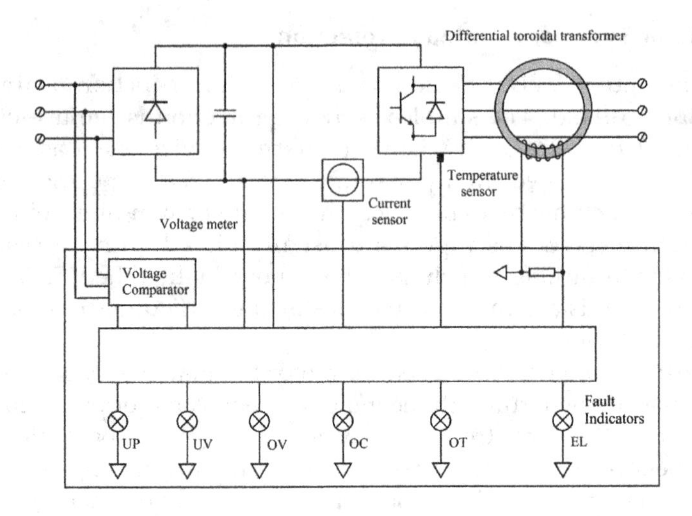

FIGURE 9.5
Built-in VFD protections.

where:

UP is the unbalanced phase
OC is the overcurrent
UV is the under-voltage
OT is the over-temperature
OV is the overvoltage
EL is the earth leakage

Many of these protection functions are implemented in software, through specific algorithms, except overcurrent protection and earth leakage that are implemented in hardware to ensure that they are sufficiently fast and safe to effectively protect the VFD.

9.3 Fault and Diagnostic Information

Actual VFDs have some type of interface module that provides access to internal data on control and status parameters during normal operation and diagnostics during fault conditions. This module is generally referred to as the human machine interface (HMI). The HMI usually has a LCD or LED display and some buttons to control the circuit. This interface with the operator can also be used to change the VFD parameters.

The VFDs also allow the transfer of these parameters through industrial networks with PLCs. When an internal or external fault occurs, the control circuit registers the fault type, which helps to identify the cause of the failure and subsequent solution of the problem. Many VFDs have a diagnostic system that monitors internal and external operating conditions and responds to any fault according to user programming. The control system records the fault information in a non-volatile memory for later fault analysis. This feature is called the VFD fault diagnosis.

The internal fault system can indicate to the operator faults that can occur in elements inside the VFD, such as communication failure, output devices, etc.

For example, power semiconductor drivers may include circuits that measure the saturation voltage, which is the voltage across the device when it is connected, for each semiconductor. This makes it possible to identify a short circuit, and the VFD can be switched off before the overcurrent protection or fuses acts. A considerable additional cost is required to implement internal fault conditions, and only a few high performance VFDs have a large number of internal fault diagnostics.

The HMI can display the protection status of the circuits indicating external faults, as described above; it also shows the internal faults that indicate the failure of specific VFD modules. Table 9.1 summarises the failures of each module and their failure parameters and diagnostics.

Failure diagnostics are essential when a fault occurs and the VFD stops, and information is needed to reduce downtime. For example, there may be a persistent overcurrent fault without the motor connection indicating a failure in the switching circuit internally.

TABLE 9.1

Summary of VFD Modules Faults and Diagnostics

Module	Parameters and Fault Diagnostics
Power supply	Power supply voltage, current and frequency
DC bus	DC bus voltage and current
Motor	Motor voltage, current, frequency, torque, etc.
Control signals	Ramp times, set-points, process variables, etc.
Status	Protection circuits, modules faults, VFD internal temperatures, swithing frequency, current limit, etc.
Fault conditions	Power supply fault, control circuit fault, controller fault, etc.

TABLE 9.2

VFD Internal Faults and Their Corresponding Protections

Protection	Internal Fault	External Fault
Overvoltage	Deceleration time too fast	Power supply voltage too high, voltage peak.
Under-voltage	Internal power supply fault	Power supply voltage too low, voltage drop in the power supply cables.
Overcurrent	Power circuit fault, control driver fault	Short circuit in the motor or cable
Thermal overload	VFD internal circuit fault	Motor overload or locked rotor
Earth fault	Internal earth leakage in VFD	Earth leakage in the cable or in the motor
Over-temperature	VFD cooling system in fault or heat sink blocked	Cooling fault in electric panel
Thermistor trip	–	Motor temperature protection (thermistor) tripped

Table 9.2 shows the most common internal and external faults that can occur in the VFD and their corresponding protections.

9.4 VFD Installation

For a good operation and performance of a VFD, it is fundamental to install it correctly. The following are some basic rules that must be taken into account:

9.4.1 Power Supply

In most cases, the VFD can be connected directly to the grid, however, there are certain conditions that must be taken into account in the installation, requiring the use of isolating transformers and/or power supply inductors called line reactors.

As the power grid suffers from frequent voltage fluctuations or power outages, an insulating transformer or line reactor can be used.

A problem related to the power supply is the existence of a non-permanently connected capacitor to correct power factor. A good alternative solution this problem is to place a line reactor. However, the disadvantage of placing it is the supply voltage decreases by approximately 2% and 3%.

It is recommended to install the line reactors to minimise transient overvoltages in the grid, reduce harmonics, improve power factor, reduce voltage

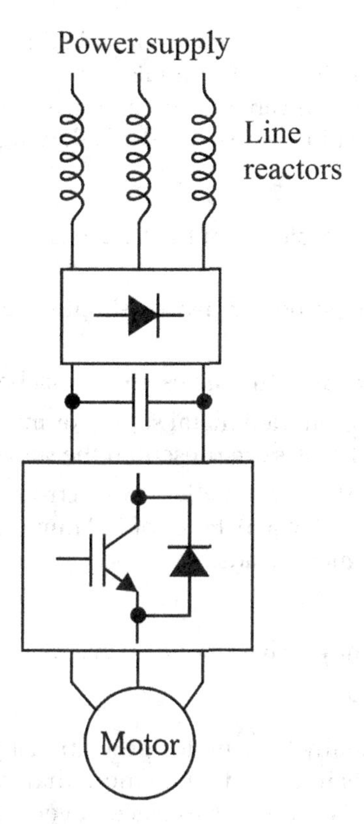

FIGURE 9.6
Line reactor application in the VFD.

distortion in the grid and, thus, increase the DC link capacitor's lifespan. Figure 9.6 illustrates the line reactor installation in a VFD.

RF filters may also be used at the VFD input to filter interference signals generated by the VFD itself, which will be transmitted over the network and may cause problems in other electronic equipment. Most VFDs already have a built-in input filter that avoids problems caused by electromagnetic interference (EMI).

9.4.2 Variable Frequency Driver Cabling

Particular care must be taken with the VFD wiring because the electrical signals transmitted by the cables can emit electromagnetic radiation and can also behave like antennas, absorbing radiation and damaging the VFD operation. To minimise this problem, shielded special cables are indicated to reduce interference. Grounding the VFD installation cabinet can also minimise this problem. If contactors are installed near it, it is important to install surge suppressors in their terminals.

Since the VFD connection cable to the motor is one of the major sources of electromagnetic radiation emission, the following precautions must be taken:

- Separate the power cables from the VFD from the signal and control cables.
- Shielded cables must be earthed or should a use grounded metal conduit.
- If it is necessary to cross the cables, they must be made at 90°.
- Insulation with a grounded metal separator must be provided if the power and control cables are present in the same cable tray.
- Place equipment that is sensitive to electromagnetic interference (PLC, controllers, etc.) at a distance of 250 mm from the motors, reactances, filters and motor leads.

9.4.3 Variable Frequency Drive Output Devices

9.4.3.1 Overload Relays

Although the VFDs usually have motor protection against overcurrent, the use of overload relays is indicated when more than one motor is driven by the same VFD, where it is necessary to place a overload relay on each motor.

Another care to be taken is to set the overload relay trip current to approximately 10% above the rated current of the motor to avoid tripping without reaching its tripping current, because the VFDs are switched to high frequencies, which may cause an occasional trip.

9.4.3.2 Load Reactor

There may be a leakage current due to the capacitive effect depending on the cable length. This effect must be considered in the VFD sizing or compensated for by an inductance in the VFD output called a load reactor.

To solve this type of problem, the use of a reactor between the VFD and the motor is indicated. It should be designed especially for high frequencies, as the output signals from the VFD can reach up to 20 kHz. Figure 9.7 shows a load reactor applied to a VFD.

With load reactor use, the leakage current is reduced, allowing the use of longer cable lengths without reducing the VFD power. Figure 9.8 shows the output reactance placed between the VFD and the motor.

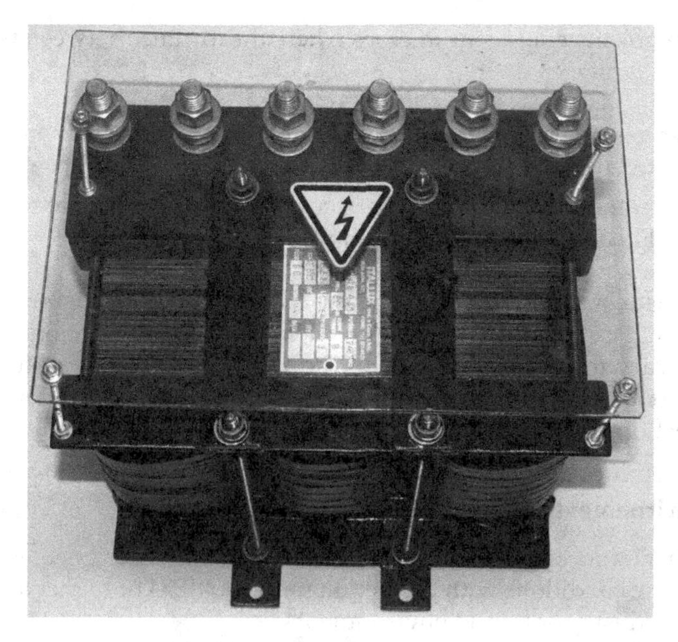

FIGURE 9.7
VFD load reactor.

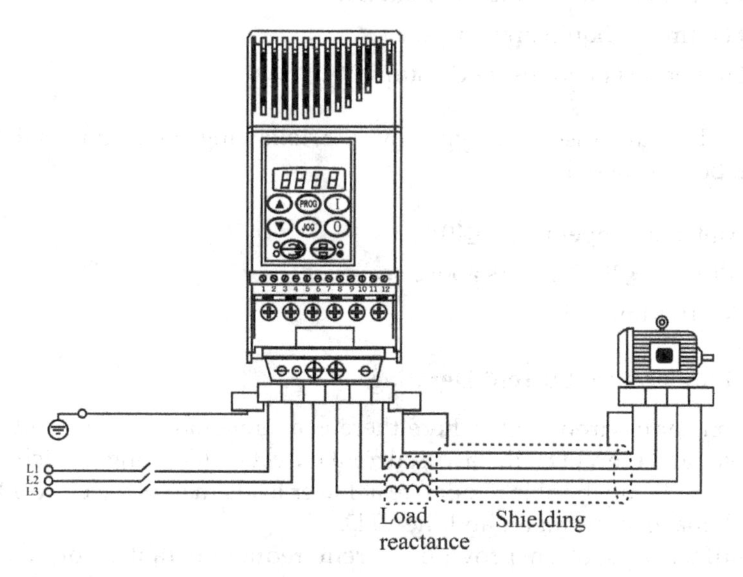

FIGURE 9.8
Output reactance placed between VFD and motor.

Other problems may occur in cases where the distance between the motor and VFD is large:

- Overvoltages in the motor produced by a phenomenon called reflected wave.
- Appearance of capacitances between the power cables that return to the VFD producing a "ground fault" effect, blocking it.

In some cases, a contactor is used before powering the VFD in order to prevent automatic starting of the motor after a power interruption. To do so, it is necessary to place a contactor in the VFD power supply or to perform some interlocking in its control. The use of the contactor also allows remote disconnection of the power supply to the VFD.

9.4.4 Environmental Conditions of the Installation

Induction motors controlled by VFDs are generally designed to operate under severe conditions with protection ratings of IP54 or higher and can be placed in industry or humid environments.

These VFDs are much more sensitive and should be placed in environments protected against:

- Powder and other abrasive materials.
- Flammable liquid and gases.
- High levels of atmospheric humidity.

When VFDs are installed, typically, the following environmental limits should be considered:

- Ambient temperature: ≤40°C
- Altitude: ≤1000 meters above sea level
- Relative humidity: ≤95%

9.4.5 Temperature Current Derating

In regions or environments where there is a temperature above 40°C, both the motor and the VFD will have a current reduction (derating), which means that they will only be able to operate at lower loads than those up to 40°C, to avoid damage to the motor and the VFD.

Manufacturers often provide current reduction tables for temperatures above 40°C. Figure 9.9 shows a typical temperature reduction table for a VFD.

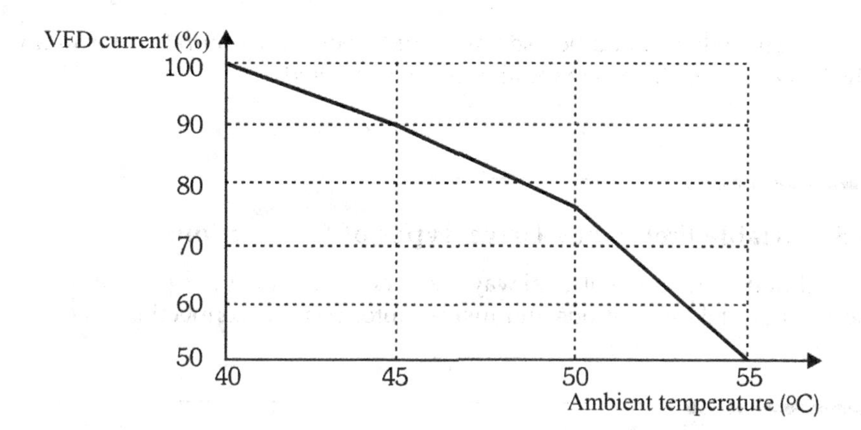

FIGURE 9.9
VFD current derating considering ambient temperature.

NOTE: This table should be used as a guide. For more accurate information, the VFD manufacturer's manual must be consulted.

9.4.6 Altitude Current Derating

At high altitudes, the electrical equipment cooling is reduced because the air becomes more rarefied, reducing the oxygen amount. According to the standards, typically VFDs are designed for altitudes up to 1,000 meters from sea level. Thus, if the VFDs are installed in environments with altitudes greater than 1,000 meters, there will be an output current reduction. Figure 9.10 presents a typical current reduction table due to altitude.

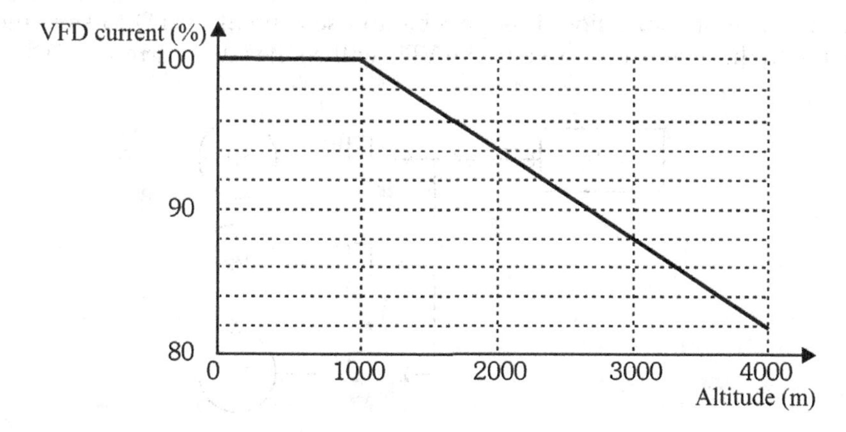

FIGURE 9.10
VFD current derating considering altitude.

NOTE: This table should be used as a guide. For more accurate information, the VFD manufacturer's manual must be consulted.

9.5 Variable Frequency Drive Types of Connections

In addition to the conventional way of connecting a VFD, it is possible to connect in parallel associations and inside motor triangle connection.

9.6 Motors in Parallel

In this type of connection, the VFD rated current must be greater than or equal to the sum of the rated currents of the motors to be controlled, that is to say:

$$I_r \text{ VFD} > I_r1 + I_r2 + \ldots I_rx \tag{9.2}$$

Care must also be taken with regard to thermal protection, with external thermal protection being preset for each of the motors. Figure 9.11 illustrates this connection.

9.7 Inside Delta Connection

In this type of connection, it is possible to use a smaller VFD to start the motor. In this way, the current in the VFD will be the rated current/1.73.

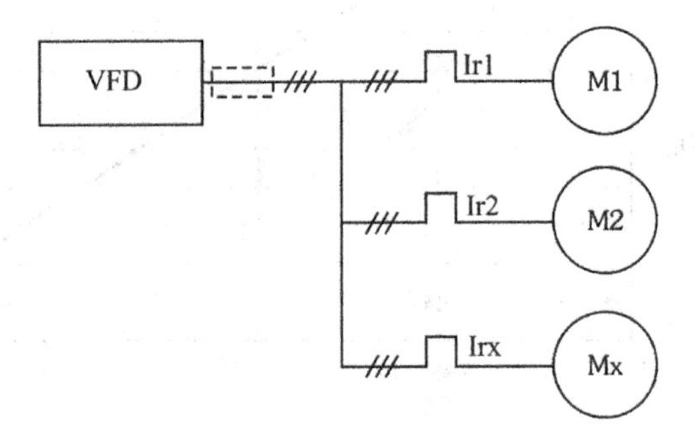

FIGURE 9.11
Motors associated in parallel with a VFD.

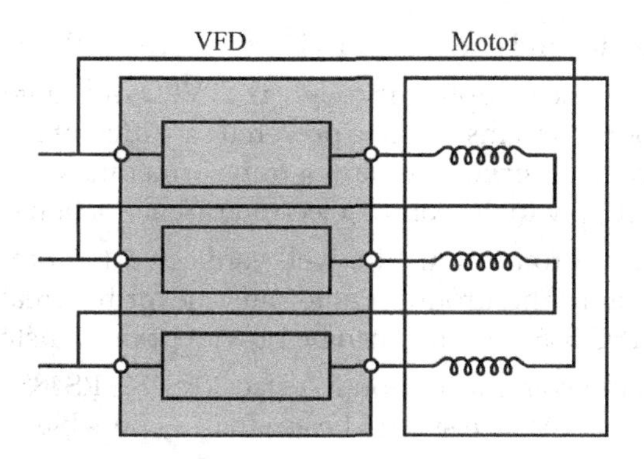

FIGURE 9.12
VFD inside delta motor connection.

Example: For a motor having 400 V and 100 kW with FLC = 195 A, it is possible to use a VFD with a rated current of 195/1.73, that is, 114 A. Figure 9.12 shows the VFD inside delta connection.

9.8 Good Practices for Installing VFDs

For the right operation of the VFD, the following basic precautions must be observed:

- All metal parts of the panel must be connected with firm contacts and wide contact areas.
- Care must be taken with the panel door, which must also be connected to the panel structure through flexible copper.
- Signal and power cables must be mounted separately, with a minimum distance of 20 cm.
- If unshielded cables (e.g. temperature sensor) of the same circuit are used, they must be twisted in order to avoid antenna effect.
- Unnecessary wiring lengths should be avoided to reduce coupling capacities and inductance.
- When backup cables are used, the backup wires must be ground at both ends.
- Interference couplings can be shortened by wiring the cables to the side of grounded plates. Due to this fact, cables should not be passed through the panel randomly, but should be close to the panel structure or to the mounting structures.

- Tachogenerator and encoder must be connected with shielded cables.
- Avoid interference coupling through grid. VFDs and automation/control electronics systems must be powered by different grids. If there is only one available grid, an isolating transformer must be provided for the power supply to the control electronics/automation system.
- The electrical ground must be well connected to both the inverter and the motor. The ground value must never be greater than 5 Ω (standard IEC 536), and this can be proven prior to installation.
- If a VFD has a communication interface (RS232, RS485 or industrial network) the cable size should be as small as possible.
- Care must be taken with the VFD cooling system which must be housed near ventilation "holes" or, if the power is too high, it must be ventilated (or exhausted), since some inverters already have a small internal exhaust fan.
- If there are contactors and coils added to the VFD operation, always use noise suppressors (RC circuits for AC coils, and diodes for DC coils). These precautions are not intended only to improve the VFD operation, but to prevent it from interfering with other equipment around. The VFD is, unfortunately, a large electromagnetic interference (EMI) generator, and if it is not installed according to the previous guidelines, it may harm the machine (or system) around it. For equipment to meet the standard certification, electromagnetic emission is required to reach very low levels.

9.9 Harmonics Generated by Variable Frequency Drives

VFDs generally use non-linear devices, such as diodes and transistors, to convert an AC signal into a DC. The rectifiers generate a non-sinusoidal current and distort the voltage in the power system, therefore causing additional losses and are the major source of electromagnetic interference. Harmonic distortion can be seen as a type of electrical pollution in a power system, which is worrying, because it can affect other equipment connected to the system. The source and intensity of harmonic distortion must be clearly understood in order to address this problem with practical results.

9.9.1 Harmonic Definition

Considering a fundamental frequency of a 60 Hz distribution system, a harmonic frequency is the sinusoidal which is a multiple of a fundamental

TABLE 9.3

Harmonics of the Fundamental Frequency (60 Hz)

Even Harmonics	Odd Harmonics
2nd harmonic = 120 Hz	3nd harmonic = 180 Hz
4th harmonic = 240 Hz	5th harmonic = 300 Hz
6th harmonic = 360 Hz	7th harmonic = 420 Hz
8th harmonic = 480 Hz	9th harmonic = 540 Hz

frequency. Harmonic frequencies can be odd or even multiples of the fundamental frequency. The fundamental frequency multiple is called the harmonic order. Examples of harmonics of the fundamental frequency of 60 Hz are in Table 9.3.

A linear load is one that provides a pure sinusoidal current when connected to a sinusoidal voltage source, such as, resistors, capacitors and inductors. Many of the traditional devices connected to a power distribution system, such as transformers, electric motors and resistive heaters, have linear characteristics.

A non-linear load is one that provides a non-sinusoidal current when connected to a sinusoidal current source, such as a diode bridge, thyristor bridge and others. Many power electronics devices, such as VFDs and rectifiers, have non-linear characteristics and result in a non-sinusoidal current waveform or distorted waveform, as shown in Figure 9.13.

9.9.2 Harmonic Distortion Analysis

The technique used to analyse the harmonic distortion level of a periodic waveform is known as Fourier analysis. This method of analysis is based

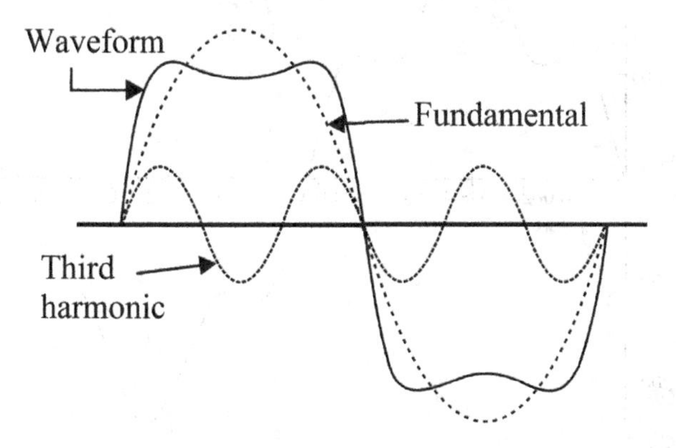

FIGURE 9.13
Distorted CA form: fundamental + 3rd harmonic.

on the principle that a periodic distorted waveform is equivalent and can be replaced by the sum of a number of sine waveforms. See below:

- A sinusoidal waveform at the fundamental frequency (60 Hz).
- A number of other sine waveforms of higher order harmonic frequencies that are multiples of the fundamental frequency.

This process of finding the frequency components of a distorted waveform is mathematically obtained by a technique known as Fourier transform. To obtain this transform, equipment employing microprocessors is used to analyse harmonics in a real-time method known as Fast Fourier Transform (FFT).

The harmonic currents are generated by the nonlinear loads connected to the grid. A load is called nonlinear when the current it absorbs does not have the same shape as the voltage that supplies it, such as a VFD.

Figure 9.14 is similar in shape to a sinusoid, though it is somewhat distorted. In fact, the waveform is composed of a set of sinusoids with different frequencies and amplitudes, as shown in Figure 9.15.

In Figure 9.15, it is possible to note a sine wave of greater amplitude, with a known fundamental, with a peak of 100 V and a frequency equal to 60 Hz.

As for the other sinusoids, one has a peak voltage of 20 V and a 300 Hz frequency (fifth harmonic), and the other has a peak of 14 V and a 420 Hz frequency (seventh harmonic). The more complex the signal, the greater the number of harmonics that form this signal. There are also even harmonics

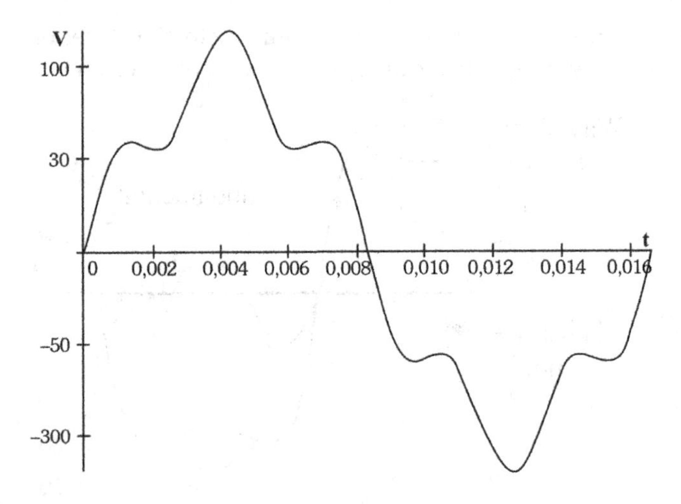

FIGURE 9.14
Example of distorted waveform.

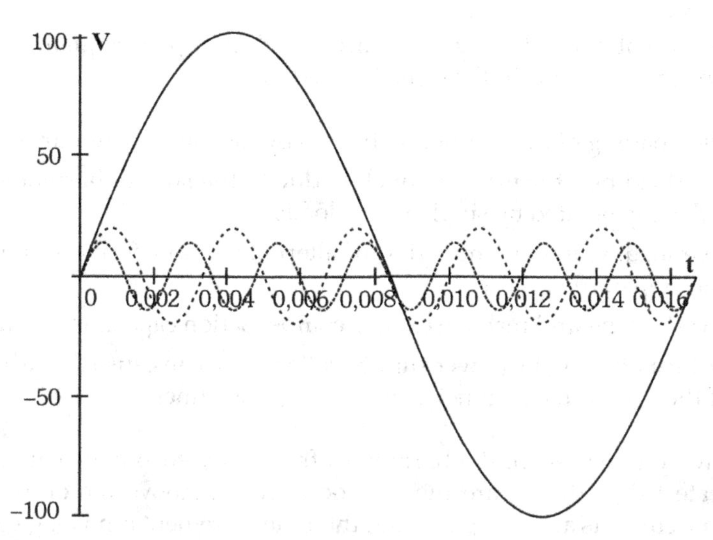

FIGURE 9.15
Composition of the wave with fundamental frequency and harmonics.

(fourth, sixth, etc.), but they appear very rarely and also have a very low value, not influencing the final value.

The presence of harmonics in voltage or current is a deformed synonymous waveform. The Fourier Theorem indicates that any non-sinusoidal periodic function can be represented as a sum of expressions (series), which is composed of a sinusoidal expression in a fundamental frequency of sinusoidal expressions whose frequencies are integral multiples of the fundamental (harmonics) and of a possible continuous component.

Rectifiers operate in a way that only transfer power to the output capacitor bank when the grid voltage is higher than the capacitor bank voltage, causing a distortion in the current.

The rectifier harmonics (η) can be calculated according to the following equation:

$$\eta = P \cdot k \pm 1 \tag{9.3}$$

where:

P is the pulse number of the rectifier bridge
k is the assumed values 1,2,3...

A three-phase six-pulse rectifier bridge has the harmonic presence of 5a, 7a, 9a etc. In the case of a 12-pulse rectifier bridge, it would not have the 5th and 7th.

Harmonics are undesired in the grid, because only the fundamental performs work, the other harmonics only cause losses in the system. The harmonics that circulate in grid reduce the electric power quality, causing the following problems in the electrical system:

- Overloading of distribution networks by increasing rms current.
- Overloading of neutral conductors due to the sum of harmonics of order 3 generated by single-phase loads.
- Overload, vibration and aging of alternators, transformers, motors, and transformer noise.
- Overloading and reactive energy compensation capacitors aging.
- Deformation of the power supply voltage that can cause disturbance of the communication networks or telephone lines.

In a three-phase system, the harmonics frequently encountered that cause the greatest disturbance are those of odd orders. Above the order 50, the harmonic currents are negligible, and their measurement is no longer significant. Attention must be paid to harmonics up to 30.

A compensation of the harmonics up to the order 13 is fundamental for a good quality power supply system, since good compensation takes into account the harmonics up to order 25.

9.9.3 Total Harmonic Distortion (THD)

Total harmonic distortion (THD) is called the global harmonic distortion rate. It is a notation widely used to describe the impact of the harmonic content of an alternating signal. The THD in voltage characterises the waveform deformation.

Thus, the THD can be defined as the relation between the value of the fundamental frequency measured at the output of a transmission system and the value of all harmonics through the following equation:

$$\mathrm{THD} = \sqrt{\frac{a_2^2 + a_3^2 + a_4^2 + \ldots + a_N^2}{a_1^2}} \times 100\% \tag{9.4}$$

where a_2, a_3, a_4, a_N are the harmonics and a_1 is the fundamental.

A value less than 5% may be considered normal. Values between 5% and 8% show a signal with a significant harmonic distortion. Being a value greater than 8%, it shows a fairly high harmonic distortion.

Care must be taken when the voltage of the power supply has a distortion rate close to 10%, as it significantly affects the life of the equipment. It is estimated that the reduction in equipment lifetime is approximately 18% for three-phase machines and approximately 5% for transformers.

The harmonics also produce overloads in the grid, making it necessary to increase the power for a given VFD—that is, oversizing the installations.

9.9.4 Effects of Harmonic Frequencies on Equipment

Harmonic currents cause voltage waveform distortion, affect the performance of other equipment and produce additional losses and heating. For example, a voltage harmonic distortion of 2.5% can cause a temperature rise of 4°C in induction motors. In cases where resonance can occur between the capacitance and reactance of the system, the voltage distortion may be even greater.

Capacitor banks (used for power factor correction) are elements that are very vulnerable to the effect of harmonics. They have a low-impedance path for high-frequency currents. This increases the dielectric losses in the capacitor bank, which can lead to overload and eventual failure.

Transformers, motors, cables, busbars, and others must be over-sized to accommodate the additional harmonic currents and the extra losses associated with harmonic frequencies.

9.9.5 Selecting the Switching Frequency

Most frequency inverters have a parameter that allows the selection of the switching frequency. If a higher output frequency is selected, the trend is a reduction of the VFD noise. The higher the frequency selected, the greater the loss of leakage currents.

Selecting the PWM switching frequency is a compromise between motor losses and VFD losses.

- When the switching frequency is low, the losses in the motor are raised because the shape is less sinusoidal.
- When the switching frequency increases, the losses in the motor are reduced, however, the losses in the inverter will be increased due to the switching increase internally in the VFD. The losses in the motor cables also increase due to the leakage current through the cable capacitance. To minimise this effect, VFD manufacturers provide a table or graph of how much the VFD should be over-sized, similar to the graph in Figure 9.16.

9.9.6 High *dv/dt* Rates in Variable Frequency Drives

High switching frequencies, using modern devices such as IGBTs in the inverter, achieve sinusoidal current with small distortion and reduce motor

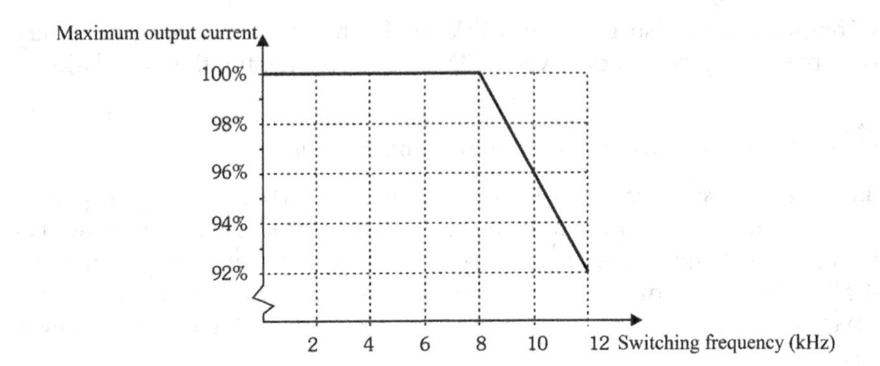

FIGURE 9.16
VFDs derating currents factors due to high switching frequencies.

losses. While the closest sinusoidal waveform reduces thermal losses and the audible noise in the motor, the fast voltage change can introduce many other problems. The voltage rise rate (*dv/dt*) may be greater than 10 kV/μsec with an IGBT. Some of the problems caused by this high *dv/dt* are:

- High electrical stress of cables and motor insulation due to high voltage growth rate (high *dv/dt*) and signal reflections at the end of the motor cable.
- High radiant electric field, due to high *dv/dt*, can exceed EMI standards implemented in Europe and the USA.
- High *dv/dt* through cable capacitance results in leakage currents on the shielded cable (if placed) or other conductive paths to ground.

Leakage currents generate additional heating in the VFD or exceed the current limit in small VFDs, which usually result in the acting of protection and its shutdown.

The most significant impact of high voltage rise rates includes the high voltage peaks that occur due to the reflected wave at the end of the motor cable.

These voltage peaks can reach up to two to two-and-a half times the DC bus voltage. This reflected wave phenomenon is widely known in communication cables operating at similar frequencies. In communication cables, the main problem is interference caused by the reflected signal. Increasing the voltage signal causes no problem, because the voltage signals are too low.

In VFDs, the high voltage caused by signal reflection in the motor at the end of the cable can damage the motor insulation and eventually lead to a short circuit.

This occurs due to the following mechanism:

The cable between the VFD and the motor represents an impedance, which has resistive, inductive and capacitive elements. The cable has its own impedance for the voltage pulses generated by the VFD PWM. If the motor impedance does not match the impedance of the cables, a partial or complete reflection occurs at the motor terminals.

The maximum reflected voltage amplitude depends on the speed of the voltage pulse, the time of the voltage rise (dv/dt) and the length of the cable between the VFD and the motor. The switching time depends on the switching device. With IGBTs, which have a short switching time (50–500 ns), the length of the cable at which the reflected voltage increases is much smaller than a BJT (0.2–2 µs) or a GTO (2–4 µs), since the latter have large switching times, allowing longer cable lengths.

The most critical cases of reflected voltage amplitude can be two to two-and-a half times the DC voltage of the inverter bus. For a nominal 415 VAC source of a VFD, the voltage on the bus will be approximately 600 V, which means that the reflected wave at the motor terminals can reach up to 1.5 kV.

The voltage peaks can be much larger than the insulation level of the electric motor, causing stress in the insulation of the dielectric and possible motor failure.

Even if the voltage peak is not large enough to damage the insulation, there will be parasitic currents in the windings, which cause hot spots that can accelerate the process of degradation of the insulation of the motor, especially in older motors.

NOTE: It is important to understand that this reflection occurs regardless of the switching device type (IGBT, TJB, MOSFET, GTO, etc.).

9.9.7 Protection of Motors against High Switching Frequency

The use of high switching frequencies with modern techniques in VFDs enables a current very close to the sinusoidal, reducing the harmonic currents and the losses in the motor, reducing the audible noise. These are desirable characteristics of modern VFDs.

New problems have arisen due to the appearance of leakage currents in the cables and insulation damage to the motors due to the high switching frequency.

From the motor point of view, the best solution is to use motors whose insulation can withstand the peak amplitudes of the reflected signal. Many motors manufacturers have recognized that motor insulation levels must be increased to operate with VFDs.

TABLE 9.4

Maximum Recommended Cable Lengths according to the Switching Device

	Maximum Voltage Allowed in Motor Windings	
Switching Device	**1.000 V (Peak)**	**1.200 V (Peak)**
IGBT (0.1 μs)	15 m	25 m
BJT (1 μs)	180 m	220 m
GTO (4 μs)	700 m	850 m

As the reflected voltage signal amplitude is dependent on the motor cable's length, it should be kept as short as possible and the installations planned to minimise the cable's length. Table 9.4 gives an idea of acceptable lengths for the different types of switching devices for insulation levels of 415 V induction motors.

If the length of motor cables needs to be longer than those recommended in Table 9.4, there are some solutions that can be used to reduce the effects of reflected voltage and, consequently, increase the lifespan of the starting device. They are:

Motor output filters: Special harmonic filters that comprise R, L and C components can also be used similarly to the output reactor described earlier to protect the cables and the motor. The filter can also be designed to reduce EMI on the motor cable. The filter achieves this by changing the impedance conditions, so the EMI is diverted to the ground and directly to the source. The filter is basically set to a low series impedance value and provides a high impedance for the high frequency, similar to the output reactor, with some additional components.

The use of shunt capacitors on the VFD is restricted because of the effect on the VFD performance. These filters have thermal losses, so the filter losses must be added to the VFD losses when determining the cooling needs in a panel. In addition, the filter must be grounded in the same grounding bar of the panel.

Terminators on the motor terminals: On the communication cables, the reflected voltages can be attenuated by the connection of a terminator at the end of the cable. A similar solution can be used with the motor cable. A terminator, comprising an RC circuit, connected to the motor terminals, may be designed to keep the voltage peak below a potentially destructive level. In comparison to output reactors and filters, the terminators occupy a small space, dissipate small

TABLE 9.5
Maximum Cable Lengths with IGBT Inverters

Protection	Maximum Motor Cable Length (meters)
No protection	10–50
Load reactor	30–100
Motor filter	60–200
Motor terminator	120–300

power and cost less than 10% of the filter value. In addition, terminators can be used on each motor in an installation that has multiple motors in the same VFD.

Table 9.5 illustrates the maximum cable lengths with converters using IGBTs with the solutions discussed above.

Changes in motor cable lengths depend on the peak voltage level that withstands insulation of the motors.

9.9.8 Conclusions about High Switching Frequencies in Variable Frequency Drives

Although the subject of reflected waves in VFDs is extremely important, there are many VFDs operating successfully without any additional protection. This does not mean that voltage reflections do not exist. Not all VFD applications suffer from this problem. Users should be aware of this potential problem that may occur in the system in a way that minimises its effects. Figure 9.17 shows the protections that can be used to improve VFD performance.

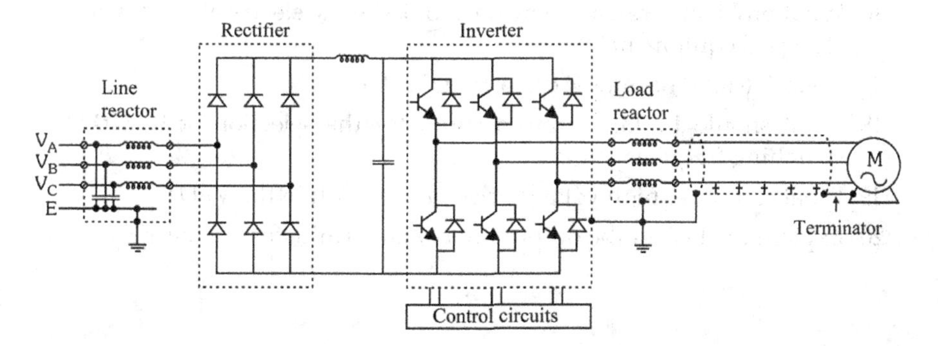

FIGURE 9.17
Protections used to improve the VFD performance.

Exercises

1. What can happen to the VFD if there is under-voltage in the power supply?
2. List the main causes of overvoltage in the VFD.
3. Why is overcurrent protection recommended to implement in hardware even in digital VFDs?
4. Describe the operation of how the earth leakage protection is implemented in a VFD.
5. How is the heat sink over-temperature protection made in VFDs?
6. What methods are used to implement overload protection in a VFD?
7. List the most common internal and external faults that can occur in VFDs and their corresponding protections.
8. Why is the installation of line reactors in VFDs recommended?
9. What precautions should be taken to reduce electromagnetic radiation emission in VFDs?
10. When is it recommended to install overload relays in a VFD?
11. What is the purpose of installing a load reactor in the VFD?
12. What is the influence of temperature and altitude on the VFD operation?
13. Is it possible to simultaneously drive more than one motor in the same VFD? How is it possible?
14. What is the advantage of inside delta motor connection?
15. Present the definition of harmonics and why the VFD is considered harmonics generator equipment when it drives a motor?
16. What problems are caused by harmonics in the electrical system and electrical equipment?
17. Conceive total harmonic distortion (THD).
18. What should be taken into account for the selection of the VFD switching frequency?
19. What can cause high voltages rise rates (dv/dt) in the VFD?
20. Explain the task of the output filters and terminators in the VFD.

10

Variable Frequency Drive Sizing and Applications

10.1 Variable Frequency Drive Sizing

Although manufacturer catalogues tried to make VFD sizing as easy as possible, there are many variables associated with selecting the motor and VFD. In most cases, considerable experience is required to make the correct VFD selection. The difficulty is finding the best cost/benefit ratio according to the following criteria:

- The need to use a safety margin in the selection procedure.
- The need to maintain an initial minimum cost by selecting the correct motor size and VFD for each application.
- Starting torque need.
- The speed range (maximum and minimum values).
- Compatibility with grid voltage.
- Ambient conditions where the motor and VFD are installed: ambient temperature, altitude, humidity, dust, etc.
- VFD and motor ventilation.
- Accuracy required for speed control.
- Dynamic response (torque and speed response needs).
- The duty cycle, including the number of starts and stops per hour.
- Control method: manual, automatic.
- VFD protection features.
- Cabling requirements for control and power.
- Maintenance and spare parts.

Besides the mentioned items, it is important to verify the characteristics that are often neglected by the designers, as described below:

- **Type of load to be driven**: It is fundamental to know the type of load to be driven, since the dynamic behaviour of these machines makes a considerable difference. For example, a compressor has a considerably different load curve than a fan.
- **Types of command**: It is fundamental to define how the VFD will be activated, for example, digital or analogue inputs, HMI, industrial network, etc.
- **Stop type**: Check the appropriate type of stop for the VFD (inertia, ramp type or braking).
- **Communication**: Check if the VFD has the type of communication available to integrate with an industrial network such as DeviceNet, ControlNet, ethernet, Modbus, etc.
- **Electromagnetic noise emission**: Verify if the VFD is able to operate within acceptable limits of electromagnetic noise, or if external filters are needed to attenuate noise.
- **Harmonics existence**: Evaluate if the value of the THD is high, if it can compromise the VFD power supply.

10.2 Basic Procedure for VFD Selection

Experience has shown that most of the problems in VFD sizing can be assigned to human errors, mainly:

- Incorrect electric motor selection.
- Incorrect VFD selection.
- Incorrect VFD parameters setting.

As with any electrical equipment, it is essential that the VFD is correctly specified for the application, taking into account all possibilities of operation.

The VFD selection must be done from a manufacturer's catalogue. In these catalogues, the VFDs are usually specified in terms of current (not kW) at a defined voltage.

For example, consider the motor with the following characteristics: 1HP/380 V and power factor of 0.8 in a industrial fan application:

The VFD current (IVFD) could be obtained from:

$$\text{IVFD} = \frac{\text{active power}}{\text{grid voltage} \times \cos\varphi}$$

$$\text{IVFD} = \frac{746 \text{ watts}}{380 \times 0.8} = 2.45\,\text{A}$$

In applications where the VFDs are very large, it is economically feasible to use higher voltages around 13.8 kV to reduce cabling costs. Although the output frequency is variable, the input frequency must be clearly determined so that there are no problems with the inductive components.

Generally, the VFD selection is emphasised, which is the highest cost part of the system, with little importance being given to the motor. However, it is first necessary to specify the motor correctly and then make the VFD selection.

10.2.1 Motor Selection

The electric motors and VFDs manufactures have developed several methods for fast selection of electric motors for a given application. Currently, the selection is made based on specific software for this purpose. However, it is important for engineers and designers to clearly understand the selection procedure. One of the best selection procedures is based on the limits of the load curves to make a basic selection of the motor size. This procedure is described below.

First, the type and size of the motor must be selected. The number of poles, which determines the base speed, must be chosen so that the motor runs at a speed as close as possible to the grid frequency, due to the following factors:

- Maintain efficient ventilation.
- Although many manufacturers offer VFDs with output frequencies up to 400 Hz, these high frequencies have few practical applications, except for very special ones. The construction of cage motors and the torque reduction in the field weakening region restrict their use to frequencies up to 100 Hz. Another important point to be highlighted is the maximum speed the motor can withstand on its axis. The noise produced by the fan also increases substantially as the motor speed increases.

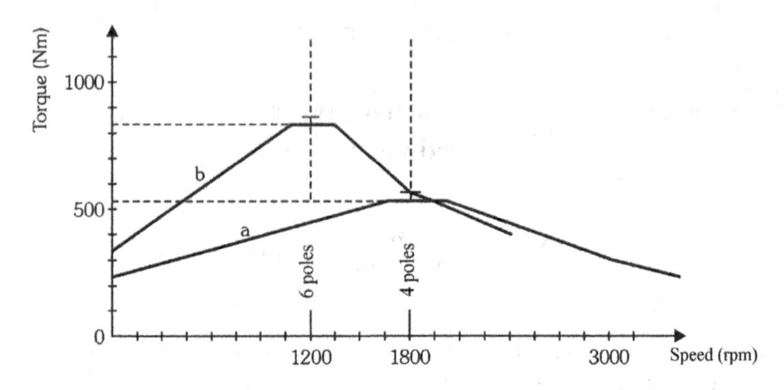

FIGURE 10.1
Comparison between speed by torque to four- and six-pole motors.

- An important issue to consider is the torque produced by the motor. Figure 10.1 shows a comparison of torque produced by a four-pole and another by six-pole motor. It may be noted in the figure that the six-pole motor has a torque greater than the four-pole torque. In this way, it is possible to conclude that the greater the number of poles of the motor, the greater its torque.

The selection of a very large motor is not advisable because the VFD must be oversized, which is developed for the highest current peak, which is the sum of the fundamental current and the harmonic currents in the motor. The higher the motor, the higher the peak currents. To prevent this peak current from exceeding the VFD limit, it should never be used with a larger motor than the one for which it was specified, since, even with large motors with low load, their harmonic currents are high.

We can summarise with the initial data the selection of a motor for the drive with a VFD, as shown below:

- Power supply voltage and frequency.
- The starting torque (N·m).
- The load torque and its change with speed.
- The speed range (rpm).
- Acceleration time.
- Moments of inertia of the motor and load.

The choice of the number of motor poles determines the motor synchronous speed and is selected according to the maximum speed requested by the application.

Motor power selection: Depending on the load torque requirements, the rated motor power can be selected from a manufacturer catalogue using the following equation:

$$\text{Power(kW)} = \frac{\text{torque(N}\cdot\text{m)} \times \text{speed(rpm)}}{9550} \tag{10.1}$$

NOTE: Consideration should be given to the motor power derating due to heating by harmonic currents, the reduction of ventilation at low speeds and the torque reduction at high speeds.

10.3 Variable Frequency Drive Selection according to the Type of Load

When an AC motor is selected for a particular application, the most important issue is to ensure that the motor will not overload or lock in all speed and load conditions, that is, across the entire speed range. To remain within the motor temperature rise limits, the load torque required for starting, accelerating and working at nominal speed must be within the rated motor torque.

For AC motors that are started direct on line, it is sufficient to ensure that the load torque is less than the motor torque at a given speed. For this type of start, this procedure is made easier, since the starting methods work only at one speed. For these methods, it is also important to ensure that the starting motor torque is greater than the torque required to accelerate the load.

In the case of VFDs that have variable speed, the load torque usually changes with speed. It is vital to check that the motor torque exceeds the load torque in all speed ranges. For example, a centrifugal pump has a variable torque characteristic, where the torque is low and increases with the speed square. For other loads, such as conveyor belts, the torque is constant for all speeds. The squirrel cage motor torque is generally less than the rated torque in the following situations:

In all speed ranges: The motor torque capacity is reduced as a result of additional heating caused by harmonic currents. This is because the current form in the VFD is not perfectly sinusoidal, even in modern devices that have switching frequencies in their PWM of approximately 10 kHz. Traditionally, a factor between 5% and 10% is used, depending on the type of motor (number of poles) and VFD type. However, it has become common practice for modern VFDs not to use reduction factors, because current motors have an internal thermal reserve that compensates for any additional heating.

Also, the mechanical loads are rarely made exactly with the torque supplied by the motor, being generally placed 20% smaller than the motor capacity. It is considered good engineering practice to allow a small margin of safety, so a safety factor of up to 5% is generally considered.

At speeds below nominal: Between 0 and 60 Hz, the motor continuous torque capacity is reduced because of the reduced stator and rotor ventilation. The output motor torque depends on the type and size of the motor; however, in the absence of a table with reduction factors, a reduction of approximately 40% of the nominal torque can be assumed.

For applications requiring constant torque, auxiliary ventilation can be used to increase ventilation in the motor and, consequently, increase its torque capacity at low speeds. Supplementary ventilation does not entirely solve the problem of torque loss. In a squirrel cage motor, the losses in the rotor are generally larger than in the stator, and losses in the rotor become difficult to dissipate at low speeds, even with supplemental ventilation. External ventilation is more efficient with open motors.

At speeds above nominal: Motor output torque capacity is reduced due to field weakening. The output torque is reduced in direct proportion to the motor speed above 60 Hz.

The load curve shown in Figure 10.2 summarises the previously mentioned elements and the solid line defines the maximum limits

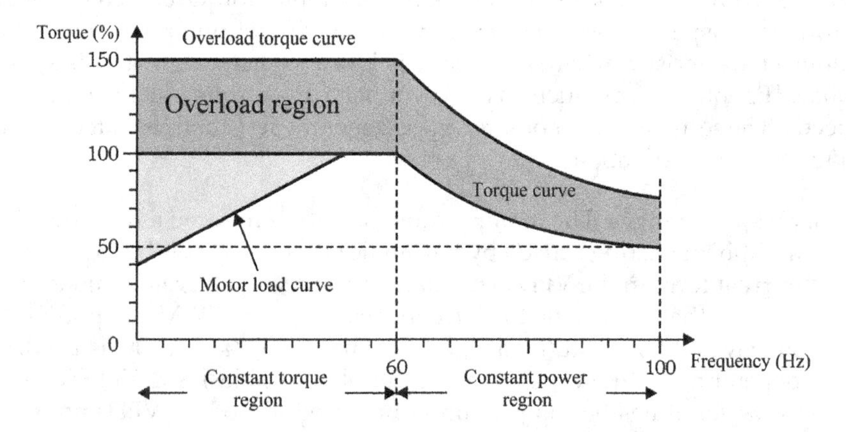

FIGURE 10.2
Torque by frequency curve for induction motor.

of continuous torque for the load. Motors driven by VFDs can drive loads continuously with torques below the load curve limit line for the motor speed range.

Motors can withstand torques larger than those allowed by the load curve for short periods of time. High torques are usually requested during start and acceleration. The duration of the overload depends on many factors, such as motor size, overload size and speed. Many VFDs have an up to 150% overload capability, up to 60 seconds, to support starting transients and sudden load variations.

The curve in Figure 10.2 shows the typical load capacity of a VFD. The curve is based on squirrel cage motors drived by VFDs according to international standards. The curves are given in percentage values, which can be applied on motors of any voltage and size. Motors below 5.5 kW have a slightly higher capacity at low speeds.

The equivalent load capacity (kW) curve is shown in Figure 10.3. In the region called the base speed (60 Hz), known as the constant torque region, the power capacity increases linearly from zero when the motor is stationary, for maximum power up to the base speed. Above the base speed, the output power cannot increase further and remains constant for speed increases. This region is called the constant power region.

As with a DC motor, the output torque is proportional to the flux and the stator current product; the VFD generates an output voltage with a constant V/F ratio to produce a constant flux in the region between zero speed and 60 Hz, which produces a constant torque characteristic in the region between zero and the base frequency and the power amplification with increasing speed.

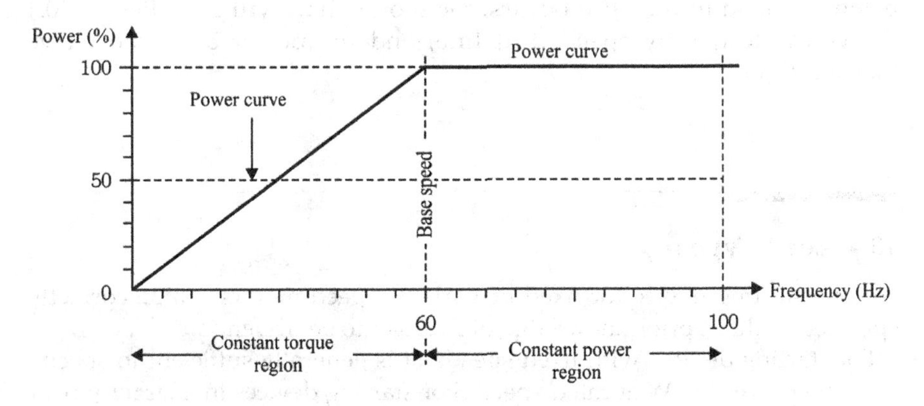

FIGURE 10.3
Power capacity of an induction motor controlled by VFDs.

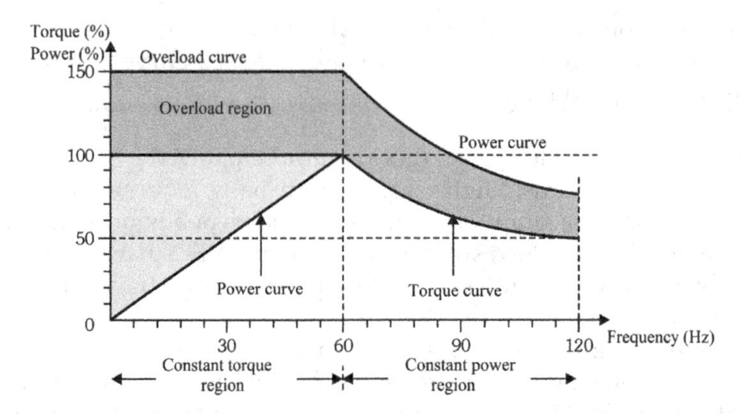

FIGURE 10.4
Torque by speed and power by speed curves for induction motors coupled to VFDs.

It is possible to increase the VFD output frequency in addition to the base frequency. On some models, you can set this frequency up to 400 Hz. At speeds above the base speed, the output voltage remains constant up to a maximum level. Consequently, the V/F ratio falls in a ratio inverse to the VFD frequency, and the motor output torque decreases at the same proportion of the flux falling. In this region, although there is a torque reduction, the output power remains constant. The power remains constant due to the fact that the power is the product of torque and speed.

The main effect of this base speed operation is the motor output torque reduction in direct proportion of the speed increase. In this region, care must be taken to ensure that the motor torque does not decrease below the load torque. If it occurs, the motor shaft will lock. Figure 10.4 shows the torque by speed (full line) and the power by speed curve (dashed line).

10.4 Load Types

No one starting device for fixed or variable speed motors can be correctly specified without prior knowledge of the load to be driven.

For starting devices with fixed speeds, it is generally sufficient to specify only the power in kW at rated speed. For starting devices in a larger power motor, the motor manufacturer usually requests more information about the load, such as moment of inertia, to ensure that the motor design meets the needs of load acceleration. In the case of starters with variable speed, more details on the characteristics of the loads are needed. The output torque of an AC motor shall be sufficient to:

- Overcome the resistant torque of a load to be driven.
- Accelerate the load from its zero speed to the desired speed with the acceleration time required by the process.
- Have a torque greater than the load by a suitable margin during continuous operation at any speed in the range of possible speeds on any conditions.
- The motor current does not exceed the thermal specifications of the electric components and remains below the load curve during continuous operation.

This selection procedure applies only to a single motor coupled to a VFD without special requirements. Multiple motor drives and other special applications require more accurate study, requiring consultation with manufacturers for better design.

There are a variety of loads commonly driven by VFDs, each with different characteristics such as: torque, inertia, etc. For most applications, it is necessary to know the following load characteristics:

- The load torque, type, amplitude and torque characteristics of the load connected to the motor shaft.
- Speed range, maximum and minimum load speeds.
- The inertia of the motor and the load connected to the motor shaft.

10.4.1 Load Torque Characteristics

The torque requested by a load determines the size of the motor, whose torque must always be greater than the torque required for the load.

A rule of thumb is that the load torque determines the cost of the motor, which is directly proportional to its output torque. The load torque is not necessarily a fixed value. It can vary according to speed, position, angle and time.

Another important aspect of the load torque is that it is supplied on the motor shaft. When couplings, such as gears and pulleys, are involved, the actual torque of the machine must be converted to the motor shaft torque.

It is fundamental to know the type of load to be driven to correctly specify the VFD. The following are some of the most commonly encountered load types in the industry.

Variable torque loads: Exhibit variable torque in their full speed range, such as centrifugal pumps and fans. The torque-speed curve for these loads is shown in Figure 10.5, where n is the speed and k is a constant that depends of the load type.

The VFD used for speed control of pumps and fans is one of the simplest applications, however, it presents some inconvenience, because the starting torque is usually very low and increases with increasing speed.

The current requested at the start is low, so the overload protection does not act or takes time to act at the starting time of the motor.

VFD manufacturers tried to reduce this effect for these applications by inserting special functions for these start up types, such as a low range for overcurrent protection performance of typically 120% for 30 seconds.

Constant torque: Have constant torque throughout the speed range, such as conveyor belts, positive displacement pumps. The torque curves of this type of load are shown in Figure 10.6.

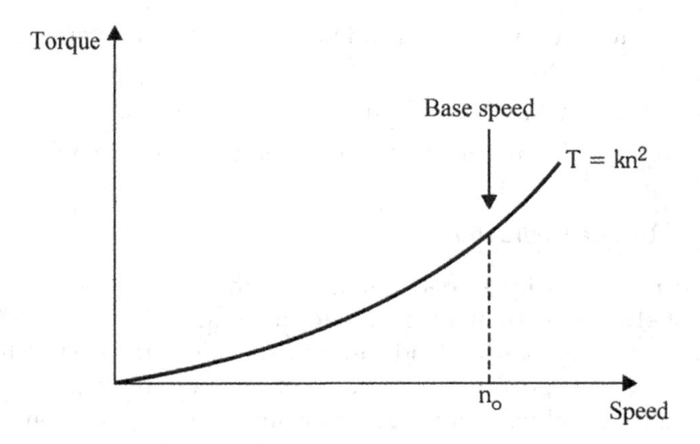

FIGURE 10.5
Characteristic torque by speed of a load with variable torque.

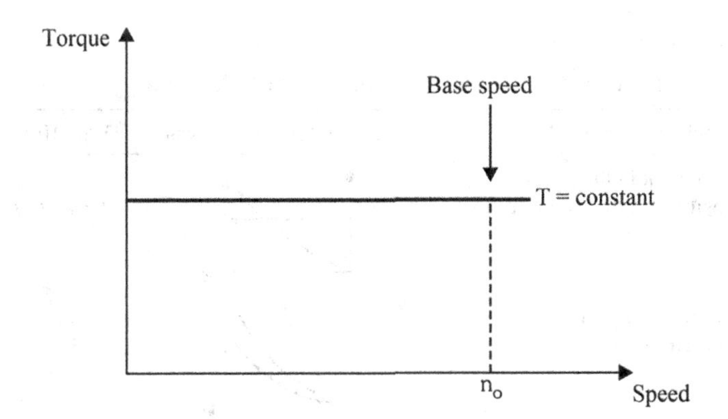

FIGURE 10.6
Characteristics of torque by speed of a load with constant torque.

For this type of load, the following problems can be presented:

- Starting torque is theoretically equal to full-load torque, but, in practice, real torque may be much higher due to greater needs to overcome inertia of load, and need for acceleration torque.
- If the motor runs for long periods at low speeds, thermal overload may occur. External ventilation may be necessary in some cases.
- Running at speeds above the motor base speed can cause excessive slip, with the possibility of the motor shaft locking.

Manufacturers have tried to overcome these problems by implementing the following features:

- High overcurrent protection capability in a short time, typically up to 150% for 60 seconds.
- Increase in voltage to compensate the voltage drop in the stator at low frequencies.
- Provides adequate protection against overload through inputs for protection thermistors placed in motor windings and advanced systems employing models that provide cooling reductions at low speeds. Table 10.1 summarises the types of loads most commonly encountered in industrial applications, along with their equations and characteristic torque and power curves.

TABLE 10.1

Typical Torque Curves of Different Types of Machine

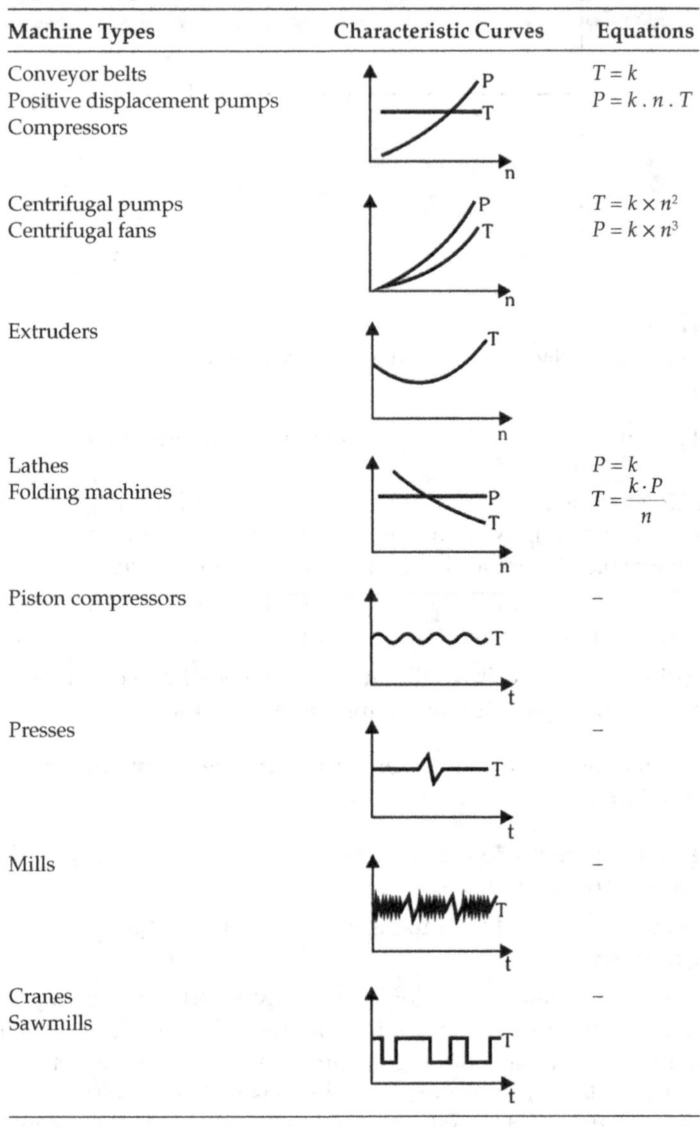

Machine Types	Characteristic Curves	Equations
Conveyor belts Positive displacement pumps Compressors		$T = k$ $P = k \cdot n \cdot T$
Centrifugal pumps Centrifugal fans		$T = k \times n^2$ $P = k \times n^3$
Extruders		
Lathes Folding machines		$P = k$ $T = \frac{k \cdot P}{n}$
Piston compressors		–
Presses		–
Mills		–
Cranes Sawmills		–

10.5 Speed Range Selection

Selecting the correct size of an electric motor for a VFD is affected by the speed range at which the motor is expected to run continuously. The important factor is that the motor must be able to drive the load continuously at any speed within the speed range, without stopping or overheating the motor—i.e., the torque and thermal capacity of the motor must be adequate at all motor speeds—within the limits of acceptable motor loads. The motor running below the base speed ($f < 60$ Hz) causes the following effects:

- Reduced motor ventilation, because the fan attached to the motor shaft rotates at a low speed, so the motor temperature rise tends to be much higher than expected.

Figure 10.7 gives the example of a torque by speed curve for a pump, operating in the speed range of 10–60 Hz. This way we will have:

- The load torque within the acceptable load limit for all speeds.
- The maximum speed below the base speed (60 Hz). The speed range must not be increased above 60 Hz, because the load torque will exceed the VFD load limit as the torque for that load will increase with the square of speed.

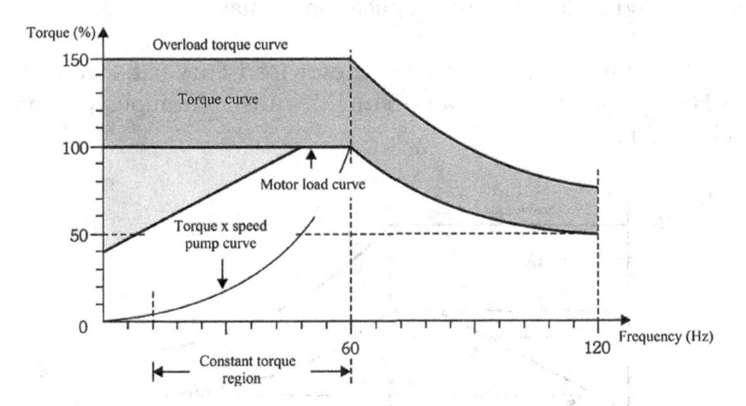

FIGURE 10.7
Speed and torque range example of a pump operating with VFD.

- The starting torque is low, so there will be no problem with the load resistant torque.
- The load acceleration torque is high, so the VFD must reach the maximum speed quickly. Rapid acceleration is desirable in pump applications.

Running above the base velocity ($f > 60$ Hz), the flux will be reduced due to the reduction of the V/f ratio. Consequently, there will be a reduction in motor torque capacity. The motor torque will be reduced in the same proportion as the frequency increase. The load torque must not be allowed to exceed the motor torque, even for a short period of time. Should this occur, the motor can lock its shaft. The following equation is practical and shows the maximum torque T_{max} allowed for frequencies above the nominal, the torque being directly frequency dependent.

$$T_{Max} \leq 0.6 T_M \frac{60}{f} \text{N} \cdot \text{m} \tag{10.2}$$

where:

T_M is the maximum motor output torque
f is the actual motor frequency
0.6 is the safety factor

Figure 10.8 shows an example of a torque by speed curve for a conveyor belt operating in a range of 15–60 Hz.

Analysing Figure 10.8, for this application we have:

- The load torque falls below the motor load curve at speeds below 28 Hz. Motor problems can occur if it runs continuously at speeds below 28 Hz.

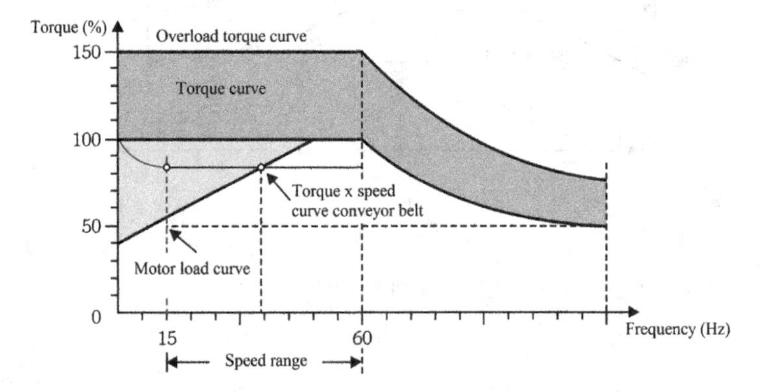

FIGURE 10.8
Example of speed and torque range of a conveyor belt that operates with frequency inverter.

- Although the maximum speed is below 60 Hz, it is possible to increase the motor speed above this value as the torque remains constant with increasing speed.

The starting torque is high due to inertia of the load, therefore, problems with the starting of this motor can occur.

10.6 VFD Applications

As the drive of machines and mechanical equipment by electric motors is a subject of great economic importance, in the field of industrial drives, it is estimated that two-thirds of the consumption of electricity is with electric motors. Figure 10.9 shows a graph with estimates of the percentage of consumption that each type of electric machine represents.

The following are examples of the use of VFDs in industrial environments.

10.6.1 Ventilation

The VFD application in ventilation processes is widely used because it has the following advantages:

- VFD and motor thermal protection.
- Automatic restart with speed recovery.
- Forced emergency operation (e.g., smoke extraction in a tunnel).
- Configuration of the types of stop in case of failure.

In addition to the advantages mentioned, another fundamental element is the energy savings that can be obtained with VFD use. To demonstrate this,

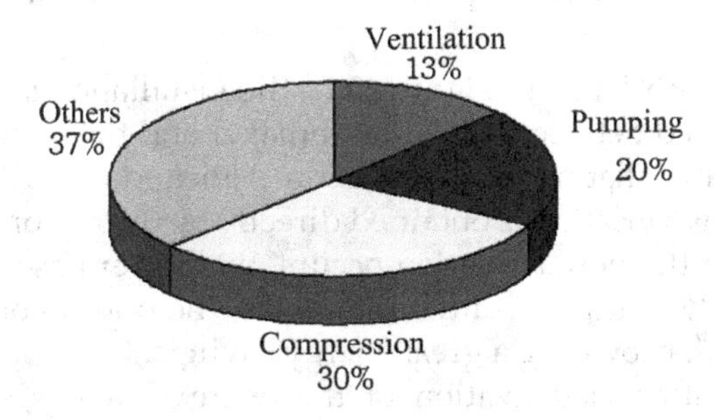

FIGURE 10.9
Estimated consumption of machines driven by electric motors.

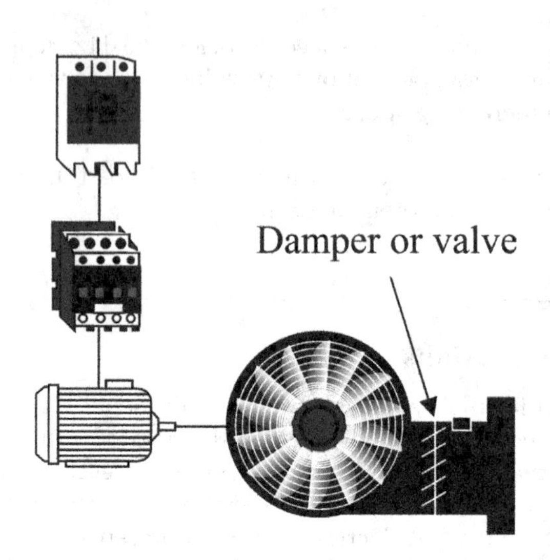

FIGURE 10.10
Motor installation without VFD.

we will make a comparison of a typical ventilation system without the use of a VFD and another using it.

Typical installation without the VFD: Figure 10.10 shows the installation without the VFD.

In this installation, the electric motor is fed directly from the grid and runs at its rated speed. To obtain the flow variation, a restriction device is used. When the flow reduction occurs, the absorbed power has a small decrease. For a fan with damper at the output, with 80% of the rated flow, the power consumed is still 95% of the rated power—that is, a great waste of energy occurs.

Installation with VFD: In Figure 10.11, the installation is with the use of VFD placed between the circuit-breaker and the motor.

In this installation, the restriction device (damper) is removed. The fan flow variation is obtained directly by the motor speed variation. When the flow reduction occurs, the power absorbed reduces considerably. With 80% of the rated flow, the power consumed will be only 50%, providing a great energy savings.

For a better visualization of the energy savings, Figure 10.12 shows a graph with the electrical power change in relation to the flow for a regulating valve and a VFD.

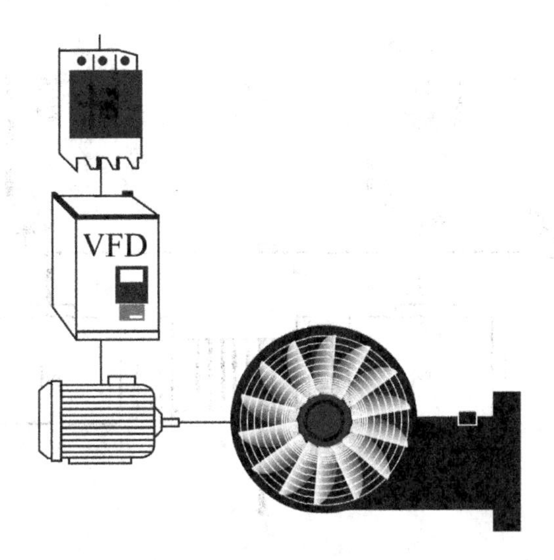

FIGURE 10.11
Motor installation with VFD.

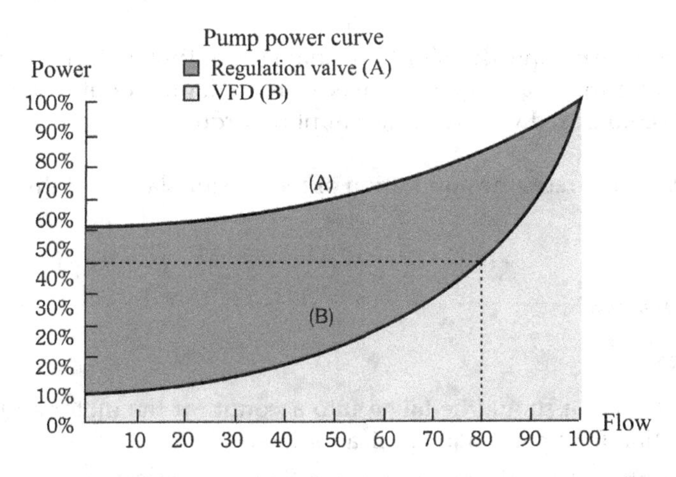

FIGURE 10.12
Comparison of the change of the electric power with respect to the flow in a motor driven by a VFD and another system with direct on line start and regulating valve.

10.6.2 Load Division (Master-Slave)

In rotary cranes, the motors are mechanically connected, therefore, they work at the same speed. When each motor is controlled individually by VFD, this function improves the torque distribution between the two motors. For this, the VFD changes the speed based on the torque, having the same slip effect.

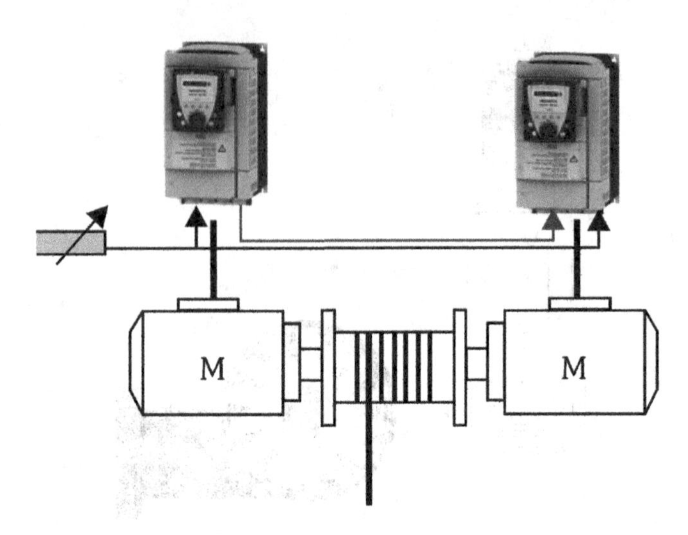

FIGURE 10.13
Application of VFD in a load division master-slave.

Master-slave (torque division): An analogue output of the master VFD is configured for torque and feeds an analogue input of the slave VFD configured with reference limit or torque.

Figure 10.13 illustrates the application of the master-slave load division.

Exercises

1. What criteria should be taken into account for the motor and VFD selection for a particular application?
2. What should be taken into consideration for the correct specification of a motor?
3. What is the VFD current for the following motor application: 20 HP/380V and power factor of 0.85?
4. Why is the region between 0 Hz and the base frequency called the constant torque region and the region above named the constant power region?
5. What are the desirable torque characteristics for an AC motor application?

6. For application of VFDs, what load characteristics should be known?
7. Describe the load characteristics of constant and variable torque.
8. What happens in a set motor/VFD when it is running below and above the base speed?
9. What is the advantage of using VFD in ventilation systems?
10. Describe the load division application with VFDs.

Appendix: Motors Wiring Diagrams

This section introduces a series of electrical wiring circuits for different applications. It makes a progression of more simple diagrams to the most complex. All should be considered as examples of studying. They cannot be applied directly in practice because they do not contain necessary signalling and safety functions (Figures A.1 through A.16).

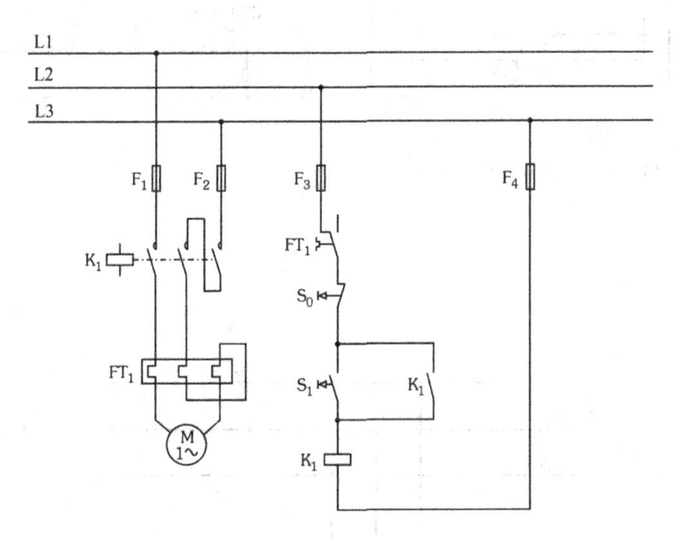

FIGURE A.1
Single-phase motor starter.

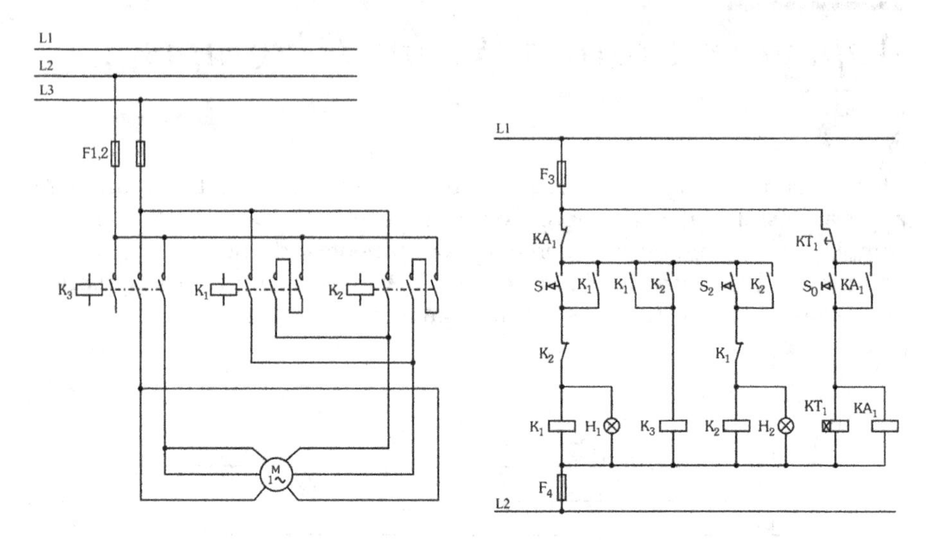

FIGURE A.2
Forward/reverse of single-phase motor.

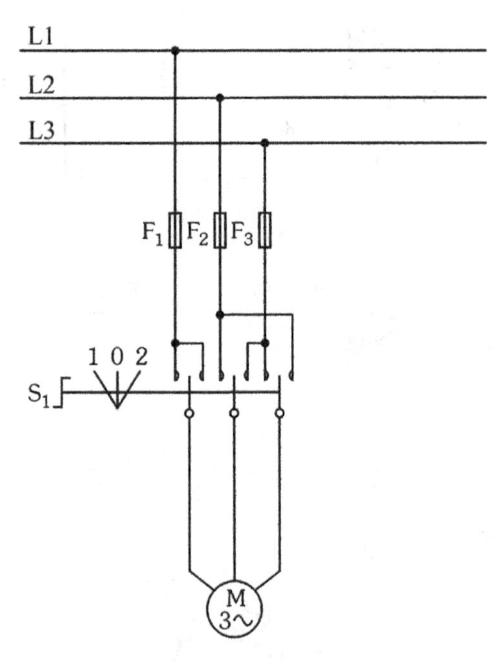

FIGURE A.3
Mechanical forward/reverse of three-phase motor.

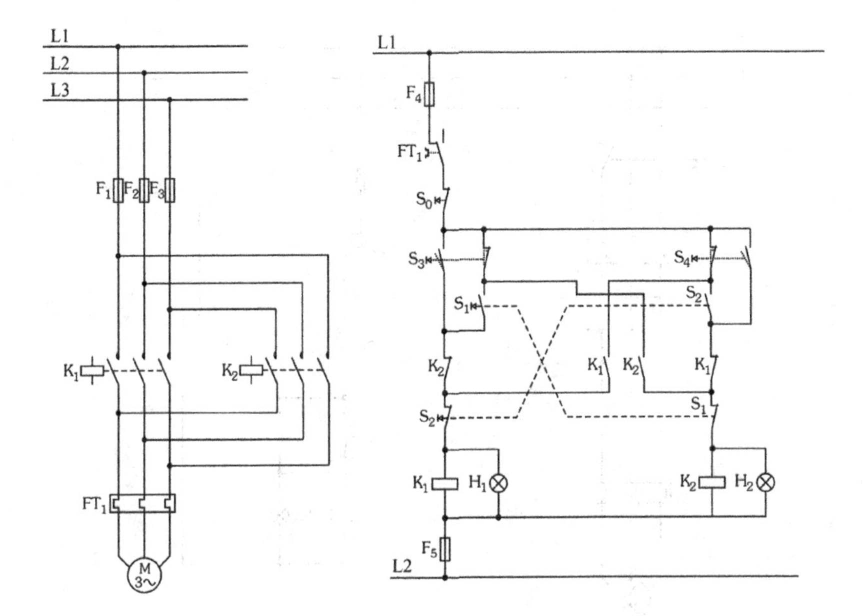

FIGURE A.4
Three-phase motor forward/reverse with limit switches.

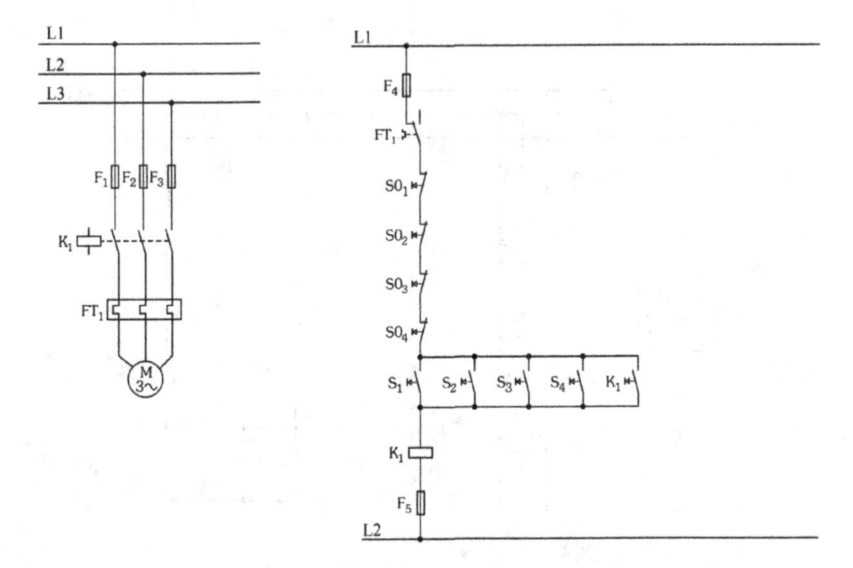

FIGURE A.5
Multi-point three-phase motor starter.

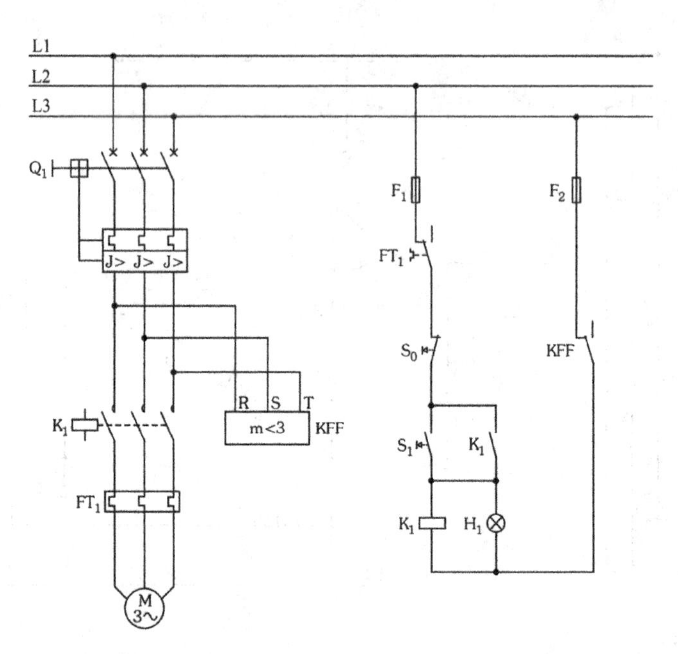

FIGURE A.6
Three-phase loss protection.

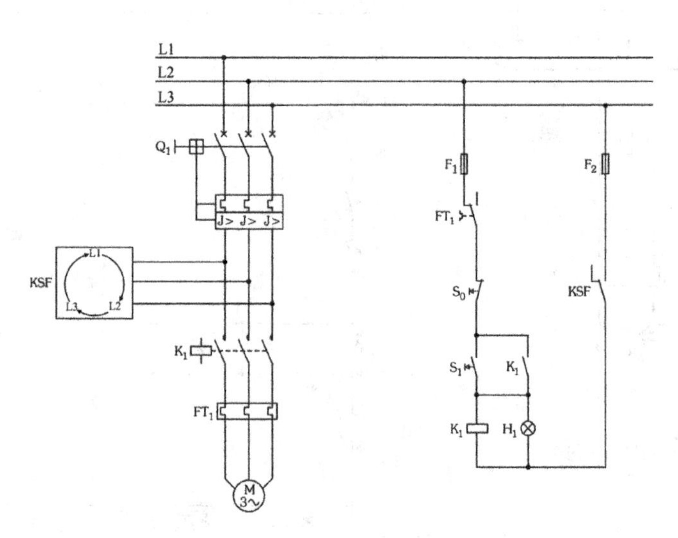

FIGURE A.7
Three-phase sequence protection.

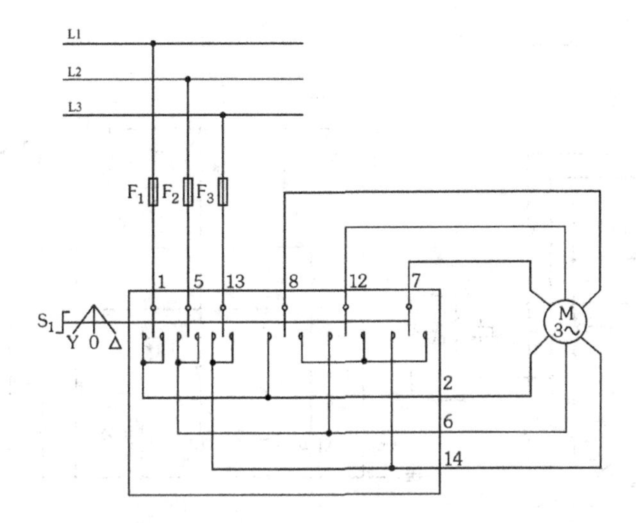

FIGURE A.8
Mechanical star-triangle starter.

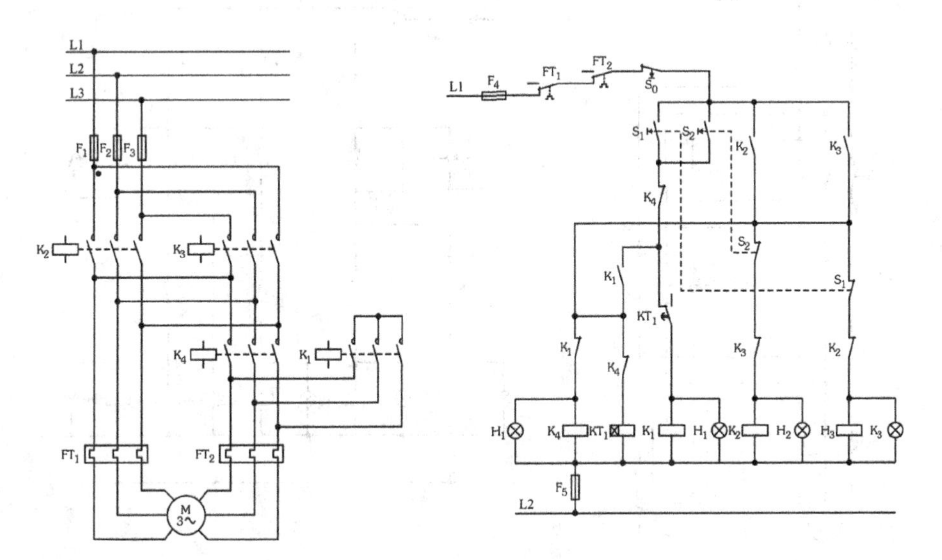

FIGURE A.9
Forward/reverse star-triangle starter.

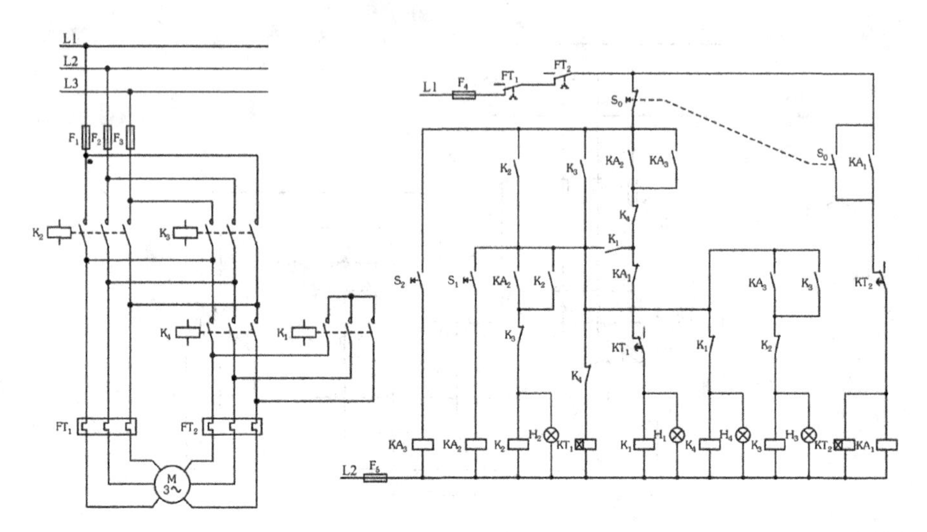

FIGURE A.10
Star-triangle starter with delay for forward/reverse.

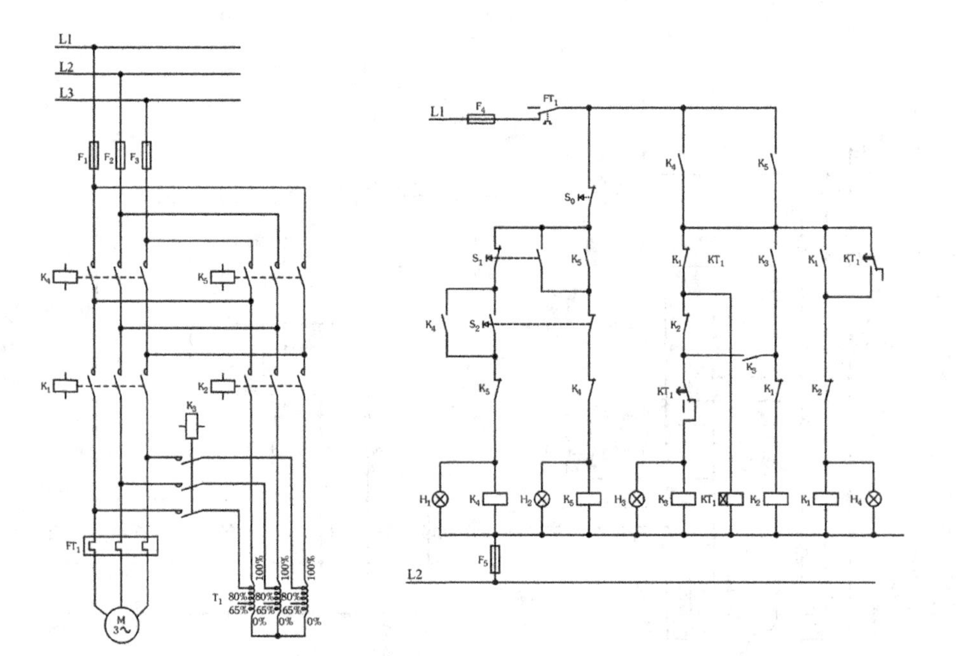

FIGURE A.11
Forward/reverse autotransformer starter.

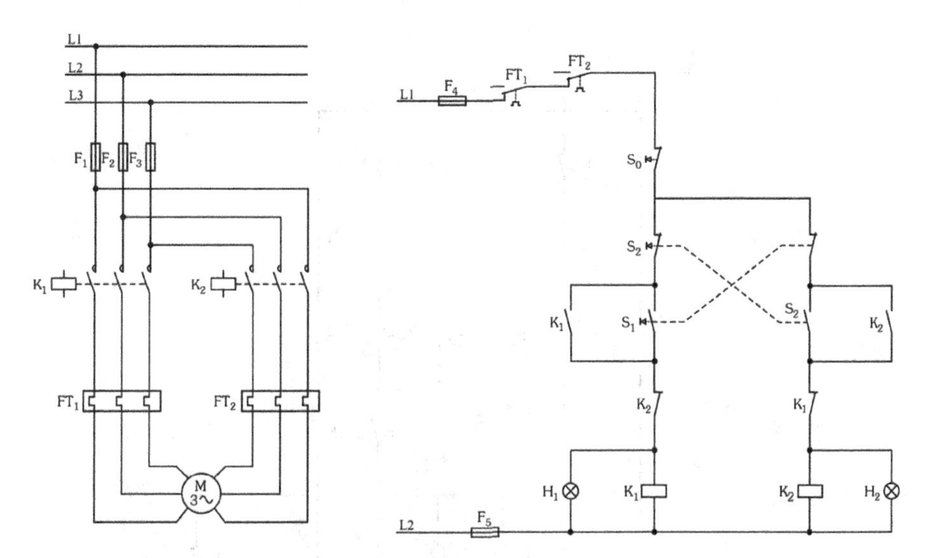

FIGURE A.12
Motor with two windings and two speeds.

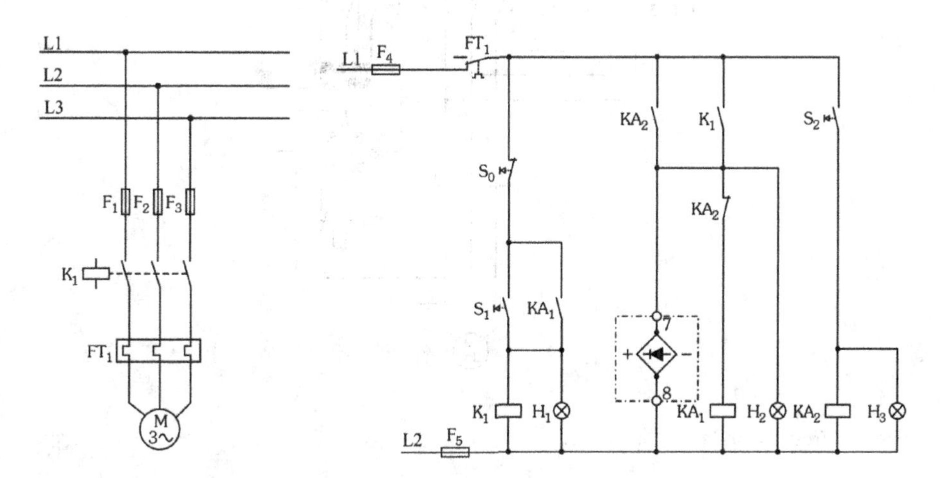

FIGURE A.13
Brake motor starter.

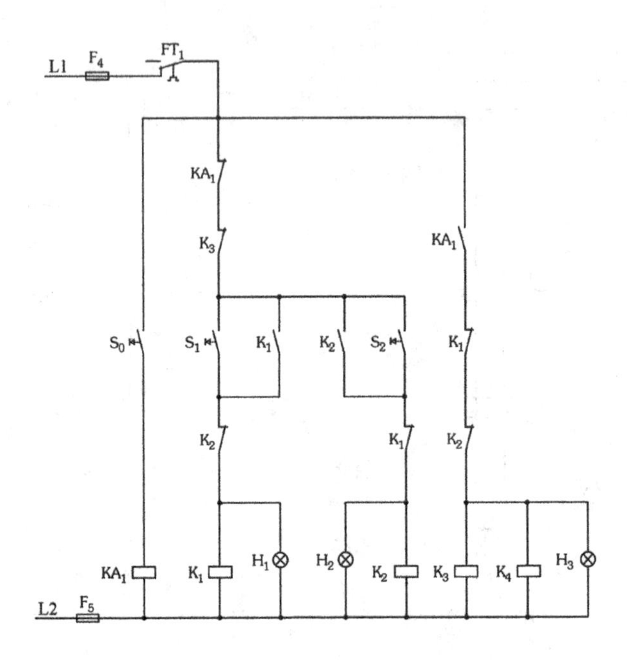

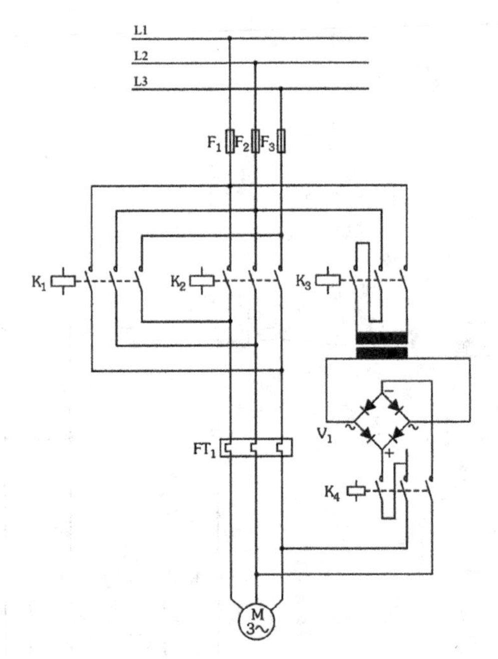

FIGURE A.14
Forward/reverse brake motor starter.

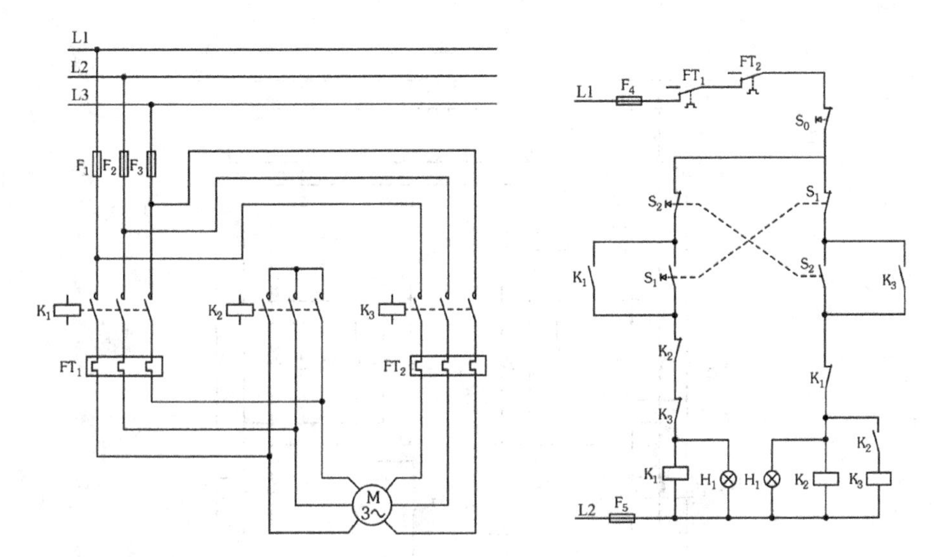

FIGURE A.15
Dahlander motor.

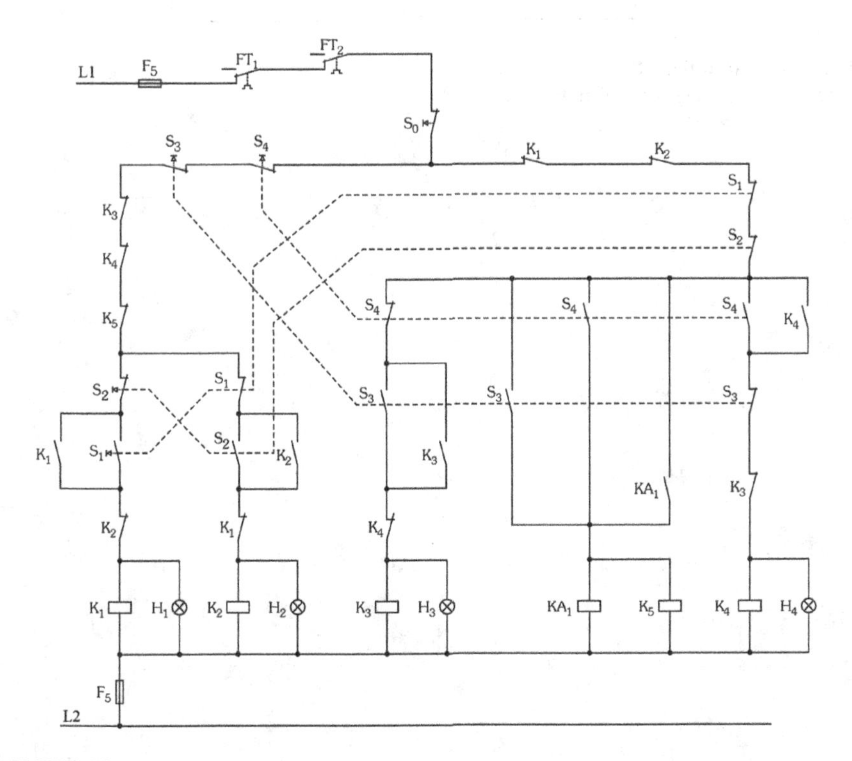

FIGURE A.16
Forward/reverse Dahlander motor. *(Continued)*

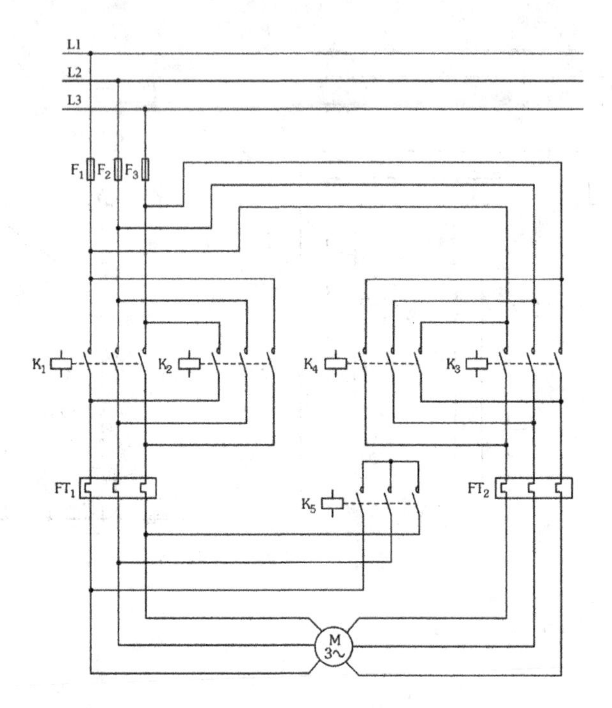

FIGURE A.16 (Continued)
Forward/reverse Dahlander motor.

Glossary

A/D converter: device that transforms analog signals into binary numbers that can be used by microprocessed systems.

AC. Alternating current: system in which the associated voltage and current alternates between positive and negative polarities, typically, at the frequency of 60 cycles per second.

Algorithm: description of the system behavior through a sequence of instructions.

Analog: signal that has characteristic of continuous values between two intervals.

Armature: the rotating part of a brush-type direct current motor or in an induction motor, the squirrel cage rotor.

Auxiliary Contacts: contacts of a switching device in addition to the power circuit contacts that act with the movement of the power contacts.

Binary: numbering system that uses only two digits: 0 and 1.

BJT. Bipolar junction transistor: three-terminal power transistor designed to withstand high currents between the collector and emitter terminals, which can be turned on and off by applying a current to the base terminal.

Breakdown torque (motor): the maximum motor torque with the rated load.

Brushes: devices that make contact with the rotating connections to the armature of a motor or generator. Usually made of carbon or metal.

Capacitance: property of a circuit to store electrical energy when a voltage is applied; your unit is farad (F).

Circuit Breaker: switch designed to protect an electrical circuit from damage caused by excess current from an overload or short circuit.

Coil: a device made by turns of insulated wire wound around a core.

Commutator: the part of a DC motor armature that causes the electrical current to be switched to various armature windings and transmit electrical current to the moving armature through brushes that ride on the commutator.

Contact NC: normally closed contact, that is, at rest allows the current flow.

Contact NO: normally open contact, that is, at rest does not allow the current flow.

Contact: element that allows or not the current flow.

Contactor: relay-like element, typically used in electrical diagram.

Control Circuit: circuit used to control the operation of some device.

CPU. Central Processing Unit: the CPU is composed of an arithmetic and logic unit, a control unit and a central (main) memory. The input and output units and auxiliary units are known as peripheral. It has, as function, to perform calculations of logical and arithmetic operations and to manage the execution of the instructions of the program.

Damper: valve used as a restriction for flow control in applications involving fluid flow.

DC. Direct current: electrical transmission system in which the voltage and current maintain a fixed polarity (positive or negative).

Dielectric: an insulating material used in a capacitor or other electrical device.

Digital: variable that admits only two states, being 1 or 0. Same as binary.

Diode (rectifier): electronic device used to convert alternating current to direct current.

Display: information display on one screen.

DSP. Digital Signal Processor: large capacity processor used in complex operations with high processing speeds.

Eddy Currents: induced currents contrary to the main currents; a loss of energy that shows up in the form of heat.

EEPROM. Electrically Erasable Programmable Read Only Memory: non-volatile memory, commonly used in microprocessors to store program data. Data in memory can be erased and updated electronically; the data is not lost when the power is removed.

Efficiency: a ratio of the input power compared to the output, typically expressed as a percentage.

Electrical Interlocking: used in control circuits in which the contacts in one circuit control another circuit.

EMI. Electromagnetic Interference: measurement of electromagnetic interference provided by an equipment.

Enclosure: mechanical, electrical and environmental protection for devices.

Encoder: electromechanical device that counts or reproduces electrical pulses from the rotational movement of its axis.

Endshield: the part of a motor that houses the bearing supporting the rotor and acts as a protective to the internal parts of the motor.

Excitation: the production of magnetic field in a motor winding by applying voltage.

Feedback: return signal from a sensor in a process control loop.

FFT. Fast Fourier Transform: equation that expresses a function in terms of sinusoidal functions, that is, as a sum or integral of sinusoidal functions multiplied by coefficients ("amplitudes"). One of its applications is the representation of a complex signal with the sum of a set of sinusoids.

Fieldbus: generic term assigned to any standard system that defines the software requirements for the connection of field devices and controllers in a control system.

Frame: standardized motor mounting and shaft dimensions as established by NEMA or IEC.

Full Load Amperes (FLA): line current (amperage) drawn by a motor when operating at rated load and voltage.

Full-Load Torque (motor): the torque necessary to produce the rated power of a motor at full-load speed.

Galvanometer: instrument that can measure electrical currents of low intensity, or the difference of electrical potential between two points.

GTO. Gate Turnoff Thyristor: thyristor that can be turned off by means of the application of a current in its gate.

Hardware: set of interdependent physical devices that make up an equipment.

Heat Sink: metal fin used to dissipate the heat of solid-state components.

HMI. Human-machine interface: interface between operator and controller that aims to provide interaction between operators and controllers applied in industrial processes. It serves as an interface for parameters insertion and process visualization.

HP: Horse Power. Equivalent to 746 watts.

Hz. Hertz: unit of measure of frequency in the metric system, being 1 hertz = 1 cycle per second.

IEC. International Electrotechnical Commission: international entity specializing in establishing standards in electrical systems.

IGBT. Insulated Gate Bipolar Transistor: electronic switching device controlled by voltage.

Impedance: combination of resistance (R) and reactance (X), measured in ohms, is conceptualized as an opposition to the current flow in an electric circuit.

Inductance: property of an electric circuit to oppose electric current; your unit of measure is Henry.

Interface: common electrical boundary between two separate devices, in which data or other electrical signals can be transmitted.

Inverter: an electronic device that converts direct current to alternating current.

Jogging (Inching): momentary equipment operation where the circuit quickly repeats its actions to start a motor from rest for the purpose of promoting small movements.

Ladder Language: language based on electrical contacts and coils. It is one of the graphic languages specified in IEC 61131-3.

LCD. Liquid Crystal Display: very light and thin monitor with no moving parts.

Leakage Current: current that flows through an unintended path.

Locked Rotor Current (motor): the steady-state current taken from the line with the rotor locked (stopped) and with the rated voltage and frequency applied to the motor.

Locked Rotor Torque (motor): the minimum torque developed by a motor at rest for all angular positions of the rotor with the rated voltage applied at a rated frequency.

Magnetic Field: space in which a magnetic force acts.

MOSFET. Metal Oxide Semiconductor Field Effect Transistor: field-effect transistor of semi-conductor of metallic oxide. It is designed to work with great powers; compared to other semiconductor devices (IGBT, SCRs, etc.); its main advantage is high switching speeds and good low-voltage efficiency.

NEC (National Electric Code): a safety code regarding the use of electricity. The NEC is sponsored by the National Fire Protection Institute. It is also used by insurance inspectors and by many government bodies regulating building codes.

NEMA (National Electrical Manufacturers Association): a non-profit trade organization, supported by manufacturers of electrical apparatus and suppliers in the United States. There are NEMA standards for motors cover frame sizes and dimensions, horsepower ratings, service factors, temperature rises and various performance characteristics.

Network: communication system in which several devices are interconnected and share the same communication channel.

Nm. Newton per meter: the unit of measurement of torque in the metric system.

Noise: unwanted electrical signal that is induced in a communication network and usually results in signal distortion or error in data transmission.

Off-Delay: timer in which the contacts change position immediately when the coil or circuit is energized, but delay returning to their normal positions when the coil or circuit is de-energized.

On-Delay: timer in which the contacts delay changing position when the coil or circuit is energized, but change back immediately to their normal positions when the coil or circuit is de-energized.

PID. Proportional Integral Derivative: control type extensively used in industrial processes.

PLC. Programmable Logical Controller: digital electronic equipment that uses a programmable memory to internally store instructions and implement specific functions such as logic, sequencing, timing, counting and arithmetic, controlling, through input and output modules, various types of machine or process.

Plugging: braking by reversing the line voltage (phase sequence).

Pole: the north or south magnetic end of a magnet; a terminal of a switch; one set of contacts for one circuit of main power.

Potentiometer: a variable resistor with two outside fixed terminals and one terminal on the center movable arm.

Power Factor: from any electrical system, which is operating on alternating current (AC); it is defined by the ratio of actual power or active power by total power or apparent power.

Profibus. Process Fieldbus: the name given to a standardized industrial network.

Programming Language: set of rules that syntactically defines valid statements.

Protocol: formal set of rules specifying all software, flow control, error detection, and time requirements for exchanging messages between two connection devices on a network.

PT100: temperature sensor built in platinum alloy that has its electrical resistance proportional to temperature. It has, by definition, the resistance of 100 Ω at 0°C.

PTC. Positive Temperature Coefficient: thermistor or semiconductor electronic component sensitive to temperature. Used for control, measurement or polarization of electronic circuits. It has a coefficient of variation of resistance that changes positively as the temperature increases, i.e., its electrical resistance increases with increasing temperature.

Pull-up Torque (motor): the minimum torque developed by the motor during the period of acceleration from rest to the speed at which breakdown occurs.

Pushbutton: a non-latching switch which causes a temporary change in the state of an electrical circuit only while the switch is physically actuated. An automatic mechanism (i.e., a spring) returns the switch to its default position immediately afterwards, restoring the initial circuit condition.

PWM. Pulse Width Modulation: modulation technique used in the inverters to control the output amplitude of a signal in alternating current and to improve the quality of the waveform of the current.

Reactance: opposition to the current passing when an alternating voltage is applied in an electric circuit, due to the inductance of the circuit or to the capacitance of the circuit.

Rectifier: electronic circuit whose purpose is to convert alternating current into direct current.

Rectifying bridge: set of semiconductors for rectification - conversion of alternating current into direct current.

Relay: electromechanical device composed of one or more coils and electrical contacts that switch when their coil is energized.

Resistance: opposition to the passage of current when a voltage is applied to an electric circuit; measured in ohms.

Rheostat: a resistor that can be adjusted to change its resistance without opening the circuit in which it may be connected.

Ripple: the component of alternating current that appears in signals of direct current.

RMS. Root Mean Square: the mean square value or effective value is a statistical measure of the magnitude of a variable quantity.

RPM. Rotation per minute: rotational speed measuring unit.

RS-232: a standard that defines the electrical and mechanical details of the physical interface between two devices, employing serial data signals.

RS-485: standard that defines the electrical details of the physical interface between two devices employing binary serial data exchange, allowing up to 32 transmitters and receivers in a network.

RTD. Resistance Temperature Detector: temperature sensor type where the temperature is directly proportional to the resistance.

SCR. Silicon Controlled Rectifier: alternative name for thyristors.

Sensor: a device capable of detecting variations of a physical variable.

Sensorless: without sensor; type of control that does not use sensor, open loop control.

Setpoint: or setpoint. Corresponds to the desired value of a given variable.

SFC. Sequential Function Chart: graphical sequencing of functions, one of the graphic programming languages specified by IEC 61131-3. It is derived directly from Grafcet.

SI: international System of Units.

Slip: the difference between the rotating magnetic field and rated rotor in an induction motor, expressed in percentage.

ST. Structured Text: high-level textual language used in PLC programming. It is similar to the Pascal language that allows structured programming techniques. It is one of the textual languages defined by IEC 61131-3.

Stator: the fixed part of an AC motor, consisting of copper windings within steel laminations.

Surge: a transient variation in the current and/or voltage in a circuit.

Tachometer: used for counting the motor speed revolutions per minute (rpm).

Temperature Rise: the amount by which a motor, operating under rated conditions, is hotter than its surrounding ambient temperature.

Terminal: a fitting attached to a circuit or device to making electrical connections.

THD. Total Harmonic Distortion: relation between the fundamental frequency power measured at the output of a transmission system and the power of all the harmonics observed at the output of the system by non-linearity when a single signal of specified power is applied at the input of the system.

Thermistor: temperature transducer that exhibits a variation of its internal electrical resistance proportional to the change of its temperature.

Thermocouple: a pair of different conductors joined to produce a thermoelectric effect and used to accurately determine temperature.

Thermostat: a sensor that is temperature-sensing with a dry contact, which is mounted on the stator winding.

Thyristor: semiconductor switch with three terminals called anode, cathode and gate, which by a direct voltage across the anode-cathode terminals and a current applied to the gate causes a current to flow from the anode to the cathode. The thyristor switches off when the anode current is reduced to zero.

Transducer: device used to convert physical parameters, such as temperature, pressure and weight, into electrical signals.

Transistor: electronic switching element made up of semiconductor type N or P chips. Bipolar can be NPN or PNP.

Transmitter: transmits sensor information over a range of voltage or current values.

Triac: electronic device used to control alternating current.

Trip: the state in which a protective device is when it is disarmed by a system fault.

TTL. Transistor-Transistor Logic: a logic family of semiconductors characterized by high switching speed and average power dissipation, the basic elements of which are bipolar transistors with multiple emitters.

Vac: Alternating current voltage.

Vdc: Direct current voltage.

Winding: typically refers to the process of wrapping coils of copper wire around a core.

Wiring Diagram: diagram that shows a pictorial representation of components with connecting wires.

Bibliography

Barnes, M. 2003. *Practical Speed Drives and Power Electronics*. London, UK: Elsevier.

Chalmers, B. J. 1998. *Electric Motor Handbook*. London, UK: Butterworths.

Doeuff, R. L. and Zaïm, M. E. H. 2010. *Rotating Electrical Machines*. Somerset, NJ: John Wiley & Sons.

Franchi, C. M. 2008. *Inversores de frequência: Teoria e Aplicações*. São Paulo, Brazil: Saraiva.

Franchi, C. M. 2014. *Acionamentos elétricos*, 4th ed. São Paulo, Brazil: Saraiva.

Herman, S. L. 2009. *Electric Motor Controls*. New York: Delmar Cengage Learning.

Herman, S. L. 2015. *Understanding Motor Controls*, 3rd ed. Boston, MA: Cengage Learning.

Hughes, A. and Drury, B. 2013. *Electric Motors and Drives: Fundamentals, Types, and Applications*, 4th ed. Oxford, UK: Newnes.

Kosow, I. T. 2007. *Electric Machinery and Transformers*. Mumbai, India: Pearson India.

Miller, R. and Miller, M. 2013. *Industrial Electricity and Motor Controls*. New York: McGraw-Hill Education.

Petruzella, F. D. 2010. *Electric Motors and Control Systems*. New York: McGraw-Hill higher education.

Senty, S. 2013. *Motor Control Fundamentals*. New York: Delmar Cengage Learning.

Toliyat, H. A. and Kliman G. B. 2004. *Handbook of Electric Motors*, 2nd ed. Boca Raton, FL: CRC Press/Taylor & Francis Group.

Veltman, A., Pulle, D. W. J. and Doncker, R. W. 2007. *Fundamentals of Electrical Drives*. Eindhoven, the Netherlands: Springer.

Vukosavic, S. N. 2013. *Electrical Machines*. London, UK: Springer.

Index

Note: Page numbers in italic and bold refer to figures and tables respectively.

9781032338620